Laser Beam Shaping Applications

OPTICAL SCIENCE AND ENGINEERING

Founding Editor
Brian J. Thompson
University of Rochester
Rochester, New York

RECENTLY PUBLISHED

*Please visit our website **www.crcpress.com** for a full list of titles*

Laser Beam Shaping Applications

Second Edition

Edited by

Fred M. Dickey
FMD Consulting, LLC
Springfield, MO, USA

Todd E. Lizotte
Pivotal Development Co. LLC
Hooksett, NH, USA

CRC Press
Taylor & Francis Group
Boca Raton London New York

CRC Press is an imprint of the
Taylor & Francis Group, an **informa** business

Cover artwork is a combination of a faceted beam integrator and a fiber optic laser beam delivery system each used within the laser processing industry. Both images were supplied by II-VI Corporation, Saxonburg, PA. The authors would like to thank Steve Rummel and John Ryan for supplying those images on behalf of II-VI Corporation.

CRC Press
Taylor & Francis Group
6000 Broken Sound Parkway NW, Suite 300
Boca Raton, FL 33487-2742

First issued in paperback 2019

© 2017 by Taylor & Francis Group, LLC
CRC Press is an imprint of Taylor & Francis Group, an Informa business

No claim to original U.S. Government works

ISBN-13: 978-1-4987-1441-9 (hbk)
ISBN-13: 978-0-367-87239-7 (pbk)

This book contains information obtained from authentic and highly regarded sources. Reasonable efforts have been made to publish reliable data and information, but the author and publisher cannot assume responsibility for the validity of all materials or the consequences of their use. The authors and publishers have attempted to trace the copyright holders of all material reproduced in this publication and apologize to copyright holders if permission to publish in this form has not been obtained. If any copyright material has not been acknowledged please write and let us know so we may rectify in any future reprint.

Library of Congress Cataloging-in-Publication Data

Names: Dickey, Fred M., 1941- editor. | Lizotte, Todd E., editor.
Title: Laser beam shaping applications / edited by Fred M. Dickey (FMD Consulting, LLC Springfield, MO, USA), Todd E. Lizotte (Pivotal Development Co. LLC Hooksett, NH, USA).
Description: Second edition. | Boca Raton, FL : CRC Press, Taylor & Francis Group, [2017] | Includes bibliographical references and index.
Identifiers: LCCN 2016035542 (print) | LCCN 2016040875 (ebook) | ISBN 9781498714419 (hardback ; alk. paper) | ISBN 1498714412 (hardback ; alk. paper) | ISBN 9781315371306
Subjects: LCSH: Laser beams. | Beam optics.
Classification: LCC TA1677 .D53 2017 (print) | LCC TA1677 (ebook) | DDC 621.36/6--dc23
LC record available at https://lccn.loc.gov/2016035542

Visit the Taylor & Francis Web site at
http://www.taylorandfrancis.com

and the CRC Press Web site at
http://www.crcpress.com

We dedicate this book to our children:
Stephen and Alan
Brent and Andrew

Contents

Preface

Since the advent of the laser, many applications have required shaping the laser beam irradiance profile. A few of the primary applications include material processing, such as welding, cutting, and drilling; and medical procedures, such as corneal surgery and cosmetic skin treatments. Other applications include laser–material interaction studies, lithography, semiconductor manufacture, graphic arts, optical data processing, and military uses. Over the past 30 years, laser beam shaping technology has matured in theory, design, fabrication techniques, and applications. A good illustration of this is that flattop beams are used in lithography and material processing, but they are also used in the LIGO system that recently detected gravity waves. In the LIGO system, flattop beams are used to increase signal-to-noise in the interferometric system.

This revised edition of *Laser Beam Shaping Applications* (first edition 2005) treats the applications and related technology of laser beam shaping. It is a companion volume to *Laser Beam Shaping: Theory and Techniques* (second edition 2014). The application of beam shaping goes beyond the physical and geometrical optics theory needed to develop beam shaping techniques; it inherently includes the physics of the beam–material interaction. The subtleties of the theory and techniques are best appreciated through the experience of application and design. The chapter authors are highly recognized leaders in the field of laser beam shaping applications. In addition to revising the chapters in the first edition, four new chapters have been added. The subject of the chapters is outlined in the introduction chapter.

This book is intended primarily for optical engineers, scientists, and students who have a need to apply laser beam shaping techniques to improve laser processes. It should be a valuable asset to someone researching, designing, procuring, or assessing the need for beam shaping with respect to a given application. Because of the broad treatment of theory and practice in the book, we think it should also appeal to scientists and engineers in other disciplines.

Although "all" applications of a technology can never be treated in a single volume, this book should provide the potential user of beam shaping techniques with the major insights, knowledge, and experience that can only be derived from applied systems development. In addition, this book provides extensive references to the literature.

The editors express their gratitude to the contributing authors; it is their efforts that made the book possible. It was a pleasure working with the staff of Taylor & Francis Group. Finally, we express our appreciation to the very helpful Ashley Gasque.

Fred M. Dickey

Todd E. Lizotte

Editors

Fred M. Dickey received his BS (1964) and MS (1965) degrees from Missouri University of Science and Technology, Rolla, Missouri, and his PhD degree (1975) from the University of Kansas, Lawrence, Kansas. A fellow of the International Optical Engineering Society (SPIE) and the Optical Society of America and a senior member of Institute of Electrical and Electronic Engineers (IEEE), he heads FMD Consulting, LLC, Springfield, Missouri. He is the author of more than 100 papers and book chapters, holds nine patents, and started and chaired the SPIE Laser Beam Shaping conference, for the first 8 years, and now which is in its eighteenth year, and is currently a committee member.

Todd E. Lizotte is an entrepreneur, inventor, technologist, and author who continues to seek technological and business opportunities through the application of laser beam shaping technology. As a cofounder of several high-tech startups over the past 26 years, Lizotte has demonstrated the critical impact that laser beam shaping has had within the industrial marketplace, by improving the quality and throughput of precision laser processes. Lizotte's laser beam shaping innovations have been applied directly to the microelectronics packaging, semiconductor, and automotive industries to name a few and lead to the acquisition of NanoVia LP a company cofounded by Lizotte, by Hitachi Ltd. Japan in 2003. For more than 10 years, Lizotte worked for Hitachi holding the executive position as Director of Emerging Technology in the United States and was fortunate enough to jointly work on various laser beam shaping projects with colleagues worldwide, developing 24+ patents covering technology that continues to be used today. Since leaving Hitachi in 2014, Lizotte has focused on business development projects at Pivotal Development Co., LLC in the United States, Europe, Africa, and China.

Contributors

Victor Arrizón
Optics Department
National Institute of Astrophysics,
   Optics and Electronics
Puebla, Mexico

T. T. Basiev (Deceased)
General Physics Institute
Russian Academy of Sciences
Moscow, Russia

Lourens Botha (Deceased)
Scientific Development and Integration
Pretoria, South Africa

T. Yu. Cherezova
Department of Physics
Moscow State University
Moscow, Russia

Michael von Dadelszen
J. P. Sercel Associates, Inc.
Hollis, New Hampshire

Fred M. Dickey
FMD Consulting, LLC
Springfield, Missouri

V. V. Fedorov
Department of Physics
University of Alabama at Birmingham
Birmingham, Alabama

Andrew Forbes
School of Physics
University of Witwatersrand
Johannesburg, South Africa

Daniel D. Haas (Retired)
Eastman Kodak Company
Rochester, New York

Scott C. Holswade
Sandia National Laboratories
Albuquerque, New Mexico

A. V. Kudryashov
Atmospheric Adaptive Optics Laboratory
Institute of Geosphere Dynamics
Russian Academy of Sciences
Moscow, Russia

Andrew F. Kurtz
Principal Scientist
IMAX Corporation
Rochester, New York

Todd E. Lizotte
Pivotal Development Co. LLC
Hooksett, New Hampshire

Paul Michaloski
Senior Optical Designer
Corning Tropel Corporation
Fairport, New York

Tom D. Milster
Optical Sciences Center
University of Arizona
Tucson, Arizona

S. B. Mirov
Department of Physics
University of Alabama at Birmingham
Birmingham, Alabama

I. S. Moskalev
Laser & Photonics Research Center
Department of Physics
University of Alabama
Birmingham, Alabama

Orest Ohar
Pivotal Development Co. LLC
Hooksett, New Hampshire

Nissim Pilossof
Principal Engineer
Kodak Canada ULC
British Columbia, Canada

Rubén Ramos-García
Optics Department
National Institute of Astrophysics,
    Optics and Electronics
Puebla, Mexico

Ulises Ruiz
Optics Department
National Institute of Astrophysics,
    Optics and Electronics
Puebla, Mexico

Jeffrey P. Sercel
J. P. Sercel Associates, Inc.
Hollis, New Hampshire

Reinhard Voelkel
SUSS MicroOptics SA
Neuchâtel, Switzerland

Edwin P. Walker
Call/Recall, Inc.
San Diego, California

P. G. Zverev
General Physics Institute
Russian Academy of Sciences
Moscow, Russia

1 Introduction

Todd E. Lizotte

The goal of technical publishing in its most basic form is to transfer knowledge. In terms of applications, it is to provide examples of how new technology can be applied within a particular industry to inspire and motivate its further use by others. As the reader goes through this book, there comes a moment where the abstract focuses to a point where ideas merge and one might ask, "Could that work for my application?" The editor's goals for this book, *Laser Beam Shaping Applications* (second edition), is threefold: (1) update the information originating in the first edition, (2) explore new applications that have come about during the time period between the two publications, and (3) inspire creativity to drive further innovation within the field.

As engineers and technologists tasked with the implementation of laser technology, we continue to seek optical solutions employing coherent monochromatic beams of light generated by lasers, delivering energy to the work surface with the goal of increasing quality, reliability, and repeatability of processes. Instead of crudely compressing the laser beam into a concentrated spot, laser beam shaping offers techniques capable of finessing and tailoring a process to specific material properties. This ability to fine-tune the process using optical techniques enables a host of material processing tasks to be achieved or improved, including milling, cutting, drilling, welding, and heat treating.

The newer, updated information contained in the second edition has emerged over the past 12 years. It should also be mentioned that the real genesis with laser beam shaping applications within the industry over the past 20 years coincides with and is reinforced by the introduction of the definitive book, *Laser Beam Shaping Theory and Techniques* (2000, first edition), written by *Fred M. Dickey and Scott C. Holswade*. It can be said with confidence that the introduction to laser beam shaping theory brought forward the adoption of the technology into the mainstream laser industry and enabled the creation of a host of laser material processes that spurred the further development of products, such as high-density electronics, handheld personal electronic devices, computer miniaturization, and flat panel displays. Most people, if not all, have a large screen flat panel TV, shifted over from flip phones to smartphones, and have acquired a tablet-sized computer. The fact is laser beam shaping has made significant contributions to the quality and capability of these technologies on the factory production line, opening the door to the visualization of and access to data by individuals to a level not seen since the introduction of the first personal computers. It is clear to many of us within the industry that laser beam shaping is no longer new and strange as it was 20–30 years ago. More people throughout the industry are now aware of its potential and familiar with the terms used to describe beam shaping. The second edition of *Laser Beam Shaping Applications* represents a continuation of the transfer of knowhow to the industry, reiterating its prominence as a means to push the boundaries of laser processing technology.

It will become apparent while reading that topics are discussed within the context of their applications in order to familiarize the reader with the solutions derived for each purpose and in some cases omit many other discussions about the specific details of the theory of laser beam shaping. Such an approach keeps the focus on the potential of the application of laser beam shaping technology and its successful implementation. If a reader wants to delve into a more complete and integrated book on theory, the book, *Beam Shaping Theory and Techniques*, which is the predecessor to the *Laser Beam Shaping Applications* is an excellent choice.

In this edition, new and additional chapters will demonstrate clearly that the field of laser beam shaping is an enabler for increasing the stability, quality, and speed of laser-based processes. What becomes evident is that, although the application may change, the concepts described and much of the discussions on the specific approaches continue to stay current and demonstrate the longevity, adaptability, and transferability of laser beam shaping. The editors are pleased to acknowledge that Chapters 2, 3, and 7 have been revised by their respective authors to include updates and further work realized since the last printing. Chapters 4, 10, 11, and 12 are included without revisions and although they do not contain updated material, each still provides a unique perspective on technologies and applications that are still relevant today. Chapter 4 addresses beam shaping for excimer lasers, covering varying integrator and prismatic-based homogenizer beam shaper designs of interest to the industry for bulk material removal, using ablation techniques and precision micromachining. Chapters 5, 8, and 9 cover entirely new areas as well as new applications, which include beam shaping using micro-optical elements for illumination in lithography work, through the use of shaped fiber optics, and the use of beam shaping for optical tweezers. Chapter 10 treats the application of deformable mirrors and explains their method of actively shaping higher powered laser beams due to the segmentation of actuators under a deformable substrate. Chapter 11 is an example of the application of spectral control in dispersive lasers. Each of these chapters provides the reader with ample opportunities to compare and contrast applications.

A great comparative example is included in the updated Chapter 2, "Illuminators in Microlithography," written by Paul Michaloski, and Chapter 3, "Laser Beam Shaping in Array-Type Laser Printing Systems," written by Andrew Kurtz, Daniel Haas, and Nissim Pilossof. Each chapter reveals that two applications that occupy either end of a product spectrum can benefit from similar laser beam shaping technology. Both chapters offer insights into the use of fly's eye uniformizers, also known as fly's eye homogenizers, or refractive integrators as well as the use of kaleidoscope or light pipe design for homogenization of laser beams. Each application describes the clear benefits from the two styles of integrator designs in relation to their respective technology areas and provides the reader the opportunity to recognize how the technology can be leveraged within the two such diverse cases.

Chapter 12, "Beam Shaping: A Review," written by Fred M. Dickey and Scott C. Holswade, continues to provide an excellent base foundation for the readers who are new to the field of laser beam shaping. What makes Chapter 12 useful to a novice and to those actually engaged in the technology is that it provides foundational material in a manner that clarifies and assures the reader that as he or she reads other chapters, he or she will have the basics to clearly and concisely understand the opportunities as well as the difficulties that are associated with beam shaping.

New to this edition, Chapter 5, "Micro-Optics Is Key Enabling Technology for Illumination Light Shaping in Photolithography," written by Reinhardt Voelkel, expands on a similar theme as Chapter 2; however, it introduces the use of monolithic wafer-based refractive micro-optics and vividly describes their use in a host of lithographic illumination techniques. Chapter 5 further expands on the natural evolution of established lens array integrators over the past years and demonstrates the importance of hybridization of established techniques and the evolution of integrators as they began to take the form of monolithic wafer-based micro-optical elements. Section 5.2.3 with the introduction of Kohler integrators that were utilized for early lithographic tools in the 1970s provides a splendid historical perspective on why the need for continued development of illumination in some cases drove the development of micro-optical fabrication techniques that transformed today's lithography industry. Chapter 8, "Applications of Diffractive Optics Elements in Optical Trapping," written by Ruben Ramos-Garcia, Victor Arrizon, and Ulises Ruiz, provides a primer for the broader role of spatial light modulators within this field and the use of diffractive optical elements. Chapter 8 serves as a basic building block for the value of laser beam shaping within the field of optical tweezers or trapping while recognizing the different degrees of technical sophistication of the intended reader. Chapter 9, "Laser Beam Shaping through Fiber Optic Beam Delivery," written by Todd Lizotte and Orest Ohar, demonstrates that light pipes adapted into smaller form factors, such as drawn square fibers, can open doors to new laser materials processing applications for precision material removal and offer laser beam shaping opportunities for more traditional robotic based fiber optic based processes. Chapter 9 provides the reader with an existing robotic example on a production floor to emphasize the concepts governing the application and to demonstrate the true nature of a product that incorporates laser beam shaping technology.

Considering these new and revised chapters, it becomes apparent that solutions in each individual industry continue to evolve as beam shaping technology and fabrication techniques improve, and/or newer technologies emerge. From the chapters added to this new edition, the reader can ponder the evolution of beam integrators over time and witness the shift from the basics of light pipes and the early days of refractive integrators to the modern versions that leverage micro-optical technology.

The theory of laser beam shaping is bound within the constraints of physics and calculations, whereas applications provide a tangible point in time to gauge the progress of a technology, within real terms of its proven limitations at that moment in time, including in some cases, the lack of computing power, access to the most efficient optical materials, and the basic state of fabrication techniques used to realize the final application. Each chapter could be considered a historical marker. As an example, take optical integrators back in the late 1970s and early 1980s, at that time cylindrical and spherical refractive lens and reflective mirror integrators began to find traction within the burgeoning laser industry. Made by stacking individual lenses (spherical or cylindrical) or faceted mirror segments, the lens array integrators were expensive and somewhat clumsy to use and replicate (Figures 1.1 and 1.2).

FIGURE 1.1   Refractive lens array homogenizer, circa 1985.

FIGURE 1.2   Reflective segmented mirror array homogenizer, circa 1990.

FIGURE 1.3   Refractive monolithic lens array homogenizer, circa 2008.

Approximately 15 years ago, such devices began to take the form of reactively ion-etched monolithic refractive lens arrays (Figure 1.3), which also came with their own limitations, including part-to-part reproducibility and alignment. Ultimately, through continued fabrication development, these monolithic micro-optical arrays reduced the cost of the beam shaper designs and allowed the technology to leap into other markets, such as printing, telecom, and laser-material processing. Each chapter provides the reader a clear vantage point to evaluate the evolution of beam shaping technology and weigh its progress. There are times, however, where even the older designs become more relevant to today's needs or industrial demands on a true production floor. Even stacked lens array integrators continue to find relevance within the latest generation of laser processing systems for large panel displays as they continue producing cutting edge products that consumers purchase every day.

It is evident in the growth of laser beam shaping that even though the solutions are specific within this book, they may be applied in different ways to other applications in the future. Even though these chapters represent some of the most understood techniques within the industry, each of them can be enhanced, modified, or combined to tackle the next challenge faced within an industry that continues to leverage laser technology.

The editors thank the authors who submitted new and revised chapters to this edition and add that as with most books, no editor or author can ever be completely right, although they all strive for that level of perfection. The editors shall be grateful for any comments, corrections, or criticisms that readers may find relevant and bring to their attention.

2 Illuminators for Microlithography

Paul Michaloski

CONTENTS

2.1 INTRODUCTION

The process of imaging a patterned reticle by steppers and scanners onto photoresist-coated silicon wafers has been the backbone of the mass production of integrated circuits for over four decades. Like photography, imaging optics allows the parallel transfer of tremendous amounts of information from the object to image, or the reticle to wafer, in a fraction of a second. The specialized projection lenses of microlithography must accomplish this image transfer while maintaining uniformity of imagery and image placement on the nanometer level over centimeter sized image fields. Although demands on the imaging lens are severe, performance cannot be obtained without a properly designed illuminator. The illuminator controls or *sculpts* the diffraction pattern in the pupil that accomplishes the imaging on the wafer and maintains the radiometric uniformity across the image field. The illuminator does not require the same fabrication precision as the projection lens but has its own set of difficult design problems. Nonetheless, the imaging lens and illuminator must be designed together as a single system to achieve the desired overall performance. This chapter examines the design considerations, difficulties, solutions, and techniques of designing these optical systems from the source to reticle.

2.2 ILLUMINATORS IN MICROLITHOGRAPHY

2.2.1 The Path of Microlithography

In 1965, Gordon Moore predicted that integrated circuits would double in performance every year based on observation.[1] In 1975, Moore revised his prediction to doubling every two years, and it has proven relatively accurate.[2] Moore's law, though it is a prediction, has served as the guideline for the path and pace of development of the semiconductor industry and its microlithographic equipment (Figure 2.1).

Most discussions on microlithography begin with the imaging equations of resolution and depth of focus (DOF):

$$CD = k_1 \frac{\lambda}{NA_{PL}} \tag{2.1}$$

$$DOF = k_2 \frac{\lambda}{NA_{PL}^2} \tag{2.2}$$

The numerical aperture (NA) of the projection lens, NA_{PL}, is $n\sin(\theta)$, where n is the refractive index of the image space and θ the semiangle of the imaging cone. The drive to produce smaller features (the critical dimension) has resulted in the use

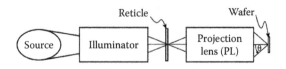

FIGURE 2.1 A simplified microlithographic system.

of shorter wavelengths and greater NAs. The former is preferred, since the DOF only decreases in proportion to the wavelength, as opposed to the inverse square of the NA. The DOF is a vital parameter to the manufacturability of the circuits. Furthermore, increasing the NA also increases the complexity of the projection lens, and the sensitivity to fabrication and alignment error. Historically, each step down in wavelength introduced new challenges to the development of the source, lens materials, photoresists, reticles, and more. Yet, many dramatic improvements in the resolution factor k_1 and the DOF factor k_2 have been made through improvements in illumination, which will be covered later.

2.2.2 DESCRIPTION OF AN EXPOSURE

The viability of expensive optical microlithographic tools is dependent on the number of exposed wafers per hour, or more precisely, the wafer area rate of exposure. These sophisticated cameras, or tools, need to move a resist-coated wafer from a clean container into the tool, align the wafer in all six degrees of freedom to the optimum aerial image of the reticle, and maintain this position during the exposure. This process is repeated as the wafer is stepped from site to site until the entire wafer area is exposed in a matrix fashion. The wafer is then shuttled to an output container and the process repeated. This crudely describes the tool known as a stepper. A full-wafer scanner, on the other hand, moves the wafer in synchronism with the moving reticle during exposure to allow the use of a smaller projection lens image field in the shape of a curved slit or rectangle. The technique that has become most prominent today is a step-and-scan system (but also called a scanner) that combines both stepping and scanning. The scan field covers a section of the wafer, and then the tool steps the wafer to the next section and scans. This allows larger wafers to be covered with less mechanical and optical demand on the tool than a stepper or full-field scanner.

2.2.2.1 Exposure

The duration of an exposure is proportional to the resist sensitivity, and the irradiance that the illuminator delivers. In the case of scanners, the speed of the scan and width of the slit also relate to the exposure duration. The combined factors of increased resist sensitivities with acceptable resolution, improved processes, and the higher photon energies of lower wavelengths have allowed shorter exposures. Even with these improvements, transmission requirements of the illuminators have remained stringent due to the economic value of high productivity.

The dimensions of printed features vary with exposure. This variation is a function of the aerial image and the contrast behavior of the resist. The so-called aerial image might be formed in water for an immersion lithography system. The acceptable range of the critical dimension variation of a feature from all sources (typically less than 10%), along with the characteristics of the aerial image and resist processes define the range of exposure allowed, which is known as the exposure latitude. The exposure latitude is one of the limiting aspects of the lithographic *process window*. For a continuous source, such as a lamp, the exposure energy is proportional to the duration. For a pulsed source, such as an excimer laser, the energy is proportional to the number of pulses summed during an exposure. Excimer lasers have increased pulse energy and pulse repetition rates from typically 100 Hz to as high as 6 kHz. This reduces sensitivity to pulse-to-pulse energy variations, and in the case of scanning tools, improves irradiance uniformity over the field. The more the pulses, the smaller the variability in the number of pulses summed over different portions of the scanned field. The choice of where to shutter in the illuminator can also vary the exposure over the field, which is discussed in Section 2.3.4.

The laser pulse plasma (LPP) extreme ultraviolet (EUV) sources generate pulses at rates that are an order of magnitude higher than the excimer laser. Tin droplets are heated by a 20 kW CO_2 (10.6 um IR) laser pulse to induce a high-temperature plasma that generate EUV photons at 13.5 nm.[3,4]

2.2.2.2 Uniformity

A very important objective in lithography is to obtain uniform control of the intended critical dimension over the whole imaging field. Yet, the uniformity of exposure duration is a just one contributor to the uniform exposure goal. A much larger contributor is the uniformity of the irradiance delivered by the illuminator through the projection lens at the wafer. Therefore, to maintain the exposure latitude, both exposure duration control and irradiance uniformity are needed. In terms of specification and measurement, uniformity for lithography is defined as the deviation of irradiance around the mean irradiance:

$$\pm U = \frac{\left(Ir_{max} - Ir_{min}\right)}{\left(Ir_{max} + Ir_{min}\right)} \tag{2.3}$$

This can be thought of as the deviation from the optimum exposure energy. The specification for the uniformity is on the order of $\pm 1\%$ for all illumination conditions. This is typically measured with an *open field* reticle that has no features present. This is important to mention, since the transmission from the reticle to the wafer varies for different features being printed. This is due to diffraction overfilling the projection lens pupil and is discussed in more detail in Sections 2.2.4 and 2.3.1.

2.2.2.3 Alignment

As stated, the microlithographic tool aligns the wafer to an optimum *aerial image* field. An aerial image is formed in the image space. An exposure in the resist is altered by the refractive index and other optical properties of the resist, including the medium above the resist. In the case of immersion lithography, this medium will alter the coupling of the light into the resist. The evaluation of the imaging performance by modeling and measurement is often based on the aerial image and not the image in the resist. The DOF of the imaging is the depth of acceptable aerial image performance perpendicular to the optical axis. Any tilt of the wafer will move a portion of the imaging field out of the DOF. Lateral and rotational errors of the position of the wafer will misalign the new image to any existing features. Integrated circuits are composed of many layers, all of which need to align to previous layers to a tolerance on the order of a fraction of the feature sizes being imaged. Also, the alignment capability allows multiple exposures with different reticles to the same field. In the case of DUV (193 nm, 248 nm) lithography, the positioning of the wafer to the image is within tens of nanometers over imaging fields of 20–30 mm². In the case of EUV (13.5 nm) lithography, the alignment needs are on the level of a few nanometers, and some day will be even less.[5] The spacing of the atoms of the silicon wafer is ~0.2 nm.

2.2.2.4 Telecentricity

In order to maintain the same magnification through the DOF, lithographic projection lenses are designed to be telecentric on the image side (Figure 2.2). A telecentric lens, in the most basic terms, is defined as one that has the chief ray normal at the image plane. This is the ray that emits from the edge of the field and passes through the center of the aperture stop. In more precise terms, it is the illuminator's angular radiance distribution at the wafer that defines the telecentricity of imaging at the wafer. Typically, the projection lens has insignificant NA variations over the imaging field, so it is the illuminator that determines the degree of telecentricity correction. This is discussed in greater detail in Sections 2.2.3 and 2.3.5. The illuminator not only maintains the magnification of the image through the DOF, but can also increase the depth by tailoring the pupil profile for the specific objects being imaged. This is discussed in more detail in Section 2.2.4. The DOF along with the exposure latitude defines the total process window. In the absence of other factors, the process window is usually illustrated as an exposure defocus (ED) window. Factors such as reticle flatness, wafer flatness, thermal drift, aberration correction of the projection lens, flare, and other characteristics can reduce the DOF. A larger ED window means a more efficient and robust lithography process.

2.2.3 OVERVIEW OF PROJECTION OPTICS AND IMAGING CONCERNS

It is helpful to review both historical and contemporary projection lenses in order to understand the requirements of illuminators.

Early lithography systems operated in the visible spectrum with the use of high-pressure mercury lamps, which have strong emission lines at convenient wavelengths. The first systems used the visible g-line (436 nm) and progressed toward the near

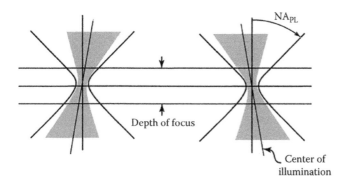

FIGURE 2.2   The center of the illumination determines the telecentricity of the imaging.

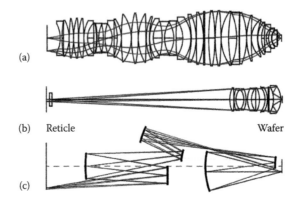

FIGURE 2.3 Three projection lenses: (a) an all-refractive single material production 248 nm, NA 0.70, 1.0 m total track,[6] (b) catadioptic small field 157 nm, NA 0.85, 0.6 m total track[7,8], and (c) an all-reflective 13 nm, NA 0.25, 1.5 m total track.[9] The three figures are not to the same scale.

UV i-line (365 nm) by the late 1980s. The illuminator filters out the entire unwanted spectrum, including all the lines and heat continuum in the infrared. The systems that use broader spectrums have higher power and throughput, but also required a higher degree of chromatic correction in the projection lens. The achromatization of all refractive designs requires lens materials of varying dispersion of their refractive index. There are many optical glasses available at the visible g-line that allow the achromatization over broad bandwidths. Optical glass manufacturers developed high transmission glasses specifically targeted for i-line microlithography. The number of acceptable glass types at i-line is much less than g-line, but enough to achromatize all-refractive designs to bandwidths of 6–10 nm at high imaging NAs. A catadioptic design, which uses mirrors along with refractive lenses, can allow even broader bandwidth correction. Some catadioptic projection lenses were achromatized for g, h, and i-lines. This is possible since the optical power of a reflector is constant over the spectrum. An example of a catadioptic lens is a Newtonian form shown in Figure 2.3b. The achromatization of this design allows both lines of an unnarrowed bandwidth 157 nm excimer laser to be used.

The next step down in wavelength was to the 248 nm KrF excimer laser source.[10] An all-refractive projection lens is shown in Figure 2.3a. Fused silica and calcium fluoride are the only practical materials at this wavelength with adequate transmission. Unfortunately, these two materials are close in refractive index and dispersion, so chromatic correction requires high-optical power lenses that are difficult to fabricate and align. The solution is to greatly narrow the spectral bandwidth of the laser rather than achromatize the lens.[11] The availability of high-purity fused silica along with its superb mechanical and optical properties also played a part in the choice of narrowing the source spectrum. The development of reliable line narrowed excimer laser sources, and sensitive, high contrast, chemically amplified resists were essential to the success of 248 nm lithography.

The next wavelength reduction was from 248 to 193 nm. Narrowing of this laser also allowed all fused silica refractive designs. The high photon energy of the 193 nm ArF_2 excimer laser introduced new problems. The high flux pulses induce optical and mechanical changes to fused silica as a function of the fluence, pulse duration, and number of pulses.[12,13] The slow change of absorption in optical materials had long been observed in visible and near UV optics. This higher power regime brought about by short-pulsed lasers introduced additional changes in density and thus in refractive index and optical surface figure. In order to reduce these effects, some excimer laser manufacturers added optical delay lines to stretch out the pulses.[14]

The wavelength reduction to 157 nm of the F_2 excimer required calcium fluoride due to the low transmission of standard fused silica. Fused silica doped with fluorine is the only amorphous material with adequate transmission to be used as a reticle and thin optical elements.[15] The challenge of polishing crystalline CaF_2 with its directional-dependent hardness was solved. The influence of the directionally dependent intrinsic birefringence on image formation was never completely solved.[16]

Instead of 157 nm lithography, the industry chose the path of 193 nm water immersion. The NA of Equations 2.1 and 2.2 is proportional to the refractive index of the imaging medium. The index of water at 193 nm is 1.43, so for CD the equivalent wavelength can be thought of as $193/1.43 = 135$ nm.[17] Unfortunately, the DOF also decreases by $1.43^2 \sim 2x$. Yet, there are advantages of coupling from the lens into the resist through a higher index medium.

The next step down in wavelength is a large step to the EUV of 13.5 nm. As described, the EUV radiation is created by focusing a high powered IR laser on to a tin (Sn) drop. This radiation is concentrated backward toward the laser and *collected* by a concave mirror similar to the elliptical reflectors that are discussed in Section 2.3.4. It took years to develop the mirror coatings for 13.5 nm to be capable of reflecting 70% of the incident radiation, which produces system transmissions of less than 1%.

In an ideal imaging system, a wavefront emanating spherically from a single point at the object converges spherically to a point at the image. Deviations in the wavefront from the converging sphere are due to portions traveling a different optical path length. These wavefront errors become significant on the order of 1/40th of a wavelength or larger. This level of correction on an all-refractive projection lens with 25 or more elements requires that the optical surfaces be spherical on an order of 12 nm. To appreciate this very small relative error, the 12 nm departure from a 200 mm diameter optical surface when scaled to the distance across the Atlantic Ocean from New York to Ireland would represent a departure equal to a 250 mm wave. This becomes even more extreme for EUV systems that require mirror surfaces with a fraction of a nanometer correction. A six-mirror system at 0.25 NA as shown in Figure 2.3c over the same span of ocean would have departures of 6 mm waves.

2.2.4 The Role of Illumination Pupil Distributions

This section covers the theory of partial coherent imaging without delving into the equations of image formation. Both the classic 1953 paper by H.H. Hopkins[18] and Alfred Kwok-Kit Wong's[19] work are valuable references on the subject.

Lithographers will often refer to the illumination pupil distribution as the *source*, which forms the incident angular distribution on any feature of the reticle that is being printed. This might follow from the desire to classify all illuminators as being either a critical or a Köhler. The actual source is imaged to the pupil of a Köhler illuminator, and considering the importance of the pupil to lithography and the use of a field and pupil stop that Köhler describes, this is the better choice of the two classifications. Yet, the section on uniformizers reveals that the actual source is not simply, or necessarily, imaged to the pupil. The term *source* might also follow from the use by H.H. Hopkins of the term *equivalent source* in his 1953 paper.

2.2.4.1 Partial Coherence

The illuminator has an aperture stop that is imaged into the projection lens' aperture. The ratio of these two aperture diameters is called the partial coherence ratio, σ. Since the NA is proportional to pupil dimensions, σ is also the ratio of the illumination NA (NA_{Illum}) and projection lenses NA (NA_{PL}).

$$\sigma = NA_{Illum}/NA_{PL} \qquad (2.4)$$

The coherence parameter σ is actually the opposite of the degree of coherence. The degree of coherence is a term of the mutual coherence function in partial coherence theory.[20] As σ approaches zero, the lateral spatial coherence increases. At zero, we have a totally coherent illuminator, and the irradiance pattern at the projection lens aperture is the Fourier transform of the reticle pattern. The intensity pattern for any other illumination pupil fill can be calculated by the convolution of the Fourier transform and the image of the illuminator's aperture stop (Figure 2.4).

Consider the imaging of a diffraction grating. As the grating pitch decreases, the spatial frequency of the diffracted orders increase, and the first orders move toward the outside of the imaging NA_{PL}. For a low σ, the contrast will drop to zero when the pitch of the grating reaches λ/NA_{PL}, since the first orders move outside of the aperture stop.

As the partial coherence factor σ is increased, some energy from the first order of these same higher frequencies can pass the pupil to interfere at the wafer and form an image. These regions of interference are shown hatched in Figure 2.5; the blocked portions of the first orders are darkened. As an example of how partial coherent illumination influences the aerial image, slices are shown in Figure 2.6 for various σ at a pitch of 1 λ/NA.

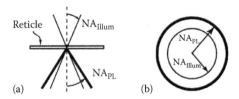

FIGURE 2.4   Illuminator and projection lens NAs at the (a) reticle and (b) aperture stop.

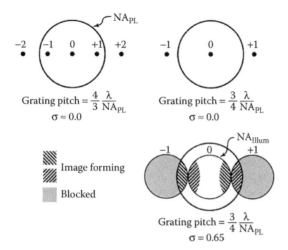

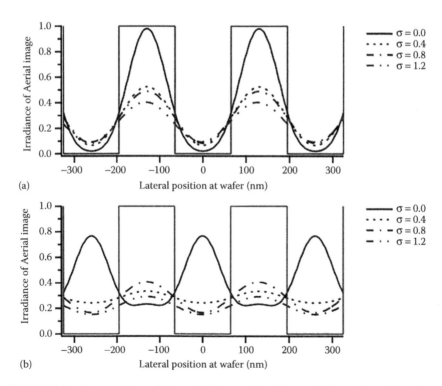

FIGURE 2.5 The pupil fill at the aperture stop of the projection lens for imaging a diffraction grating demonstrates the dependence on pitch and sigma.

FIGURE 2.6 Cross sections of the aerial image of a 260 nm pitch grating at best focus (a) and 200 nm out of focus (b) for a 193 nm 0.75 NA projection lens. At σ = 0.0, the out of focus profile has a phase reversal.

The peak contrast is less of a concern for imaging into the resist than the DOF. This is due to the high contrast or *threshold* characteristic of the resist and the importance of DOF in the total process window latitude as discussed in Section 2.2.2. The DOF for dense lines can be defined by the range of defocus positions that meet some value of image modulation. This modulation value can be one that matches empirically measured DOF for a similar lithographic system. Another merit value, which also works for isolated features, is the image's normalized log-slope (NILS). This is the log of the slope of the image's irradiance profile at the edge of the target width and multiplied by the target width. This might seem convoluted, but it does generate a unitless value that pertains to the edges of the image that influences lithographic performance more than the peak and valleys. For now, we will use modulation, which is still useful. The system modeled for these figures is a 193 nm 0.75 NA projection lens with no wavefront aberration and random polarization. The partial coherence factor σ describes only the perimeter of the illumination cone at the reticle, whereas the distribution within that perimeter also influences the performance of the imaging. The 260 nm pitch in Figure 2.7 is 1 λ/NA and 300 nm is a 15% longer pitch, and thus 15% less spatial frequency. The Gaussian source distribution is at a relative height of 0.3 at the illuminator's pupil. As shown, the optimum σ is higher for the Gaussian fill; there is less energy at the outer NAs.

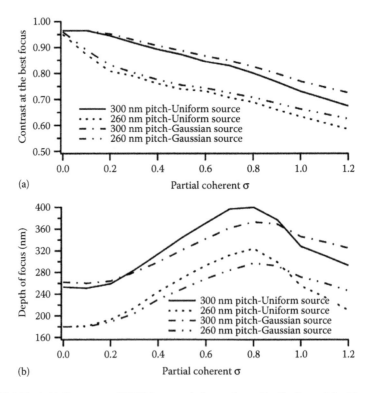

FIGURE 2.7   (a,b) Contrast and DOF for two pitches and two distributions of the illumination pupil versus σ.

Results will vary with many parameters, but these calculations provide some insight into the sensitivity of imagery to the illumination pupil distribution and its perimeter. As an example, the allowable ellipticity of the illumination pupil's perimeter can be extracted from these curves as a change in σ in orthogonal direction.

Along with the intensity of this angular distribution at the reticle, the polarization of this distribution also has great influence.[21] If two wavefronts are polarized at orthogonal polarizations, they will not interfere. In the case of imaging a grating, the interference of two diffracted orders of transverse magnetic (TM) or P polarized light will diminish as they approach 45° incidences in the resist, whereas transverse electric (TE) or S polarized light will interfere at all angles of incidence. These two polarization conditions are shown graphically in Figure 2.8. Also shown is a vector diagram of an azimuthally polarized electric field in the illumination aperture that will have interference for any NA or orientation of dense lines being imaged.

The importance of polarization is more evident with immersion lithography due to the higher angles of interference obtained in the resist.[22] Immersion also reduces the index change at the resist, which improves the coupling of the electric fields into the resist. If light does not interfere at the image, then it is background or stray light that lowers the contrast of the image.

2.2.4.2 Types of Lithographic Features

For diagnostic purposes, patterns printed in microlithography are classified as dense lines, isolated lines, and contacts. Dense lines approximate diffraction gratings and can be found in wiring patterns and memory chips. Isolated lines are considered sparse interconnections when the line separations are greater than three to five times the line width. Contacts are square or circular interconnecting points between layers and are generally surrounded by gaps larger than the contact. Each of these feature types forms a distinctive far field diffraction pattern in the pupil of the projection lens, which can be modified by the pupil distribution of the illuminator.

Figure 2.9 is a scanning electron microscope image of a lithography tool test site that has all three types of features. Unfortunately the optimum illumination pupil distribution varies with feature types. In the earlier days of lithography, simple pupil filling was employed, as shown in Figure 2.4, with partial coherence typically in the range of 0.4–0.7. As lithography progressed, the illumination pupil

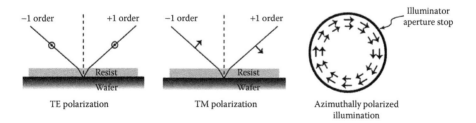

FIGURE 2.8 The electric field vector is pointing out of the page for the two beams polarized TE and forming an image in the resist. If the pupil of the illuminator is polarized azimuthally, then TE is obtained for all orientations of the first orders.

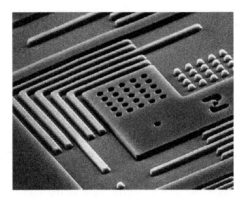

FIGURE 2.9 SEM image of printed features, showing dense, isolated, and contacts. (Courtesy of Rudolph Technologies, Flanders, NJ.)

distribution (source) was optimized for feature type that was being printed. Now, the source and the reticle are optimized simultaneously for printing critical features, which will be discussed further.

2.2.4.3 Off-Axis Illumination

The advantage of illuminators that can vary the illumination source distributions became clear in the early 1990s with the introduction of off-axis illumination techniques.[23–26] Some of these techniques were known decades earlier in the field of microscopy.[27] Consider a diffraction grating illuminated by a single point off-axis as shown in Figure 2.10. The zero order is shifted in the pupil by the illumination's angle of incidence at the reticle.

If the angle of the off-axis source is set so that the zero- and first-order points have the same angle of incidence magnitude at the wafer, then the optical paths along the zero- and first-order path to the wafer are equal at focus and out of focus. Although

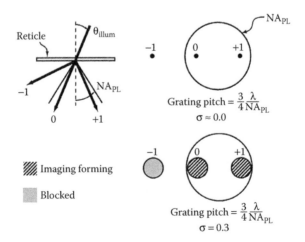

FIGURE 2.10 Pupil fills for off-axis illumination.

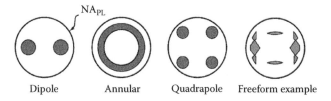

Dipole   Annular   Quadrapole   Freeform example

FIGURE 2.11   Off-axis illumination distributions: dipole, annular, quadrapole, and a freeform example.[28]

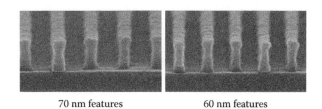

70 nm features   60 nm features

FIGURE 2.12   Features printed with an alternating PSM reticle at $\sigma = 0.3$. These were printed with the 157 nm NA 0.85 projection lens as shown in Figure 2.3b. (SEM courtesy of Selete.)

this method extends the DOF, it has the disadvantage that the zero order has more light than the first. This results in an asymmetric image when slightly out of focus. This is corrected by using another off-axis source point on the opposite side to obtain balance. This is known as dipole illumination, and is ideal for printing a grating (Figure 2.11).

The quadrapole is the optimal illumination for the printing of dense features in two orthogonal directions, known as Manhattan geometries. For features that are not orthogonally aligned to the dipole or quadrapole, the DOF is actually degraded. An annular illumination pupil has no orientation preference and allows more of the illumination pupil to be used. Maintaining uniform irradiance becomes more difficult for the illuminator as the area of the pupil is decreased.[29] This is discussed further in Section 2.3.3. Conventional illumination, unlike off-axis illumination, tends to have more of the zero-order light *not* participating in image formation (Figure 2.12). This light contributes to background and stray light or flare in the image formation. Off-axis illumination can also reduce or increase the influences of aberration in the projection lens.[30,31]

2.2.4.4   Source Mask Optimization

Phase shift masks (PSM) and optical proximity correction (OPC) are modifications of a standard reticle that can increase the process window. The term mask and reticle are interchangeable. The PSM have phase features that do no block, but alter the phase of the wavefront. OPC are unresolvable features near the main feature that diffract in a way to increase the NILS through focus. The performance of these techniques can be enhanced or degraded by the illumination conditions.[28] The PSM masks are often improved by low-sigma illumination and OPC by polarization. The previous discussion was how off-axis illumination can improve the depth of focus of certain feature types, which is the optimization of the source for a feature of the mask. What has developed is the optimization of the source and the mask simultaneously for improving greatly the process window of printing of critical patterns.[32-34] This is known as source mask optimization (SMO). These computer intensive calculations simulate the nonscalar imaging in the resist and the Fourier transform space of the reticle at the aperture stop of the projection lens. The simulations account for many factors such as the aberrations of the projection lens and illuminator relay, polarization characteristics of both, and the limitations of the illumination pupil distribution in intensity and polarization. Telecentric imaging requires that the energy centroid of the freeform source (example in Figure 2.10) be centered. The exposure and the performance of the uniformizer require a minimum etendué of this source. These limitations are why the mask and illuminator pupil distribution are co-optimized. How the illuminators switch between the different source patterns is discussed in Section 2.3.4.

2.3 DESIGN OF ILLUMINATORS

An illuminator is an optical system that collects and shapes the energy from a source, and delivers the target irradiance distributions and polarization to the wafer and pupil of the projection lens. The illuminator may filter the spectrum of the radiation and control the irradiance for exposures. The previous section provided some historical background and a description of projection lithography with an emphasis on the illuminators. While the previous sections defined the design goals, the next sections cover the design techniques and the challenges involved in meeting these goals.

2.3.1 CHARACTERIZING THE PROJECTION OPTICS

2.3.1.1 Optical Invariant, Etendué

The first value calculated is the Lagrange invariant or optical invariant for the illumination, also known as etendué:

$$H_{\text{illumW}} = \left(\sigma_{\max} \text{NA}_{\text{PL}} \phi \right)^2 \tag{2.5}$$

Illuminators with adjustable apertures have varying sigma and the maximum sigma is $\sigma_{\max}$ in Equation 2.5; ϕ is the diameter of the image field. If the product of the solid angle (NA^2) and area are considered a volume, then the etendué of the maximum sigma describes the instantaneous volume of light at the wafer that the illuminator needs to fill. The optical invariant can easily be calculated at the image and pupil conjugate planes. In the case of a pupil conjugate, such as the aperture stop, the NA is defined by the sine of the angle subtended by the chief ray and the optical axis. This is the ray that emanates from the edge of the field and passes through the center of the stop. If the sigma is unity, then Equation 2.5 also defines the imaging invariant.

The imaging invariant is proportional to the space bandwidth product. This product is the limit of information, or the total number of features, that can be imaged from the reticle to the wafer simultaneously or in parallel.

$$N = \pi \left(\mathrm{NA_{PL}} \phi / \lambda \right)^2 \tag{2.6}$$

For example, the capability of lithographic systems operating at 193 nm with 0.75 NA over a 28 mm diameter field can transfer 37×10^{10} features from the reticle to the wafer per exposure. A rotationally symmetric projection lens provides a circular imaging field, but the image area is typically rectangular in microlithography. The regions outside the rectangle may be needed for printing features used to align overlaying images, or for dose detection. For scanning systems, the instantaneous field is smaller than the final image, which is elongated by the scan length and bounded by the reticle size. The key advantage of a scan system is a smaller optical invariant lens. The trade-off is the added complexities of moving both the reticle and wafer during an exposure. The optical invariant of a lens is a crude measure of the difficulty of fabricating it.

2.3.1.2   Transmission Characterization

The illuminator is designed to produce a uniform irradiance field at the wafer, not at the reticle. The image overlay demands requires the projection lens be highly corrected for distortion, which is a variation of magnification over the field. This high correction means the contribution of geometric aberration to nonuniform mapping from reticle to wafer is insignificant. A possible exception is pupil vignetting or obscurations that vary across the imaging field. Each image point at the wafer has an incident angular profile that is subtended ($\mathrm{NA_{PL}}$) by the aperture stop. Ideally, this is a cone whose axis is normal to the wafer and subtends the same NA across the field. This will not be achieved if a surface other than the aperture stop is limiting the subtended NA. A central obscuration in a catadioptic system that is not at the pupil produces similar results (Figure 2.13).

If pupil vignetting is small and outside of the illumination NA, then it has no effect in an open-frame exposure. This is an exposure where there are no features on the reticle, and consequently there is no imaging. If the reticle has features whose dimensions are small enough for diffraction to overfill the imaging pupil, then the transmission across the field varies. This form of nonuniformity to imaging is not included in the specification of an illuminator, since the uniformity is commonly measured with an open frame. The degree of this nonuniformity will vary with sigma and the feature type, size, and orientation. The negative influence of vignetting is not limited to transmission, but may also affect imaging due to spatial frequency filtering.

Nongeometric contributors to the objective transmission profile are either surface or optical medium factors. The transmission from a field point at the reticle to its image point on the wafer is obtained by taking the weighted average of transmission values from a bundle of rays evenly distributed over the pupil.

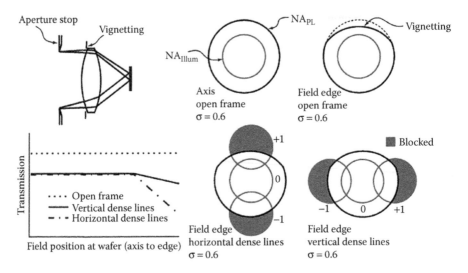

FIGURE 2.13 A graphical representation of vignetting and its influence on transmission. Note that more of the +1 order is blocked for the horizontal than the vertical dense lines, and none is blocked for the open frame. The transmission curves are illustrative.

$$T_f = \frac{\sum_{R=1}^{n}\left[W_R \prod_{S=0}^{m}\left[T_{\text{coating}}T_{\text{medium}}\right]\right]}{\sum_{R=1}^{n}W_R} \qquad (2.7)$$

T_f is the transmission based on a bundle of n rays traced from that field point f, and W_R is the weighting of the ray based on the pupil distribution. T_{coating} is a function of the coating design, coating taper, angle of incidence, and radial position of the ray, absorption, and scatter of the coating. T_{medium} is a function of the path length of the ray in the medium, absorption, and scatter of the medium. T_{coating} and T_{medium} vary over the m surfaces. The weighting function is defined by either the illuminator's angular profile for modeling an open frame, or by the imaging energy pupil distribution, which takes into account the diffraction of the features.

The thin film coatings used as antireflection, mirrors, or beam splitters vary in performance as a function of the light's polarization and angle of incidence on the surface. The polarization state along the ray's path will vary from diattenuation and retardance at the coatings. This is the difference in attenuation and propagating phase between P and S polarization. The performance also varies as a function of the thickness of the layers, which generally are not uniform in thickness over a surface due to coating deposition geometries. As the wavelengths used became shorter, both coating absorption and scatter has had more influence on performance. The more one's optical design software can model these variations, the better. What is left over, the optical designer will be left to cross reference ray information with other models usually based on empirical measurements to calculate each ray's transmission.

Loss of transmission through an optical material is from fluorescence, absorption, and scatter. Fluorescence is only a concern if it is within the actinic bandwidth of the resist. Absorption can heat the elements, changing the index and surface shapes; scatter will lower the image contrast. These factors on transmission are not limited to the optical element material, but also, to a lesser degree, to the medium between the elements. The fluid at the wafer for immersion lithography can have a very significant effect on the transmission profile. Any residual oxygen in the high purity nitrogen needed for 157 nm systems will absorb. Most molecules in the required vacuum of EUV will absorb.

2.3.1.3 Telecentricity Characterization

Telecentricity keeps the magnification of the image constant through focus, which is needed for the overlay demands of microlithography. It is the illuminator that determines how telecentric the imaging is at the wafer.

There is a direction of a ray emitting from a point on the reticle that will be normal (telecentric) at the wafer. This telecentric ray direction can be determined by tracing a ray, normal to the wafer, backward through the projection lens to the point at the reticle (Figure 2.14). There is also a direction of a different ray that passes through the center of the projection lens' aperture stop. This is the chief ray at the edge of the field, and we can call the direction of these rays at any point in the field as the stop-ray direction. The illuminator is designed to have the centroid of the angular distribution along the telecentric-ray direction at any point on the reticle. If this is different than the stop-ray direction, then the centroid will be decentered in the aperture stop of the projection lens.

If the projection lens is removed from the system, the image of the illuminator's aperture stop is formed at the entrance pupil of the projection lens. The entrance pupil is the virtual image of the projection lens aperture stop in the object space, or reticle space. For a rotationally symmetric system, the telecentric entrance pupil distance from the reticle is defined by Equation 2.8, where r_{ret} is the radial distance of the field point at the reticle from the optical axis, and θ_{inc} is the telecentric ray's angle of incidence at the reticle.

$$Z_{ep}\left(r_{ret}\right) = r_{ret} / \tan\left(\theta_{inc}\right) \tag{2.8}$$

The range of Z_{ep} over the reticle is known as the longitudinal pupil spherical of the projection lens. Unlike standard spherical aberration in the imaging of the reticle, this spherical aberration is not detrimental to imaging. The term pupil spherical

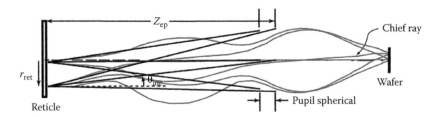

FIGURE 2.14 The lighter rays provide the real path through the projection lens. The darker rays are extended from the object space to the entrance pupil positions.

aberration is used since the image of the pupil shifts with annulus diameter of the field, as opposed to standard spherical aberration, where the image of the field shifts with annulus diameter of the pupil. The illuminator is designed to have its exit pupil to match the pupil spherical of the projection lens entrance pupil. In other words, the image of the illuminators aperture stop must match this focus shift versus field position in order to maintain telecentricity at the wafer.

2.3.2 LAYOUT OF AN ILLUMINATOR

The last section describes the first task in the illuminator design: characterizing a particular projection lens. The designer at this point is faced with the formidable task of providing a layout of the illuminator. As in all large design efforts, it is best to organize by dividing the work into modular sections and list out the many options. Most of the components of Table 2.1 will be covered in detail but are listed here to characterize the decisions needed at the beginning of a design process (Figure 2.15).

TABLE 2.1
Some Design Considerations of an Illuminator

- Pupil Distribution
  Aperture stop: Fixed, iris, removable masks, or polarizers
  Beam shaping: Fixed or zoom system, axicons, freeform sources by switchable engineered diffusers, or flexible mirror arrays
- Spatial Distribution
  Field masking: Adjustable or fixed field stops
  Uniformizer: Kaleidoscope tunnel or rod, fly's eye, and corrective adjustment techniques for uniformity
- Spectral Filtering
  Filter: Narrow band coatings, broadband mirror dumps, and correction techniques
- Power/Dose Control
  Attenuation: Continuous or variable step
  Dose metering: Detectors, sampling percentage, and sampling position
  Shutter: Position
- Interference Concerns
  Speckle busting
  Avoidance of interference fringes
- Polarization Concerns
  Strategy: Minimizing, tailoring, or variable
  Optics: Polarizers, wave plates, mirrors, other
- Mechanical Concerns
  Envelope constraints
  Reticle clearance
  Heat source locations
  Natural frequency
  Center of Gravity

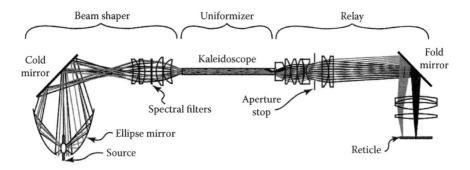

FIGURE 2.15 Shown is a lamp illuminator demonstrating the three sections. The rays are traced backward from two field points at the reticle, showing how the scrambling of the light pipe varies the sampling of the lamp.

In an effort to organize the design process, the illuminator can be broken down into three distinct modules or sections: the beam shaper, the uniformizer, and the relay. The beam shaper collects and shapes the light from the source to optimally fill the uniformizer, and sculpt the light distributions in the pupil of the illuminator. The uniformizer integrates the light into a uniform plane. The relay images this plane to the reticle and controls the final shaping of the incident angular distributions at the reticle. In the case of laser sources, a subsection of the beam shaper is the beam delivery system (BDS), which transfers the source from the laser to the input of the illuminator. This allows the laser to be isolated from the rest of the exposure tool.

One of the first steps is to consider the placement of the components of Table 2.1 within the three sections of the illuminator. For example, a bandpass filter for line narrowing is best placed in the beam shaping section. The spectral transmission of these filters varies with angles of incidence. The integrating function of the uniformizer will ensure that every point at the reticle has the same distribution of spectrum.

An important consideration of the layout is the change of the optical invariant, or etendué, as the light progresses through the system. The printable field at the reticle needs to be overfilled. This not only accommodates some misalignments, but also excludes the penumbra at the edge of the field created by the uniformizer. This requires an illumination field with a greater etendué than the printable field at the reticle.

$$H_{\text{IllumRet}} = H_{\text{IllumWaf}} \left(\frac{\phi_{\text{IllumRet}}}{\phi_{\text{ImgRet}}} \right)^2 \tag{2.9}$$

In Equations 2.9 and 2.10, ϕ is the diameter of the fields. Overfills decrease the irradiance at the reticle and wafer, requiring longer exposures. Simply, any light that overfills is lost. There is also an overfilling of the illuminator's aperture stop. So likewise, the invariant at the output of the uniformizer is larger by a corresponding diameter squared.

$$H_{\text{IllumUniformizer}} = H_{\text{IllumRet}} \left(\frac{\phi_{\text{IllumStop}}}{\phi_{\text{ImgStop}}} \right)^2 \qquad (2.10)$$

The beam shaper does not need to overfill or completely fill the uniformizer's invariant, but that the illuminator's performance is influenced by how this invariant is filled.

2.3.3 THE DESIGN OF UNIFORMIZERS

A multitude of incoherent sources at a sufficient distance from a plane will produce a uniform field of irradiance. A rough approximation to this principle is observing a floor in a room with an array of many lights on the ceiling. The floor will tend to have uniform irradiance. Each point on the floor sees light from multiple sources, and multiple angles, overlapping, summing, and thus averaging on the floor. This is why uniformizers are also referred to as integrators. This simple analogy breaks down with highly directional sources (i.e., spotlights), or when sources with a common angular distribution are aimed to the center of the floor and sum to produce only that common distribution at the floor. If not careful, similar flaws can result with a uniformizer. Kaleidoscopes and fly's eyes are used to create a field of multiple sources. The degree of uniformization resulting from this field is highly dependent on the design of the uniformizer.

2.3.3.1 Kaleidoscopes Uniformizers

The term kaleidoscope is used for a tunnel or a rod. The difference is that a tunnel uses mirrors to reflect the light and a rod uses total internal reflection (TIR). Other common terms for the kaleidoscope include light pipe, light tunnel, and the integrating rod.

$$N_{\text{sources}} \propto \left(\frac{\text{NA}_{\text{IllumUniformizer}}}{n_{\text{Kal}}} \frac{\text{Length}}{\text{Width}} \right)^2 \qquad (2.11)$$

A kaleidoscope creates a new source with each reflection within the tunnel or rod (Figure 2.16). One advantage of a kaleidoscope is that simply increasing the aspect ratio of length to width increases the number of sources. Increasing the aspect ratio by reducing the width requires a larger NA to achieve the same etendué. A smaller width

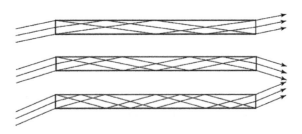

FIGURE 2.16 The angle of incidence is increased to produce 2, 3, and 4 reflections in the 3 kaleidoscopes. Note that the sign of the exit angle switches with each reflection.

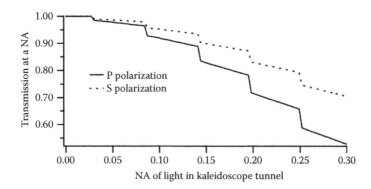

FIGURE 2.17 An example of angular attenuation of a tunnel as a function of polarization and NA. The discrete changes in transmission occur at additional reflections.

forces a higher magnification for the relay section; higher magnification is higher angles of incidence and a more complex optical design. In the case of rods, a smaller width means a higher irradiance in the glass. Increasing the length may be an issue for fabrication and possible space constraints. A high aspect ratio tunnel is more easily fabricated than a rod for structural reasons. In addition, since the index of the medium in the tunnel is unity, more reflections are obtained than for a rod of the same dimensions. An additional advantage of a tunnel, particularly in the deep ultraviolet (DUV) or vacuum ultraviolet (VUV), is the lack of absorption or higher irradiance issues associated with bulk materials. Yet, a significant disadvantage of tunnels is that mirror coatings vary in reflectivity for P and S polarization and with angle of incidence (Figure 2.17).

Attenuation in the reflections of the tunnel alters the pupil distribution for both intensity and polarization. If unpolarized light enters the tunnel, then it is partially polarized by the P and S diattenuation. A linearly polarized source aligned to a rectangular tunnel face is attenuated more in the P direction. This produces an asymmetric transmission function and creates an elliptical source in the pupil of the illuminator. The reflections in the rod do not change the transmission or degree of polarization. Yet, the phase retardance of the TIR and orientation flips produce a variation of polarization states across the pupil with the exception of when linearly polarized light is aligned to the flats of a rectangular rod.

Historically, the degree of polarization was kept low to avoid variations of performance over the field or with feature types. The influence of partial polarization in the illuminator could be characterized with Stokes maps in partial coherent image simulators.[*] As the precision improved and NAs increased, the benefits and possible degradations from polarized illumination could not be ignored. As an example, the projection lens' retardance and change of aberrations with polarization states (Jones-Zernikes) are some of the factors used to simulate and optimize the polarization of the source.[35] If polarizing optics are used to purify the polarization in the relay section, the performance of the uniformizer will be altered by this downstream filtering.

---

[*] Code V Manual, Optical Research Associates, Partial Coherence Calculation PAR.

Rectangular kaleidoscopes are used in lithography without illuminating the full possible imaging field of rotationally symmetric projection lenses. If a complete fill of the projection lens' circular field is desired, then a hexagonal kaleidoscope is used to reduce overfill. The hexagonal kaleidoscope permits more flexibility in the printed field size and shape, but it requires a larger etendué illuminator, which depending on the source may mean less irradiance at the wafer.

A circular cross section kaleidoscope or cylinder provides the least overfill for covering a circular field. Unfortunately, a cylinder rod does not produce a uniform field.[36] A kaleidoscope with flat faces will segment the broad input angular distribution into a multitude of smaller output angular distributions without altering the individual smaller distributions. Reflection off a cylindrical face will alter the smaller distributions since there is focusing in the transverse plane. This is illustrated graphically in Figure 2.27. These images are the results of a powerful means of analyzing uniformizers. In this method, a view is generated looking backward from a single point at the output face of the kaleidoscope over the used NA of the kaleidoscope. The failure of the cylinder rod is clearly demonstrated by the change in the angular distribution or *kaleidoscope pattern* between two points at the exit plane: one on the optical axis and one at the edge of the field (Figure 2.18).

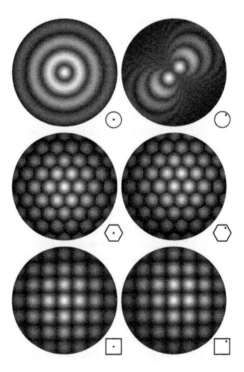

FIGURE 2.18 Pupil maps for two field points and three different types of kaleidoscopes. The position of the field point is indicated on the respective cross sections of the output face. All have the same Gaussian angular and spatial input. The variation in the distributions from center to edge of the cylindrical graphically demonstrates its inability to uniformize.

The generation of these kaleidoscope patterns is very useful in the illuminator design process. They are representative of the final angular distribution, as seen from a point at the wafer plane looking back toward the exit pupil. The perimeter is defined by the aperture stop of the relay and is the collection NA of the uniformizer. Light outside this collection NA influences the uniformity at the kaleidoscope's exit face, but not at the reticle. This perimeter is also the limit of integration for determining the irradiance and centroid shift at the field point.

$$I_r(x_2, y_2) = \int\limits_{NA} A_1(x_1, y_1, \theta_1, \phi_1)\, S_1(x_1, y_1, \theta_1, \phi_1)\, d\theta_2\, d\phi_2 \qquad (2.12)$$

$$C_S(x_2, y_2) = RSS \left[\frac{\displaystyle\int\limits_{NA} A_1(...)\, S_1(...)\, \theta_2 d\theta_2\, d\phi_2}{I_r(x_2, y_2)}, \frac{\displaystyle\int\limits_{NA} A_1(...)\, S_1(...)\, \phi_2\, d\theta_2\, d\phi_2}{I_r(x_2, y_2)} \right] \qquad (2.13)$$

The angular and spatial distributions at the input of the kaleidoscope are denoted by the functions A_1 and S_1. The directional coordinates θ and ϕ are relative to the central ray of the collection NA and therefore also the angular energy centroid shift C_S. The energy centroid of the angular distribution at the wafer determines the image placement through focus. It is the pupil energy centroid that determines telecentricity at a field point. This has been verified empirically, but also confirmed in a modeling study on asymmetric illumination pupil distributions.[37] The relationship between the coordinates at the entrance face (subscript 1) and those at the exit pupil (subscript 2) depend on the cross section type and dimensions of the kaleidoscope. With the many flips due to reflections, particularly in the case of a hexagonal cross section, the difficult analytical solution to these equations makes the use of numerical integration attractive. The mapping and integration can be performed in most lens design programs with nonsequential surface ray tracing. The tracing of the pupil does not need to start or end at the faces of the uniformizer but can incorporate the optical systems anywhere from the wafer to the source.

The needed aspect ratio of the kaleidoscope, or number of reflections, to meet the uniformity requirements is dependent on the angular and spatial distribution at the input face, and also the position of the relay's entrance pupil. This position is where the central rays of the NA cones for all the field points at the exit face of the kaleidoscope converge. If the entrance pupil is at infinity, then the relay lens is telecentric at the exit plane and all central rays are normal to the exit face (Figure 2.19).

The significant difference between the two entrance pupil positions is demonstrated by the slice profiles of Figure 2.20. There is no difference in the angular distribution as seen at the center of the output face for the two entrance pupil positions (Figure 2.20a). As the field point moves from the center (off-axis), the angular distributions vary. The enveloping function is a low spatial frequency fit to the local maximums of the angular distribution or kaleidoscope pattern. If the entrance pupil is at the input face, the enveloping function shifts, and the local maximums are fixed

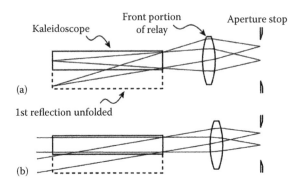

FIGURE 2.19 (a) The input face of the kaleidoscope is imaged to the aperture stop by the relay; in the case of a tunnel, the input face is the entrance pupil to the relay and (b) the entrance pupil is at infinity and rays at a common angle to the face of the kaleidoscope come to focus at the aperture stop.

(Figure 2.20b). If the entrance pupil is at infinity, the local maximums shift, and the enveloping function is fixed (Figure 2.20c).

The enveloping function is defined by the angular distribution entering the kaleidoscope and the collection NA. If the entrance pupil is at the input face as in Figure 2.19a, then the collection NA shifts with field position across the output face. The centroid shift will never be outside the central source in the pattern, where the light passes straight through without reflecting. So, as more reflections are used (i.e., the kaleidoscope increases in length), then the shift in relative NA decreases. The amount of centroid shift is also dependent on the input angular distribution and shifts less with a more uniform angular fill of the tunnel. In the case of the entrance pupil at infinity, the collection NA is telecentric at the output and does not change with field position (Figure 2.21).

Figure 2.22 presents the kaleidoscope performance as calculated by numerical integration of Equations 2.12 and 2.13 for the two different entrance pupil positions and for the different fill distributions shown in Figure 2.21.

In contemporary lithography, telecentricity requires that the centroid be normal to the wafer within a fraction of a percent of the illumination NA. The relay lens can compensate for the centroid shift at the wafer, which is discussed further in Section 2.3.5.3.

If the entrance pupil is at the input, the change in uniformity versus kaleidoscope length is from the variation of the enveloping profile. If the entrance pupil is at infinity, the change in uniformity is from the variation of the kaleidoscope pattern under a stable enveloping function. There is nothing to prevent the designer from placing the entrance pupil somewhere between infinity and the input of the kaleidoscope in order to find a more favorable balance between uniformity and telecentricity.

The advent of off-axis illumination techniques expanded the range of pupil distributions that the uniformizer needs to accommodate. The advantage of the entrance

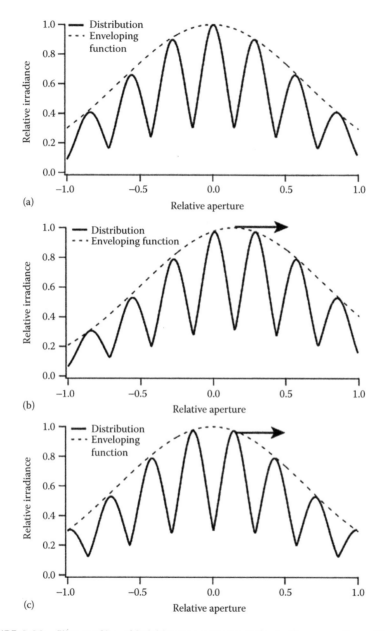

FIGURE 2.20 Slice profiles of kaleidoscope patterns at the pupil as seen from different field points, where (a) is the pattern as seen at center of the output face, (b) is the edge of the output face for the entrance pupil at the input face, and (c) is the edge of the output face for the entrance pupil at infinity.

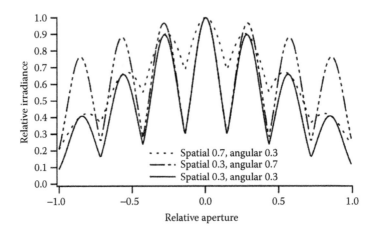

FIGURE 2.21 The angular profiles for the on-axis field point are shown for a variation of angular and spatial fill of the kaleidoscope. All are Gaussians fills defined by the relative height at the edge of the fill.

pupil at the input face is that the kaleidoscope pattern does not shift, and uniformity is less dependent on the spatial fill and the alignment of that fill at the input of the kaleidoscope. Even with the rather smooth fills being analyzed, changing the sampling of the collection NA with aperture stops in the relay can have drastic effects on uniformity performance as shown in Figure 2.23.

Note that the scale of uniformity has changed from the previous figures and that an illuminator needs to achieve uniformity on the order of 1%. The uniformity needs for the various fills and stop conditions require longer kaleidoscopes. This places difficulties on fabrication and space constraints. One solution is to use multiple kaleidoscopes. These can be connected by a relay lens or coupling with a prism. The prism coupling only works for rods; small air gaps are needed between the prism and rod faces to maintain TIR and avoid loss at the interfaces.

2.3.3.2 Fly's Eye Uniformizers

Though a fly's eye uniformizer has many similarities to a kaleidoscope, examining the differences provides useful insight. Instead of each reflection producing an individual source, each lenslet in a lens array produces one. A fly's eye array needs a transform lens, or a field lens, to produce a uniform field at the opposite conjugate of the array of sources (Figure 2.24). A kaleidoscope has its length to transform between the array sources at the unfolded input, and the uniform plane at the output.

The fly's eye is producing a uniform field by overlaying the portions of the segmented input spatial distribution. For a fly's eye, it is the apertures of the lenslet entrance faces that segment the incident spatial distribution. For a kaleidoscope, it is the reflections that segment the input angular distribution. The reflections in the kaleidoscope not only segment, but also flip the segmented distribution with alternating reflections. The alternating flips make the kaleidoscope less sensitive to an

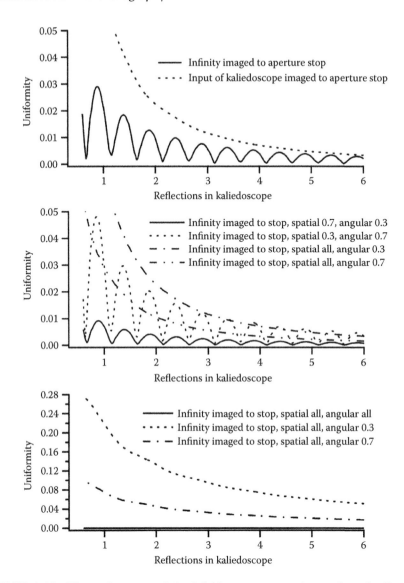

FIGURE 2.22 The performance of the kaleidoscopes versus the number of reflections (length) used. The curves are for the two entrance pupil positions and the different fill conditions exhibited in Figure 2.21.

asymmetric input angular distribution caused by misalignments.[38] A fly's eye array limits the NA by the edge of the lens array, or by an aperture stop placed at the output face of the lens array. The NA of a kaleidoscope is only limited by the edge of the input angular distribution before or an aperture stop in the relay after (Figure 2.25).

$$\theta_f = \frac{n}{2}\frac{A}{T} \tag{2.14}$$

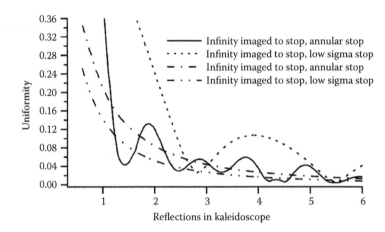

FIGURE 2.23 The uniformity versus reflections when using a low sigma or annular aperture in the relay portion of the illuminator.

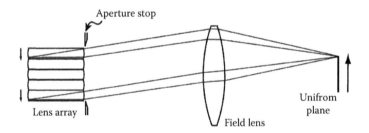

FIGURE 2.24 The general layout of a fly's eye uniformizer.

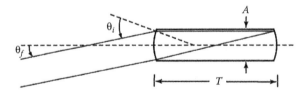

FIGURE 2.25 The design parameters of a single lenslet of the lens array.

$$\theta_i = \left(\frac{n}{n-1} \right) \theta_f \approx 3\theta_f \tag{2.15}$$

The angle of incidence to the fly's eye lenslet is limited by the aspect ratio and index. Any light beyond this angle strikes the side wall of the lenslet, and will be scattered, absorbed or pass through as *cross talk* to another lenslet. All of which are a loss of light or a source of uniformity degradation. A kaleidoscope tunnel has no limitation to the angular input, nor does a rod if the refractive index is greater than $\sqrt{2}$.

The lenslet alone is a transform lens as is the field lens. This means that the incident far field pattern is the image on the output face of each lenslet by the input face's curvature, or the first curvature. Similarly, the second curvature is a transform lens, where the pattern on the input face is imaged to the far field after the lens array. The first and second curvatures are shown in Figures 2.24 and 2.25 on the same piece of glass, as a single lens, but are often broken into two lenses with air space in between. In other words, the single lens array shown can be replaced by two arrays of the same focal length, and separated by that focal length.

Two problems might arise from nearly collimated light being fed into the fly's eye. One is high flux in the glass. The other is degraded uniformity from an interference pattern at the uniform plane from the array of point sources at the opposite conjugate. The same occurs with a focused point on a kaleidoscope input face. A *chirped* fly's eye uses a lens array of varying focal lengths to alter the regular pattern of these spots. This is described in Chapter 5.[39] Just as the finesse of a diffraction grating increases with the number of effective sources, the contrast of the interference pattern will increase with the number of lenslets used. The contrast will decrease with a larger angular fill, which increases the size of the sources in the array.

There is some speckle reduction in a kaleidoscope. The optical path of sources with many reflections is longer than those near the axis. This delay is insignificant for the long coherence lengths of narrowed excimer lasers, but it is significant for lamps and unnarrowed lasers. Efforts have been made to reduce this interference in some fly's eye systems.[40,41]

The field lens is typically two to five elements and has the difficult task of overlaying the images at the uniform plane of each of the lenslet input faces. Since the first curvature of the lenslet is segmenting and creating a secondary source, why is the second curvature needed? The answer is that the second curvature allows a much tighter overlay of input faces. The same edge point of each lenslet input face is collimated and pointed in the same direction by the curvature on the output face. The field lens is a transform lens and will ideally focus all of these lenslet points to one point at the edge of the uniform field. However, the aberration of a lenslet induces an imperfect collimation for its edge point. This aberration is segmented in the pupil of the field lens by all the lenslets. The field lens cannot correct for each individually but can find a balanced correction for the whole array. One means to evaluate the performance of a field lens design, with regard to these aberrations, is to examine the width of the penumbra of the overlaid images at the uniform plane (Figure 2.26).

The kaleidoscope requires the relay and shaping sections to work at higher NAs than in a fly's eye illuminator, so the fabrication and alignment of the optics are more difficult. It also means higher irradiance levels on the optics. The fly's eye uniformizer preserves polarization. There is no flipping due to reflections and the only influence on the polarization is the influence seen by any refractive optic. Unlike a kaleidoscope, a fly's eye can produce a uniform circular field. This is of limited advantage since the packing ratio of circular cross section lenslets loses roughly the same portion of light as hexagonal lenslets will lose in overfill at the reticle. The fact that each additional generated source requires an additional lenslet is a significant disadvantage to the use of fly's eye array. As shown in Figure 2.23, a great

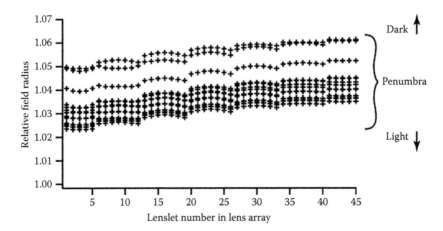

FIGURE 2.26 An analysis of the imaging performance of a field lens and the individual lenslets of the array, by the use of the ray intercepts at the uniform plane.

number of sources are needed to obtain the uniformity. One solution is to use arrays of cylindrical lenses crossed at 90° to each other to approximate spherical lenses. This approximation results in a reduced usable field. It is also possible to create a fly's eye uniformizer with diffractive or refractive microlens arrays.[42] The difficulty with this is zero-order leakage from diffractive lenses or interference patterns from repetitive refractive arrays.

Instead of two refractive lens arrays, a fly's eye can be two reflective arrays, which is advantageous for EUV systems.[43–45] The first reflective array has the freedom to shape each small mirror to alter the shape of the beam for the next array. This allows the EUV systems to produce the arc imaging field and the circular pupil with different shaped mirrors in the arrays. It is also easy to vary the tilt of each of the mirrors so that the beams can be scrambled from the input to the output arrays.

One last note: more than one uniformizer can be used in an illuminator. The benefits of a more uniform angular or spatial input in the final uniformizer were demonstrated with the study of the kaleidoscope.

2.3.4 FROM SOURCE TO UNIFORMIZER: BEAM SHAPING

The brightness of a source is the amount of energy within a unit of etendué. This can be thought of as the number of photons within a *volume* of area times solid angle. As this light travels through an optical system, the brightness can be decreased by such actions as scatter and absorption but cannot be increased; this is the brightness theorem. The term etendué also includes the light's spectrum and polarization. If a source is polarized, the energy within a unit of etendué can be increased by combining light of a secondary source of the orthogonal polarization with a polarizing beamsplitter. Similarly, sources of different spectrums can be combined with chromatic beamsplitters.

The effort to collect light and shape it for the entrance of the uniformizer is very different between lamp and laser sources. A lamp has a high etendué and the uniformizer can only sample a portion of it, which becomes an even smaller when sculpting off-axis distributions. As the laser has a low etendué, so the effort is to shape the optimum source distribution while also filling the uniformizer in a manner that produces a uniform plane.

2.3.4.1 Lamp Sources

The lamp designs of high-pressure mercury lamps have specifically been optimized for use in microlithography. The single spectral lines 436 nm (g-line), 365 nm (i-line), and 248 nm (254 nm low-pressure line) have been targeted by narrow bandwidth projection lens designs. The broader bandwidth designs, such as those used in flat panel printing, cover the g, h (405 nm), and i-lines. These are lower NA systems that are easier to achromatize, or higher NA catadioptic systems. The anode–cathode axis is aligned vertically, otherwise the arc becomes unstable from gravitational pull. The collectable solid angle, in terms of latitudinal range from the equator of the arc, is limited by the shadowing of the anode and cathode. Typically, the light is collected with an ellipsoidal or nearly ellipsoidal shape reflector with the arc at one focus. An elliptical coated metal reflector is made by either plating on to a mandrel or by diamond turning. Another type of reflector is a glass ellipse that is coated with a cold mirror, a dielectric coating that passes the infrared radiation. In terms of thin film coatings, it is preferable to remove the extensive bandwidth of infrared radiation found in mercury lamps by reflecting the shorter usable bandwidth. A common method is with a cold mirror acting as a fold between the ellipse's foci as shown in Figure 2.15.

The simplest lamp illuminator design has the second focus of the ellipse at the input face of the uniformizer. Downsides of this design are the limited shaping of the pupil distribution and the higher angles of incidence at the spectral filters. The system in Figure 2.15 relays the focus of the ellipse to the kaleidoscope, which provides an optimum space for spectral filters. The relay can reshape the angular distribution at the kaleidoscope from the one at the focus of the ellipse by designing each doublet to have different spherical aberration.

A nearly collimated space permits the addition of an axicon or pair of zooming axicons.[46] An axicon is an optical element with a conical shaped surface.

As the axicons separate, the light is shifted away from the optical axis toward the outer edge. The second doublet converts this spatial separation to an angular one at the input of the kaleidoscope. This angular distribution is the enveloping distribution at the illuminator's aperture stop. The light at the edge of the aperture is optimal for an annular or quadrapole aperture stop in the relay section. If not stopped, a soft annular source distribution is created. When the axicons are together, the two elements act as a window, and the light is concentrated at the center of the aperture stop for low sigma applications. One variation of axicons is the use of prisms in a circle to direct light to the four points in the pupil for quadrapole illumination.[47]

Another, more traditional use of axicons is to collapse the central obscuration created by the anode or cathode of the bulb. For this to be accomplished, the first doublet images the ellipse to the collimated region. Element(s) with two axiconic surfaces flipped to those shown in Figure 2.27 will collapse the beam to a smaller aperture. Collapsing the beam with axicons alters what portion of the sources' etendué is

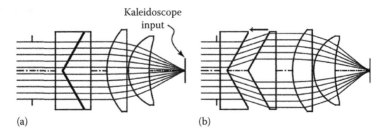

FIGURE 2.27 A zooming axicon pair. When the two are close as in (a) the shape of the distribution in the pupil is unaltered. When the two are separated as in (b) an annular distribution is produced, putting more light at higher NAs.

sampled by the uniformizer. If a pair of axicon elements is used, then the optimum separation can be found empirically when the system is installed.

A radiant modeling of the lamp, based on measurements, is used to optimize the collection from the lamp and the shaping of the distribution. Lamp manufactures provide measured radiance angular profiles as function of latitude from the equatorial plane of the lamp and spatial profiles as a function of position between the cathode and anode. These profiles of the lamp's distributions can be used in conjunction with forward ray tracing to simulate the lamp source in standard optical design programs. A trick to get around most programs' limitation of angular range of launching rays is to trace from infinity and reflect off a dummy parabolic reflector. Altering the ray position relative to the axis of the parabolic reflector allows a ray to launch over the broad range of angles needed. Altering the position of the focus of the parabola varies the spatial origin of the ray. Tracing such rays provides insightful information. The disadvantage of this method is the inaccuracies of the source model; angular exitance variations across the arc are not accounted for. The advantages are the ability to optimize and evaluate designs with a few rays.

Another technique for radiance modeling the source, is to map out the spatial and angular distributions at a plane after the elliptical reflector, and spectral filters.[48] This is based on spatial measurements by scanning a point detector for spatial characteristics and a simple pinhole camera, using an array detector (i.e., CCD) for the angular distributions.

The kaleidoscope pattern of Figure 2.28 is generated by tracing a grid of rays backward from a single point at the exit face of a hexagonal kaleidoscope rod to the plane of spatial and angular profile measurements. This technique does not help with the design of the optics from the lamp to the measurement plane, but does allow the optimization of the second half of the beam shaping section, and the characterization of the rest of the illuminator.

A more stringent method of modeling is to obtain a characterization of the source from hundreds to thousands of CCD measurements made on a goniometer, such as the one that is done by the Radiant Imaging Company.[49,50] The goniometer rotates cameras around the source, obtaining images from all directions. This is used to generate a spatial and angular radiance model. These models are incorporated into a number of illumination modeling programs that trace from the source to defined

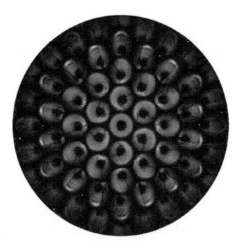

FIGURE 2.28 Combining measurements and ray tracing to generate pupil maps provide a means of optimizing an illuminator. The ellipse and cables of a lamp are imaged through ray tracing of a hexagonal kaleidoscope.

detector surface(s). These programs launch rays from a source in a random manner similar to a Monte Carlo calculation. Millions of rays are traced to generate irradiance distributions anywhere in the optical system.[51–53] The starting coordinates, position and angles, of rays are randomly generated within the volume of the source. The probability distribution of the ray's coordinates or the radiometric weighting of the ray are based on the source measurements. Each ray is traced until they miss all optical surfaces or diminish in radiometric value below some chosen limit. The detector surface is broken into a grid and irradiance profile is generated by summing the radiometric values of the rays within the cells of the grid.

These illumination modeling programs use nonsequential ray tracing.[54] An algorithm is used to determine which surface a ray will strike next, if any. This method of ray tracing allows the flexibility to track scatter, ghost images, and other complicated surfaces. Yet, this method also makes the tracing of millions of rays a slow calculation. Along with speed, the inherent noises of these statistical-based calculations have limited the use of optimization algorithms used in standard optical design program. Tracing only the edge rays is one trick to help.[55] The use of multithreading, or parallel computing, has greatly increased the use of these programs over the past 10 years; it is now possible to even perform Monte Carlo tolerancing calculations using these programs.[56] As computer power and programs improve, these tools will become even more useful in the design process. Improving computers improves designs for microlithography, which improves computers and so on.

Another consideration for lamp illuminators is the optimum position of a shutter for controlling the exposure duration or dose. The advantage of placing the shutter before the uniformizer is that the shuttering duration will be more uniform across the wafer. The importance of this is discussed in Section 2.2.2, concerning exposure latitude. A smaller cross section of the optical beam means a faster and simpler shutter. Some possible positions are the focus of the ellipse and just before a kaleidoscope.

2.3.4.2 Laser Sources

The advantage of a lamp over an excimer is simply cost; lamps are still the mainstay of less demanding lithography. Excimers are the most powerful deep ultraviolet lasers presently available.[57] These pulsed lasers have high gain efficiency in the excitation by electrical discharge over a broad volume. This large volume and relatively few passes through the cavity produce a source with low spatial coherence. The advantage to the lithographer is lower speckle. This can be thought of as hundreds of independent beams where the speckle contrast is reduced by roughly one over the square root of the number of beams. Speckle is reduced further because full exposure requires tens to hundreds of pulses that add incoherently.

Characterizing the excimer beam profiles and modeling propagation requires detailed information from the laser manufacturer, such as spatial and angular irradiance distributions at one or more planes of varying distances from the output of the laser. The measurement of angular distributions or divergence by focusing the beam with a transform lens provides no information on the variation of the divergence of the beam over the spatial distribution. Typical spatial and angular distribution profiles are shown in Figure 2.29.

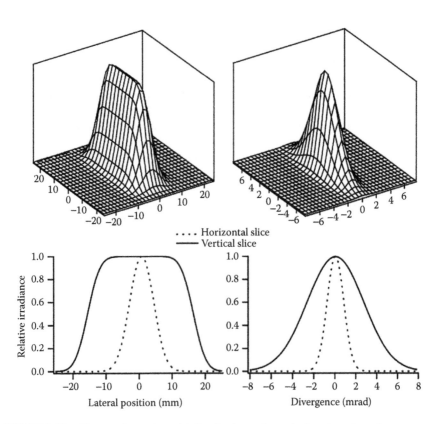

FIGURE 2.29 The angular and spatial distributions of a typical excimer laser beam.

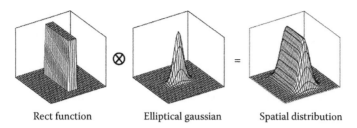

Rect function   Elliptical gaussian   Spatial distribution

FIGURE 2.30   Generation of the spatial distribution by a convolution.

The spatial distribution can be generated for modeling by the convolution of a rectangular distribution and a Gaussian distribution. Assuming that the angular or divergence profile of the beam is constant over the spatial profile, then the spread of the beam due to propagation can also be calculated by a convolution (Figure 2.30).

One means of determining the variation of divergence over the beam is to measure the irradiance distributions at multiple planes that step through a focus of the beam with a slow lens. The divergence distributions at the face of the lens are determined by an iterative recursion modeling of divergences that reproduce the measured distributions at these multiple planes. The divergence variation can also be calculated from a spatial coherence measurement over the whole beam. The van Cittert–Zernike theorem of partial coherence states that the degree of coherence is proportional to the Fourier transform of the intensity distribution of the source, which is the divergence in this case. The spatial coherence over the whole beam can be measured by the contrast of fringes in a shearing interferometer.[58,59] An excimer beam is essentially the summation of many coherent sources that can vary in elliptical shape.

Although a convolution calculation can be used for propagating in free space, a Monte Carlo ray generation is used is to calculate distributions through optical systems. The same convolution used to simulate the spatial profile can be used to randomly generate starting positions for rays in the design.

A BDS is used to relay the laser to the lithography tool, allowing them to be in separate spaces. Relaying the beam with lenses or mirrors that have optical power can often reduce the effective distance of free propagation. The BDS can shape the highly asymmetrical beam to be more symmetrical or square. Most projection lenses require a symmetrical beam in the pupil and some at the reticle conjugate. In order to make a rectangular beam square in cross section, asymmetrical optics or cylinder optics are needed. The divergence of the beam is high in the same direction as the spatial beam is long. Squaring the spatial beam can only increase the aspect ratio of the horizontal and vertical divergence.

Generally, the BDS maintains the etendué of the laser beam, whereas the rest of the beam shaping portion of the illuminator will generally increase the etendué of the beam to sculpt the desired distributions for the uniformizer and aperture stop of the illuminator. Uniformizers perform better when they are filled spatially and angularly. Better yet is to be overfilled to desensitize the alignment of the illuminator. This is contrary to the needs of efficiently producing off-axis source distributions at the aperture stop, so there must be compromise.

One optic that increases the etendué is the zooming axicon. If a nearly collimated beam is incident on the axicon, the beam is expanded around a dark center covering more space, although having little impact on the angular distribution. The enveloping etendué is larger with a dark center. A lens array produces an array of foci from a single beam. The enveloping cross section of this array of foci covers nearly the same area as the initial beam, but the angular spread has increased from the divergence to the NA of the lenslets, thus filling a larger enveloping etendué with dark regions in the spatial extent. A faceted reflector or refractor also redirects light paths to different angles to fill a larger invariant. This can be thought of as a lens array with varying wedges, instead of a constant optical power. One such facet reflector is a micromirror array (MMA). A kaleidoscope tunnel can be thought of as a faceted reflector that does not divide by spatial aperture, but by angle and multiple reflections. If a beam is focused at the input of the kaleidoscope, the output has the same enveloping angular distribution as the input, but now there are dark regions in the angular extent, but spatially the output is uniform. All of these devices increased the enveloping etendué by producing dark regions within it.

Figure 2.31 is an image of a random refractive lens array often referred to as a diffuser screen or simply a diffuser. This is an array of microlenses of random apertures and focal lengths. Etching a ground surface created this one, yet they can be created by more deterministic fabrication methods as well. When coated, the loss in transmission is at most a percent or two. This is due to this diffuser being predominantly a refractive device, with very little diffraction at the edges of the lenslets. Yet, a disadvantage is that a Gaussian angular distribution is generated and there is little latitude to change the angular width of it. The distribution being Gaussian is another example of the central limit theorem found in nature.[60] Some aperture downstream will truncate the Gaussian distribution. If the truncation and Gaussian are symmetric, then the transmission is 1-R, where R is the relative irradiance of the Gaussian at the point of truncation. A smaller R allows greater transmission to trade-off for a less uniform beam. Previously, in the study on kaleidoscopes it was shown that more uniform fills produced better performance. Arguably, all devices that fill a larger

FIGURE 2.31 Microscope image of a refractive diffuser surface.

invariant do so by creating dark regions. This refractive diffuser produces a random array of foci, with dark regions in between. Refractive diffusers may also be fabricated by lithographic processes. One technique developed is to generate precise NAs and apertures to sculpt a targeted far field or angular pattern. The positions and sizes of the lenses are set in a controlled random manner to avoid interference effects.[61] An alternate approach is to generate arrays of varying wedges or prisms to direct the light into the targeted pattern.[62]

Another category of diffusers available is diffractive in nature.[*] These diffusers do not divide the beam into multiple apertures, but increase the invariant by spreading the beam into many diffraction orders to generate an angular pattern. The divergence of the beam incident on these diffusers is convolved with the array of orders to produce a smooth profile. Printing and etching a single-phase level on a substrate can create a two level diffractive optic. The efficiency of such a diffractive is 85% at best, where light is lost to unwanted higher orders. Using more phase levels can improve on this, but decreases the robustness of the fabrication process due to the need to align multiple layers and other concerns. Yet, another technique is to produce *gray* diffusers, where the depth of diffractive features is shaped by various fabrication techniques. A more thorough discussion of this technology is provided in the predecessor to this book.[63]

One issue with diffractive diffusers is the difficulty of suppressing the zero order. This is due to light that has not interfered as intended, which results in a sharp intensity spike at the center of the angular profile. A strong zero order is often the result of fabrication errors in the depth or shape of the features on the diffusers. Refractive diffusers are not sensitive to zero order, but such fabrication errors can alter the dimensions of the angular pattern from the targeted pattern. One issue with refractive diffusers is that the faceted apertures can be a source of speckle. An interesting application of engineered diffusers is the squaring of the excimer beam spatially and angularly. This is accomplished by having one of the diffusers in the transform space of the other, but unfortunately it does increase the etendué.

The beam shaper shown in Figure 2.32 is a double telecentric canonical processor. The darker rays trace the path of the laser beam, as if no diffuser were in place.

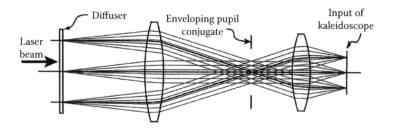

FIGURE 2.32 A laser beam shaper, where the dark rays trace the beam path, if no diffuser was in place.

---

[*] At this time, some fabricators of these various types of diffusers include Diffraction Limited, Digital Optics Corporation (DOC), Heptagon, MEMS Optical, and RPC Photonics.

The lighter rays show the path of light deviated by the diffuser. The first lens transforms the common angles emitting off the diffuser to the spatial distribution at the plane labeled *enveloping pupil conjugate*. This distribution is the convolution of the pattern without the diffuser and the far field pattern of the diffuser. The diffuser's far field pattern is the angular pattern generated from a collimated incident beam. The second lens transforms this spatial profile to an angular profile entering the kaleidoscope, which defines the enveloping function of the kaleidoscope pattern observed at the pupil of the illuminator. The spatial profile of the beam incident on the diffuser is imaged to the input of the tunnel by the reduction defined by the ratio of the focal lengths of the two transform lenses. This reduction increases the NA at the kaleidoscope from that at the diffuser, which reduces the difficulty of the diffuser fabrication, and lowers the irradiance on the optics. Imaging the diffuser to the input of the kaleidoscope allows the diffuser to be switched to another diffuser without changing the spatial distribution at the input of the kaleidoscope. Another modification that is possible is to place a second diffuser at the enveloping pupil conjugate. This will increase and smooth the spatial distribution at the input of the kaleidoscope, which improves the uniformization as demonstrated earlier, but does not alter the enveloping profile in the illuminator's pupil.

A modification of the beam shaper shown in Figure 2.32 is to move the first transform lens in front of the diffuser as shown in Figure 2.33. This places the diffuser in a converging beam. Sliding the diffuser axially alters the laser beam size on the diffuser. This allows a variation in the size of the enveloping pupil distribution without altering the transmission of the system. The range of angles from the diffuser is constant, but the spatial fill is changing, which changes the value of the etendué that is being filled.[64] The defocus of the image at the tunnel from the shift has insignificant effects on transmission, especially if a second diffuser is used at the enveloping pupil

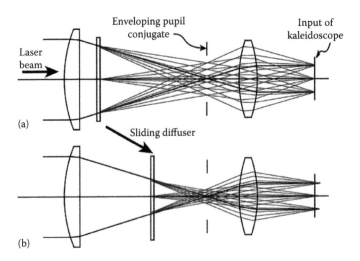

FIGURE 2.33 The technique of sliding a diffuser in a converging beam. The diffuser's position in (a) produces a larger angular fill of the kaleidoscope than in (b).

conjugate. This simple setup allows the pupil fill of the illuminator or the effective source to change without changing the transmission.

A similar change in pupil fill or partial coherence can be accomplished by using diffusers with different angular distributions.[65] Likewise, the diffusers can be generated to produce annular, quadrapole, or freeform source distributions. These can be switched with use of slides or wheels.

Many laser optical systems take advantage of two separated mirrors in the train. Tilting the first mirror sets the beams position on the second, and tilting the second sets the angle, and thus the beam can be aligned to any desired position and angle. ASML and Zeiss took advantage of this with the FlexRay© system by using two separated MMA.[66,67] There are thousands of micromirrors in an array that are individually tilted. The first MMA sculpts the spatial distribution on the second MMA, which sculpts the angular distribution. Having the two arrays allows control over both pupil and field conjugates of the uniformizer. There is a close loop mirror-tilting optimization that converges on the targeted free-form source distribution. A similar scaled down technique is used in ASML's EUV systems, but where the fly's eye mirrors are individually tilted to change the illumination to a limited set of off-axis illumination distrbutions.[68]

In this section we have covered methods of backward tracing, forward tracing, and Monte Carlo ray tracing in and out of standard optical design software. Independent of design method, the beam shaper section of the illuminator allows some creative approaches and the use of novel elements to shape and fill the uniformizer and sculpt the final pupil distribution.

2.3.5   THE DESIGN OF RELAYS

The previous two sections (Sections 2.3.3 and 2.3.4) covered the forward direction from the source through the uniformizer. The characterization backward from the wafer to the reticle was covered in Section 2.3.1, which now leaves the last portion of the illuminator to link them together.

It is the relay section that tailors the final angular (pupil) and spatial (field) distributions at the wafer (Figure 2.34). It is true that the beam shaper sculpts the off-axis and

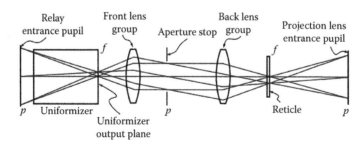

FIGURE 2.34   A diagram of a relay with the pupil and field conjugates labeled p and f, respectively. The design of the relay requires attention to the imaging to the aperture stop, reticle, and projection lens entrance pupil.

free form source distributions, but it is the relay that delivers and varies these pupil distributions over the field of the reticle, and thus the wafer. The stringent requirements at both field and pupil conjugates require the relay design be optimized at both conjugates simultaneously. In the field conjugate, the uniformizer exit face is imaged to the reticle. In the pupil conjugate, the relay's entrance pupil is imaged to its aperture stop and the aperture stop is imaged to the projection lens entrance pupil. Along with this dual imaging, the relay needs to satisfy other design considerations listed in Table 2.1.

2.3.5.1 Uniformization Plane to Reticle

Only the edges of the kaleidoscope are being imaged to the reticle. For a fly's eye system, only the penumbra of Figures 2.25 and 2.26 is being imaged. The exception is when field blades or field stops are used. Field blades provide the ability to print only a portion of the imaging field of the projection lens. The design choice is either to image the blades to the reticle as in Figure 2.35, or to place them in close proximity to the reticle. In the case of a fly's eye uniformizer, there is an accessible uniformization plane that is directly imaged onto the reticle. In the case of a kaleidoscope, blades can be close to the output of rod or tunnel and imaged to the reticle. However, the kaleidoscope needs to be oversized more than required by the reticle invariant in order to fill the blade area uniformly. The other option is to have the blades in close proximity to the reticle. The penumbra of the blades at the reticle is generally larger than with the imaging option, but there is no distortion of the blades from imaging. Pupil aberration increases the penumbra of the image of the blades. Distortion warps the image of the straight blades and alters the shape of the perimeter of the stopped field. If there are no focus blades, the field curvature is only a concern in the outer third of the field. This relieves the design of additional negative lenses to flatten the field. If there are no focus blades, the penumbra is only a concern outside of the field of use. A larger penumbra can always be countered by increasing the mag of the imaging as a trade-off in throughput from the increased overfills at the reticle. Spherical aberration, while influencing the penumbra, also alters the linearity between different aperture stop diameters and NA, or distorts kaleidoscope pattern in the radial directions.

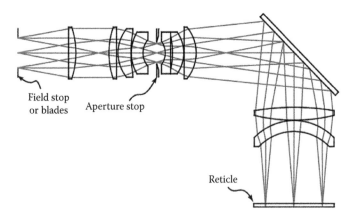

FIGURE 2.35   A double telecentric relay for a fly's eye uniformizer.

Both are acceptable, if the spherical aberration does not vary over the field. Coma and astigmatism, on the other hand, are aberrations that vary over the field and are asymmetric in the pupil. If severe enough, the aperture stop image and kaleidoscope pattern, as seen in the NA space at the wafer, can vary over the field and influence the effective source shape, position, and field uniformity.

The technique of calculating the transmission profile was covered in Section 2.3.1 on the characterization of the projection lens. This provides a target irradiance distribution at the reticle and one of the primary functions of the relay is to match this from the uniform plane of the uniformizer. This is accomplished by deliberate design for distortion in the relay's image of the exit face of the uniformizer. Distortion is a change of local magnification over the imaging field. Barrel distortion, for example, maps an array of equal area squares at the uniformizer to an array of increasingly smaller area shapes from the center to the edge of the reticle. The amount of energy in each square is the same, which means the irradiance is higher at the edge of the field than the center. If the relay is designed backward from the direction of the light, which is often a useful technique in optical design, the designer needs to recognize that distortion switches sign and pincushion distortion will produce hotter irradiance at the edge of the reticle.

Distortion is calculated by ratios of localized object and image heights or by localized object and image NAs. The latter is preferable for an illumination system since the calculation is less sensitive to the quality of the imaging. Also, it allows easier weighting of the pupil distribution that may not be uniform. If the exit pupil is uniform spatially and is Lambertian angularly, then Rimmer[69] points out that mapping the NA in direction cosine space allows a simple calculation of relative irradiance that includes cosine effects for obliquity. The final plane of interest for uniformity is the wafer, which is telecentric, and has no obliquity. The fall off due to obliquity at the reticle only needs to be taken into account when it is included in the calculation of the projection lens transmission profile.

The relative irradiance in a relay can be calculated from the perimeter of the aperture stop as mapped in NA space of both the object and image (Figure 2.36).

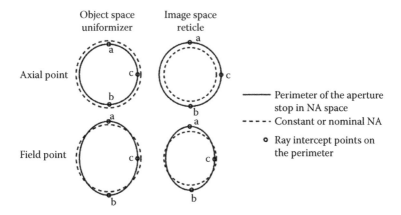

FIGURE 2.36 The perimeter of the aperture stop in NA space as seen in the object and image space.

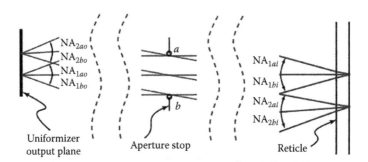

FIGURE 2.37   Rays defining the solid angles in image and object space. The rays to point c are out of the page as shown in Figure 2.36.

The area of the aperture stop in the NA space subtended from a point on the reticle corresponds to the solid angle of the illumination. The ratio of the solid angles in image and object space for an imaged point provides a value of localized irradiance without taking obliquity into account. Relative irradiance or uniformity is calculated by comparing an off-axis to the on-axis ratios. Six rays can provide a useful estimation of the relative irradiance at an off-axis field point for a relay that is rotationally symmetric about the optical axis. The area of an ellipse in the NA is calculated from the tangential rays (a) and (b) and the one sagittal ray (c) shown in Figure 2.37.

$$Ir_{\text{relative}} = \frac{(NA_a NA_b NA_c)_{\text{field-img}}(NA_a NA_b NA_c)_{\text{axis-obj}}}{(NA_a NA_b NA_c)_{\text{axis-img}}(NA_a NA_b NA_c)_{\text{field-obj}}} \qquad (2.16)$$

The accuracy of this calculation improves by tracing rays to multiple aperture stop diameters, and weighting each stop diameter by the pupil distribution. This simple evaluation is easy to incorporate into optimization routines. Targeting distortion in the optimization will alter the tangential NA (a,b) more than the sagittal NA (c), since the field elements have more leverage in altering the tangential direction. Ellipticity of the pupil in NA space alters the partial coherence in the directions defined by the ellipse. The sensitivity to imaging due to the asymmetry of partial coherence is discussed in Section 2.2.4. This asymmetry needs to be monitored and possibly constrained in the design process.

Monte Carlo ray tracing is a very useful means of calculating the geometric irradiance distribution. Under the appropriate circumstances, this method can be used in optimization as discussed in the section on beam shaping. The disadvantages relate to the speed and noise in the calculation, especially when trying to calculate uniformity on a scale of tenths of a percent. A Monte Carlo evaluation is straightforward for the imaging of the uniform plane to the reticle or wafer, and works with any amount of aberration in the imaging. It is good practice to check an optical design by multiple calculation methods.

As with the projection lens, the evaluations of nongeometric uniformity contributions of the relay section need to be accounted for in reaching the target uniformity distribution at the wafer. These are the transmissions of the optical surfaces and optical materials of the relay.

There is a limit to the efficiency of tailoring geometrical contributions by designing in distortion. Variations of the NA across the field, after all, are variations in the partial coherence and effective source. If further shaping of the uniformity is needed, then nongeometric techniques are required. This can be achieved by thin film coatings designed to absorb or reflect light differently over the field. These coatings either vary with the angle of incidence on surfaces near the pupil, or vary spatially for surfaces near the field.

2.3.5.2 Relay Entrance Pupil to Aperture Stop

The section on the uniformizer covers how the entrance pupil position of the relay influences both the uniformization and telecentricity of the system. Designing with the entrance pupil at infinity (telecentric at the uniformizer output face) is optimum for telecentricity at the wafer since the centroid of the pupil distribution remains centered in the aperture stop of the relay for all points in the field. However, placing the entrance pupil at the input face of a kaleidoscope (for a fly's eye, the equivalent would be the output face of the array of lenslets) is advantageous for high uniformity. The centroid shift in the aperture stop versus field position can be accounted for by shifting the relay aperture stop in the opposite direction in the NA space at the wafer (Figure 2.38).

There are two potential flaws with this approach. One is that the center of the diffraction pattern is aligned to the centroid and so the spatial filtering of the aperture stop is not the same in all radial directions across the field. A second flaw to this strategy is that the magnitude of the centroid shift across the field is a function of the pupil distribution. Changing stops or energy distributions will alter the shift and alter the telecentricity at the wafer. This problem can be solved by having more sources in the uniformizers, which requires longer kaleidoscope or more fly's eye lenslets. As the need for illuminators to work with a larger variety of pupil distributions has developed, more secondary sources in the uniformizer have been added to maintain telecentricity and uniformity. Another possible solution is to alter the feed into the uniformizer at both conjugates actively, which can be done with the dual MMA approach of ASML and Zeiss.

Examining the pupil maps of kaleidoscope patterns in the aperture stop is a useful way to analyze the imaging requirements of the entrance pupil to the aperture stop.

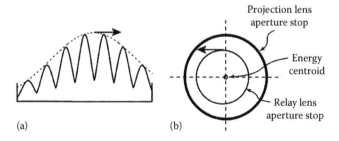

(a)   (b)

FIGURE 2.38 Designing the relay such that the aperture stop (a) shifts in the image NA space with field position can compensate for the centroid shift of the uniformizer (b). Both are shown for a point at the edge of the field.

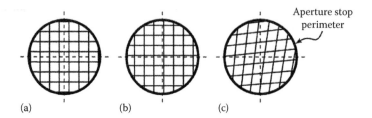

FIGURE 2.39 Kaleidoscope patterns in the relay aperture stop as seen from a single field point. The pattern (a) is for the on-axis field point with no aberration, (b) at the edge of the field for the entrance pupil at infinity and aberration free, or entrance pupil at the input of the kaleidoscope and pupil aberration present, and (c) at the edge of the field with field aberration present.

This is illustrated in Figure 2.19. If the imaging of the input of the kaleidoscope to the aperture stop is aberration free, then the kaleidoscope pattern is centered and identical for all field points. If there is aberration, then the pattern shifts, changes in size, and distorts for different field points along the field. With defocus only, the pattern shifts in proportion to the field point's displacement from the axis but will not distort or change size. A change of the kaleidoscope pattern with field position can alter uniformity by having sources fall in and out of the kaleidoscope pattern (Figure 2.39) with field position.

In the case of imaging the entrance pupil at infinity to the aperture stop, the image of the input of the kaleidoscope is defocused from the aperture stop. In this case, the kaleidoscope pattern in the aperture stop will shift one subpattern for a field point at the edge of the field. Also, sources are falling in and out of the aperture stop with shift in field position, but preferably in a compensatory manner.

2.3.5.3 Relay Aperture Stop to the Projection Lens Entrance Pupil

The aperture stop of the illuminator is imaged to the entrance pupil of the projection lens. This imaging defines the shape and placement of the effective source in the NA space at the wafer as a function of field position at the reticle. An excellent discussion of this topic has been given by D.W. Goodman.[70]

In Section 2.3.1.3, the mapping of the telecentric rays at the wafer to the angle of incidence at the reticle was described. This defines the needed spherical aberration in this imaging to obtain telecentricity at the wafer. This matching of pupil spherical aberration of the illuminator and projection lens can be done by fitting Z_{ep} of Equation 2.8 to a simple power polynomial as a function of r_{ret} in order to target ray intercepts in optimization. The source of these rays should be from the energy centroid in the aperture stop as a function of field position at the reticle. Telecentricity is commonly specified as a magnification change through focus, yet mismatching of the illuminator's aperture stop in the entrance pupil of the projection lens can have some detrimental effects as illustrated in Figure 2.40.[71]

A simpler technique is to design the relay by tracing rays backward through the projection lens. The rays originate at field points at the wafer-filling telecentric cones. If this method is used, there is still a benefit to characterizing the pupil spherical for

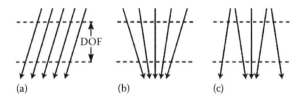

FIGURE 2.40 Representations of telecentricity in the wafer space due to mismatch of the relay stop's image in the projection lens entrance pupil. A lateral decentration of the stop image (a) produces a shift in image position though focus known as *focus walk*, an axial misplacement (b) produces a magnification change through focus, and a mismatch of the pupil spherical (c) between the two optical systems can produce altering distortion through focus.

determining the sign and magnitude of pupil spherical aberration that the illuminator needs to match. Since the projection lens has a high degree of aberration correction, the rays are traced through the projection lens solely to match the low order pupil spherical of the two systems. Whether the backward tracing method is used for optimization or not, it is needed for one of the most complete means of evaluating the illuminator design. This is to create NA or pupil radiance maps for points across the field at the wafer. The energy centroid of a map is the telecentricity. The integration provides the irradiance at the field point. The perimeter of the illuminator aperture stop is the perimeter of the effective source. Calculations from these maps can be used for targeting adjustments to the different imaging configurations of the relay.

2.3.5.4 Other Requirements of the Relay

One of the first concerns for laying out a relay design is achieving the space and clearances required. The space constraints of a lithographic tool often define the position of folds and perhaps even the choice of uniformizer. As discussed in Section 2.3.4, a kaleidoscope needs a larger magnification at the reticle than a fly's eye and therefore requires a longer relay. Near the reticle, clearance is needed for reticle handling hardware that positions and aligns the reticle. An aperture stop may need clearance for either an iris or a means of changing the aperture, such as a slide or wheel. The same means can be used for changing the polarization in portions of the pupil with wave plates and/or polarizers. We discussed methods to shape the source distribution in the beam shaper section. This can also be done with apertures, but only by removing light, which is inefficient.

The wafer is the ideal position for a dose detector, but it is a crowded space. The next best choice is the reticle. The only downside is that the variation of transmission from diffraction overfilling the projection lens aperture stop is not accounted for. This can be calibrated with a separate detector at the wafer on a per reticle basis. The uniform irradiance at the reticle is beneficial to performance and alignment of a dose detector. Every point on a detector at the reticle is collecting light from every point across the aperture stop. This means that the irradiance measured at the detector tracks that of the wafer, independent of the illuminator's pupil distribution. A dose detector can be positioned to sample the overfilled portion of the field. The reticle is also a crowded place, but a simple fold mirror might allow more room to get this sample. Another location for the dose detector is following a leaky fold mirror. The

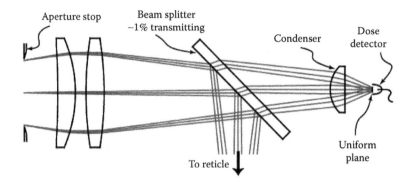

FIGURE 2.41 An example of a dose detector layout, that is, under sampling a secondary image of the uniform field created by a beamsplitter.

mirror in this case is a beam splitter that typically allows around 1% of the light to transmit. An additional optic can create the transform of the pupil (i.e., image of the exit face of uniformizer) at a smaller magnification, which increases the signal strength for the detector (Figure 2.41).

A complex beamsplitter coating might alter the polarization and transmission of the reflected light as a function of angle of incidence. One technique to avoid this is to use a full mirror with a grid of holes in the coating. This is a spatial beamsplitter, as opposed to a dielectric coated beamsplitter. These minute holes pass the light collected by the dose detector. If the mirror is between the pupil and field conjugates, then the shadows of the holes can be insignificant to the illumination uniformity. Disruptive diffraction patterns can be avoided by randomly spacing the holes on the mirror surface.

Designing in adjustments that will be made after the illuminator is installed in the lithographic tool will improve and possibly extend the performance. One method is to change the coatings on specific surfaces, even tailoring them based on measurements made on the tool. A plate window can be inserted as an apodizer substrate. A plate is insensitive to alignment and easily replaced in the field. The uniformity can also be adjusted by designing in lens groups that can be shifted axially (zoom) to adjust distortion.[72] Similarly, telecentricity can be adjusted with moving lens groups, but should not change uniformity.

An illuminator relay is a challenging design problem for a lens designer. Some aberrations are needed and targeted, whereas others are insignificant. The lack of diffraction-limited imaging is countered by the difficulty of finding a balance to the many constraints imposed at the cooptimized pupil and field conjugates.

2.4 SUMMARY

The dominating influence on the process window is the control and uniformity of exposure over the imaging field, along with achieving the maximum depth of focus. The performance of the illuminator with respect to both of these characteristics has been demonstrated. The design of this optical system requires understanding both the performance demands and practical considerations. The needs of obtaining a uniform field and controlling the effective source across the field are not limited to the printing of integrating circuits. The techniques developed for microlithography are applicable to other forms of imaging, such as the inspection of printed wafers and reticles.

ACKNOWLEDGMENTS AND DEDICATION

I am thankful to all the people who helped, but in particular I thank Dawn Revette for the help with the second edition, John H. Bruning, and the late Douglas S. Goodman for their advice and editing of the first edition. Furthermore, I dedicate the second edition to Doug, who was a friend and mentor of mine for many years.

REFERENCES

1. Moore, G.E., Cramming more components onto integrated circuits, *Electronics*, 39(8), 114–117, 1965.
2. Moore, G.E., Progress in digital integrated electronics, *IEDM Tech.*, *Digest*, 11–13, 1975.
3. Mizoguchi H. et al., LPP-EUV light source development for high volume manufacturing lithography, *Proc. SPIE*, 8679, 86790A, 2013.
4. Brandt C. et al., LPP EUV source readiness for NXE 3300B, *Proc. SPIE*, 9048, 90480C, 2014.
5. Peeters R. et al., EUV lithography: NXE platform performance overview, *Proc. SPIE*, 9048, 2014.
6. Schuster, K.H. and Belerf, H., Mikrolithographisches Reduktionsobjektiv, Projektionsbelichtungsanlage und–Verfahren, EP Patent 1 006 389 A2, Carl Zeiss (2000).
7. Phillips, A. and Michaloski, P., Imaging system for deep ultraviolet lithography, US Patent 5,650,877, Tropel Corp. 1997.
8. Webb, J.E., Rich, T., Phillips, A., and Cornell, J., Performance enhancement of 157 nm Newtonian Catadioptric objectives, Proc. SPIE, 4691, 626–636, 2002.
9. Mann, H., Ulrich, W., and Hudyma, R., Reflective projection lens for EUVphotolithography, US Patent 7,199,922, Carl Zeiss, 2007.
10. Pol, V., Bennewitz, J., Escher, G., Feldman, M., Firtion, V., Jewell, T., Wilcomb, B., and Clemens, J., Excimer laser-based lithography: A deep ultraviolet wafer stepper, *Proc. SPIE*, 633, 6–16, 1986.
11. Bruning, J.H., Deep-UV Lithography, US Patent 4,883,352, (1989).
12. Borrelli, N.F., Smith, C., Allan, D.C., and Seward, T.P., Densification of fused silica under 193-nm excitation, *JOSA B*, 14, 1606, 1997.
13. Smith, C.M., Borrelli, N.F., and Araujo, R.J., Transient absorption in excimer-exposed silica, *Appl. Opt.*, 39, 5778, 2000.
14. Burkert, A. et al., Pulse stretcher with variable pulse length for excimer laser applications, *Rev. Sci. Instrum.*, 81, 033104, 2010.
15. Moore, L.A. and Smith C.M., Properties of fused silica for 157-nm photomasks, *Proc. SPIE*, 3873, 392–401, 1999.
16. Burnett, J.H., Levine, Z.H., Shirley, E.L., and Bruning, J.H., Symmetry of spatial-dispersion-induced birefringence and its implications for CaF_2 ultraviolet optics, *J. Microlith. Microfab. Microsys.*, 1(3), 213–224, 2002.
17. Kameyama, M., McCallum, M., and Owa, S., Evolution of wavelength shrinkage in lithography, *Proc. SPIE*, 7281, 2009.

18. Hopkins, H.H., On the diffraction theory of optical images, *Pro. Royal Soc. (London) A*, 217, 408–432, 1953.
19. Wong, A., *Resolution Enhancement Techniques in Optical Lithography,* Tutorial texts in optical engineering; v. TT47, SPIE, 2001.
20. Born, M. and Wolf, E., *Principles of Optics*, Pergamon Press, 1980, Chapter 10.
21. Smith, B.W., Zavyalova, L., and Estroff, A., Benefiting from polarization effects on high-NA imaging, *Proc. SPIE*, 5377, 68–79, 2004.
22. Lin, B.J., Immersion lithography and its impact on semiconductor manufacturing, *Proc. SPIE*, 5377, 46–67, 2004.
23. Reynolds, G., A concept for a high resolution optical lithographic system for producing one-half micron linewidths, *Proc. SPIE*, 0633, 228–238, 1986.
24. Asai, S., Hanyu, I., and Hikosaka, K., Improving projection lithography image illumination by using sources far from the optical axis, *J. Vac. Sci Technol. B*, 9(6), 2788–2791, 1991.
25. Kamon, K. et al., Photolithography system using annular illumination, *Jpn. J. Appl. Phys.*, 30(11B), 3021–3029, 1991.
26. Partlo, W., Tompkins, P., Dewa, P., and Michaloski, P., Depth of focus and resolutions enhancement for i-line and deep-UV lithography using annular illumination, *Proc. SPIE*, 1927, 137–143, 1993.
27. Spitta, E., *Microscopy, the Construction, Theory and use of the Microscope*, John Murry, London, pp. 191–201, 1920.
28. Rosenbluth, A. et al., Optimum mask and source patterns to print a given shape, *Proc. SPIE*, 4346, 486–502, 2001.
29. Yen, A., Partlo, W., Palmer, S., Hanratty, M., and Tipton, M., Quarter-micron lithography using a deep-UV stepper with modified illumination, *Proc. SPIE*, 1927, 158–166, 1993.
30. Partlo, W., Olson, S., Sparkes, C., and Connors, J., Optimizing NA and sigma for sub half-micrometer lithography, *Proc. SPIE*, 1927, 320–332, 1993.
31. Yan, P., Qian, O., and Langston, J., Effect of lens aberration on oblique-illumination stepper system, *Proc. SPIE*, 1927, 167–180, 1993.
32. Tyminski, J. and Renwick, S., The impact of illuminator signatures on optical proximity effects, *Proc. SPIE*, 7140, 71402A, 2008.
33. Socha, R., Shi, X., and HeHoty, D., Simultaneous source mask optimization (SMO), *Proc. SPIE*, 5(853), 180–193, 2005.
34. Matsuyama, T., Nakashima, T., and Noda, T., A study of source and mask optimization for ArF scanners, *Proc. SPIE*, 7274, 727408, 2009.
35. Totzeck, M. et al., How to describe polarization influence on imaging (Invited Paper), *Proc. SPIE*, 5754, 23–37, 2005.
36. Dickey, F., Holswade, S., and Shealy, D., eds., *Laser Beam Shaping Applications,* 1st edn, Taylor & Francis Group, chapt.8, sect. 4, 2006.
37. Stagaman, G., Eakin, R., Sardella, J., Johnson, J., and Spinner, C., Effects of complex illumination on lithography performance, *Proc. SPIE*, 2726, 146–157, 1996.
38. Goodman, D.S., Comments on homogenizers for excimer lasers, *IBM Research Internal Study*, 1991.
39. Dickey, F., and Lizotte, T., eds., *Laser Beam Shaping Applications,* 2nd edn, Taylor & Francis Group, chapter 5, R. Voelkel, "Illumination Light Shaping in Optical Lithography"
40. Ichihara, Y. and Kudo, Y., Illumination optical arrangement, US Patent 5,253,110, (1993).
41. Alyer, A., Illumination source and method for fabrication, US Patent 5,453,814, (1995).
42. Dickey, F. and Holswade, S., eds., *Laser Beam Shaping Theory and Technique,* 2nd edn, Taylor & Francis Group, chapt. 10, D.M. Brown, F. Dickey, L. Weichman, 2014.

43. Komatsuda, H., Novel illumination system for EUVL, *Proc. SPIE*, 3997, 765–776, 2000.
44. Murakami, K., Development of optics for EUV lithography tools, *Proc. SPIE*, 6517, 65170J, 2007.
45. Lowish, M. et al., Optics for EUV production, *Proc. SPIE*, 7636, 763603, 2010.
46. Dewa, P., Michaloski, P., Tompkins, P., and Partlo, W., Variable annular illuminator for photolithographic projection imager, US Patent 5,452,054, (1995).
47. Ogawa, T., Uematsu, M., Ishimaru, T., Kimura, M., and Tsumori, T., Effective light source optimization with the modified beam for depth-of-focus enhancements, *Proc. SPIE*, 2197, 19–30, 1994.
48. Michaloski, P. and Tompkins, P., Design and analysis of illumination systems that use integrating rods or lens arrays, in *International Optical Design Conference*, Rochester, 1994.
49. Rykowski, R. and Wooley, C.B., Source modeling for illumination design, *Proc. SPIE*, 3130, 204–208, 1997.
50. Cassarly, W., Jenkins, D., and Holger, M., Accurate illumination system predictions using measured spatial luminance distributions, *Proc. SPIE*, 4769, 93–100, 2002.
51. Hayford, M. and David, S., Characterization of illumination systems using LightTools, *Proc. SPIE*, 3130, 209–220, 1997.
52. Stevenson, M., Campillo, C., and Jenkins, D., Advanced optical system simulation in a coupled CAD/optical analysis package, *Proc. SPIE*, 3634, 112–122, 1999.
53. Lungershausen, A., Eckhardt, S., and Holcomb, J., Light engine design: The software dilemma, *Proc. SPIE*, 3296, 53–61, 1998.
54. Freniere, E. and Tourtellott, J., Brief history of generalized ray tracing, *Proc. SPIE*, 3130, 170–178, 1997.
55. Cassarly, W. and Hayford, M., Illumination optimization: The revolution has begun, *Proc. SPIE*, 4832, 258–269, 2002.
56. Reimer, C., Monte Carlo tolerancing tool using nonsequential ray tracing on a computer cluster, *Proc. SPIE*, 7652, 76522M, 2010.
57. Jain, K., *Excimer Laser Lithography*, SPIE Opt. Eng. Press, chapt. 5, (1990)
58. Kawata, S. et al., Spatial coherence of KrF excimer lasers, *Appl. Opt.*, 31, 387–396, 1992.
59. Schreiber, H. et al., Applications of a grating shearing interferometer at 157 nm, *Proc. SPIE*, 4346, 1095, 2001.
60. Frieden, B.R., *Probability, Statistical Optics, and Data Testing*, Springer-Verlag, chapter 4, 1983.
61. Sales, T.R.M., Random microlens arrays for beam shaping and homogenization, *Diffractive Optics & Micro-Optics*, 2002 TOPS Vol. 75.
62. Brown, D.R., Highly divergent homogenizers for UV and deep UV, *SPIE*, 4095, 133–139, 2000.
63. Dickey, F. and Holswade, S., eds., *Laser Beam Shaping Theory and Technique*, 2nd edn, chapter 8, J.D. Brown, and D.R. Brown, Taylor & Francis Group, 2014.
64. Michaloski, P. and Partlo, W., Partial coherence varier for microlithographic system, US Patent 5,383,000, (1995).
65. Himel, M., Hutchins, R., Colvin, J., Poutous, M., Kathman, A., and Fedor, A., "Design and fabrication of customized illumination patterns for low-k1 lithography: a diffractive approach", *Proc. SPIE*, 4346, 1436–1442, 2001.
66. Mulder, M., et al., Performance of a programmable illuminator for generation of freeform sources on high NA immersion systems, *Proc. SPIE*, 7520, 2009.
67. Herkommer, A.M., Evolution of illumination systems in microlithography: A retrospective, *Proc. SPIE*, 7652, IODC, 2010.

68. Lowish, M. et al., Optics for ASML's NXE:3300B platform, *Proc. SPIE*, 8679, 86791H, 2013.

69. Rimmer, M., Relative illumination calculations, *Proc. SPIE*, 655, 99–104, 1986.

70. Goodman, D.S., Condenser Aberrations in Köhler Illumination, *Proc. SPIE*, 922, 108–134, 1988.

71. Lee, C., Kim, J., Hur, I., Ham, Y., Choi, S., Seo, Y., and Ashkenaz, S., Overlay and lens distortion in a modified illumination stepper, *Proc. SPIE*, 2197, 2–8, 1994.

72. Michaloski, P., Spatial uniformity varier for microlithographic illuminator, US Patent 5,461,456, (1995).

3 Laser Beam Shaping in Array-Type Laser Printing Systems

Andrew F. Kurtz, Daniel D. Haas and Nissim Pilossof

CONTENTS

3.1 INTRODUCTION

In a variety of applications, including longitudinal solid-state laser pumping,[1–4] fiber coupling for fiber lasers, and line printers,[5–12] laser diode arrays have proven to be very effective light sources when combined with the appropriate beam shaping optics. As laser arrays can be designed in numerous ways, such as with phase-coupled single-mode emitters, uncoupled single-mode emitters, or uncoupled multimode emitters, the output properties of both the individual beams and the ensemble of beams vary dramatically. Inherently, many beam properties, including the output power level, beam profiles and beam propagation properties, beam coherence effects, and the overall device layout, are dependent upon the emitter structure. The optical systems that have been designed to work with these highly nontraditional light sources, which are anamorphic in both physical layout and beam properties, are themselves nontraditional, and typically employ a variety of modern micro-optical components. Within that context, a variety of unique systems[13–15] have been developed for high-power, high-throughput printing applications, where the laser light is transformed into a linear arrangement of individually modulated beams and imaged onto a light-sensitive media. These systems typically combine the design and analysis techniques from classical imaging optics, illumination optics, and Gaussian beam optics into integral wholes. Certainly, many of the design concepts that have evolved to support the laser thermal printing application are applicable to other endeavors, of which laser projection is an example.

Traditionally, in a flying spot laser printer,[16] the emitted laser light is shaped into a beam, swept through space by a deflector (polygon or galvo), and focused onto a media plane by an objective lens (often an F-theta lens). The focused light creates a written spot or pixel (the smallest image element), as input by some modulation means, to create the correct density of each spot, pixel by pixel. The modulation may be applied either directly to the laser *via* a control circuit or indirectly with an external device such as an acousto-optic modulator (AOM). As the laser spot is swept in the line scan (fast scan) direction to produce a line of image data, the media may be moved in the page scan (slow scan) direction to create a two-dimensional (2D) image.

Although flying spot printers can produce high-resolution images in high-throughput applications, when the trinity of requirements for high-throughput, high-resolution, and an insensitive media are encountered, other optical design approaches are required. Efficient printer designs typically use one or more high-power lasers, with the emitted light configured into a series of writing spots. These printers are configured like lathes, where the page scan (slow scan) motion is obtained by rotating a drum (at speeds up to 3000–4000 r/min), which holds the media, and line scan (fast scan) printing is achieved by translating a multitude (12–250) of writing beams in a direction parallel to the axis of rotation of the drum. For example, laser thermal printers have been produced for the graphic arts markets, which simultaneously deliver high resolution (2400 dots/in; dpi) and high throughput (15 proofs/h) while printing on insensitive threshold-effect media (0.2–0.5 J/cm^2). Typically, thermal media works by a dye sublimation or ablation process to create an impression when heat (light) is applied locally. Other exposure mechanisms that are employed by laser thermal media include polymer cross-linking and polymer phase changing within a coated layer. All of these media can provide image resolution as high as 3000 dpi, which is finer than high-quality AgX (silver halide) photographic media.

Multiple channel print-heads are rarely used in flying spot printers because of the complexity of the optical system. More typically, a print-head is moved in a swath-like manner over a rotating drum such that the entire media can be printed on. The media may be held on a flat bed, in an internal drum or external drum. The internal drum printer holds the image-recording media stationary, while the beam is deflected through the large field of a rapidly rotating monocentric optical system. By comparison, in an external drum printer, rotation of the drum carrying the image-recording media provides fast scan operation in one direction, while the print-head moves perpendicularly along the slow scan direction. The laser light propagates along the z direction, perpendicular to both the fast scan and the slow scan directions.

A multichannel print-head for a swath-type printer can be constructed in a variety of ways. For example, the print-head can employ a multitude of laser sources, each coupled into an associated optical fiber, with these fibers arranged linearly across the length of the drum, with each channel responsible for printing a portion of the media.[17] Alternatively, an integrated print-head can be provided,[18] wherein a multitude of laser sources is individually coupled into optical fibers, and the fibers are brought together into a small print-head to form a linear array of sources. Rather than creating an array of writing spots with a series of fiber-coupled lasers, a monolithic laser array can be used as the light source, with the linear arrangement of emitters imaged directly onto the media.[6] In this case, each of the lasing elements must be individually addressable to provide the desired modulation to obtain the various pixel densities. Such systems possess both a lower unit cost and higher light efficiency, as compared with the systems that couple numerous lasers to optical fibers. However, when the individual laser emitters are imaged directly at the media, the printers are susceptible to the failure of even one element in the array, because a pattern error results. Furthermore, these printers are also sensitive to image artifacts caused by thermal and electrical crosstalk within the diode laser array package.

Alternately, the monolithic diode array source can be constructed with each lasing element split into an array of subarray laser sources. Light from each of the lasing elements of a given subarray is combined into beams (one beam per subarray), and the beams are directed onto the media.[7] Each of the subarrays is directly and individually modulated to provide the image data input. While this approach reduces the sensitivity to thermal crosstalk and desensitizes the printer to the failure of lasing elements within a subarray, the optical design is both complicated and constrained to a limited number of channels by the laser structure.

More commonly, printing systems have been designed wherein a laser or laser diode array is used only as a continuous wave (CW)-driven light source with the light incident on a spatial light modulator array, either as individual beams[19,20] or as flood illumination.[9,11,21] The pixels of the modulator array are imaged onto the media as an array of printing spots. The laser source is greatly simplified as it operates at full power without direct modulation. In addition, these printing systems, in which the modulator array is flood-illuminated by light from a laser array, are desensitized to laser emitter failure. In such systems, each emitter is imaged at high magnification to illuminate the entire length of the modulator array. However, as the array direction light emission profiles provided by typical laser diodes have both macro- and micrononuniformities, the resulting illumination can be significantly nonuniform. Without correction or compensation, this illumination nonuniformity at the modulator plane will translate into a variable pixel density with fields on the printed image, usually resulting in an objectionable level of image artifact. Thus, it can become necessary to incorporate uniformizing optics, such as a mirror assembly,[9] a fly's eye integrator,[11] or a light pipe[21] within the system design, such that the spatial light modulator array is properly illuminated.

Computer-to-plate (CTP) laser thermal printers, platesetters, or newsetters, are presently offered by a variety of companies, including Eastman Kodak Company (Rochester, NY), Presstek LLC (Hudson, NH), and Agfa Graphics NV (Belgium), as well as Dainippon Screen Manufacturing Co. Ltd (Kyoto, JP) and Hangzhou CRON Machinery & Electronics Co., Ltd (China). CREO Inc. (Burnaby, BC, CA) acquired Scitex Graphics (Israel) in 2000, and CREO was in turn acquired by Eastman Kodak in 2005.

3.2 PRINT-HEAD/MEDIA INTERACTION

As noted previously, the laser can interact with a medium to form an image in a variety of ways: (1) dye transfer or sublimation from a donor sheet, (2) ablation of dye from a support, (3) polymer cross-linking, and (4) phase change. Laser thermal printing can be used to produce directly viewable images on paper or transparencies. The principal use of laser thermal printing is the production of intermediates in the preparation of viewed images printed by a high-speed press with multicolor inks. Some of these intermediates are: (1) the laser thermal proof for the buyer of the press run to approve the image content and colors, (2) contact films for exposing photoresist-coated conventional printing plates, and (3) directly laser-written printing plates. In such cases, the print quality is dependent upon not only the quality of the laser thermal process but also the surface properties of the exposed and unexposed areas on the laser-written printing plate, ink properties, and *dot gain* of the ink from the laser-written spot.

A laser thermal-written image can be generated in a variety of ways according to the properties of the thermal media and the printer, as well as the application requirements. For example, laser thermal printers have been designed for both continuous-tone, referred to as *contone*, and half-tone printing. Continuous-tone laser thermal transfer printing, which has been previously discussed in considerable detail,[22] is summarized here to provide a better understanding of head–media interactions.

3.2.1 LASER THERMAL MEDIA

One exemplary type of a laser thermal medium is the donor media. The laser thermal donor suited for continuous-tone applications is capable of transferring intermediate amounts of its coating of visible dye to a receiver when stimulated by the incident laser radiation. In particular, low-molecular-weight dyes are vaporized from the donor and transferred to a receiver when a donor is heated to temperatures higher than the dye's vaporization temperature. This process is initiated by absorption of the incident focused light which then induces a large, rapid temperature rise within a small volume of the donor. Controlling the beam irradiance of the laser during raster scanning across the donor can produce well-defined images. No further chemical processing is needed, and all material handling can be performed in room light.

Figure 3.1 depicts a vertical section of a layer of infrared (IR) and visible dyes coated in a binder on the bottom surface of a donor sheet. Matte beads hold the donor at a fixed gap from the receiver and minimize sticking of the heat-softened dye layer to the receiver's surface. The sequence of panels illustrates a laser, scanning from left to right, at four successive times. The dye layer is cool when the laser is first activated in the left panel. Accumulation of the laser energy heats the dye in the second panel, with fastest heating occurring at the beam's focus, where laser irradiance is greatest. Dye nearest the laser captures more heat than dye farther along the beam's propagation path as a result of the dye's attenuation of the light. Heat builds up at the donor's lower surface in the third panel

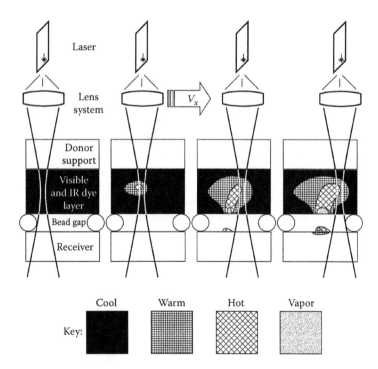

FIGURE 3.1 Schematic representation of laser heating and transfer of absorbent donor.

because insignificant thermal conductivity or convection at the donor–receiver gap causes the donor's lower surface to act as a thermal insulator. If dye at its vaporization temperature acquires its heat of vaporization, then this additional energy transforms the hot molten dye into free gas molecules, or possibly into an aerosol, which can condense onto the cooler surfaces of the receiver and surrounding donor. The fourth panel shows that as dye molecules transferred from the surface are removed, greater depths of the dye layer are uncovered with continued laser irradiance. Thicker deposits of dye can thus be formed on the receiver, providing the potential for modulated image density. Dye condensing on the receiver imparts some of its heat to that surface, warming parts of the receiver. Material in the laser's heat trail on the donor cools back to ambient temperature by diffusion of heat to the surrounding unexposed donor.

The thermal profile within the dye layer can be analytically modeled[22] as a static approximation and, more completely, as a dynamic phenomenon. In the absence of thermal diffusion, a scanning source generates a uniform temperature profile throughout the dye layer, which can be calculated by simply accounting for energy deposition. Multiplying the resulting static temperature by a proportionality factor representing the fraction of heat escaping by thermal diffusion provides an approximation for the temperature in the presence of thermal diffusion. The absorbed light energy elevates the temperature of the donor material above the ambient temperature in proportion to the amount and heat capacity per unit volume $\{\rho c_p\}$ of material capturing that exposure. As part of the static temperature analysis, a uniform temperature approximation can be made, even though the Beer–Lambert law indicates that more heat is deposited on the side through which the light enters than on the side through which the light departs. The uniform temperature approximation is relevant for light incident through the support because, at the outer surface of the dye layer, a mild vacuum is provided, which has an extremely small thermal conductivity compared with the donor materials. This outer surface acts as a thermal insulator, causing heat to build up in the nearby material. This buildup of heat at the outer surface occurs when the beam scans slowly enough so that the time for the beam's center to traverse the beam's full width at half maximum (FWHM) is longer than the characteristic time for heat to diffuse through the thickness of the dye layer. By implication, the dye layer is *thermally thin*. Heat diffuses into the cooler plastic support and away from its location of most intense deposition at its entrance into the dye layer, further leveling the temperature profile across the thickness of the heated dye layer.

The typical laser thermal dye transfer medium is a threshold medium, meaning that the dye is transferred only from regions that retain enough energy from the absorbed laser light to convert the visible dye to its vapor phase.[23] The exposure in excess of threshold constitutes the energy available for transferring the dye molecules from the surface of the donor. In general, less exposure is required by a thinner dye layer than by a thicker dye layer to attain the same image density, because the thin donor contains less material to be heated in its dye layer. Only a small fraction (~3%) of the energy beyond that needed to heat the visible dye to its vaporization point seems to be devoted to transferring the visible dye from the donor to the receiver. The rest of that energy presumably vaporizes volatile molecules, further heats nonvolatiles, decomposes some constituents, and mechanically deforms the donor.

The laser thermal dye transfer process can be considered to obey additive density in the slow scan direction so that the density profiles transferred by successive scan lines add if the donor is permitted to cool substantially to ambient temperature between fast scans.[23,24] This basic property of the laser thermal imaging is a significant departure from the *additive exposure* mechanism of silver halide imaging,[25,26] in which the exposure uniquely determines the image density, regardless of the time sequence for depositing that exposure onto photosensitive materials.

The hottest location in the donor inevitably trails the instantaneous beam center because the location in the donor at the beam center has only received light from the leading half of the beam. The trailing half of the beam subsequently attempts to make that location twice as hot. However, some of the heat deposited by the leading half of the beam has diffused away, reducing the maximum temperature attained. At slow-to-moderate scanning speeds, the hottest temperature occurs at the exposed outer interface of the dye layer as a result of the insulating character of that interface, and in spite of the initial deposition of more heat at the internal interface of the dye layer where it contacts the cooler support.

It should be noted that there are types of laser thermal media other than dye transfer media. In particular, many laser thermal media are fabricated using pigment layers. Laser thermal media can also operate by an *eruptive* process, in which an internal laser-heated layer expands abruptly through the over-layers on a localized basis.

3.2.2 LASER/MEDIA INTERACTION—DEPTH OF FOCUS

The necessity of heating the visible dye to high temperature in order to induce transfer requires that the laser beam be tightly focused. Broadening of the beam, caused by movement of the dye layer away from the plane of best focus, reduces the transferred image density. A depth of focus criterion can be determined by recognizing that the plane of best focus as the location that produces highest uniform density, and not necessarily the plane that exhibits the narrowest waist. In effect, the depth of focus becomes the distance from best focus that causes the image density to drop a specific amount,[27] instead of using the more conventional size of the aerial irradiance distribution[28] (e.g., the Rayleigh range). A representative curve of density versus distance from best focus is shown in Figure 3.2, which indicates that to hold a density variation on the receiver to less than 0.1 D requires the dye layer of the donor to be maintained within a 50 μm range about the plane of best focus throughout the image. This demand justifies the use of a mechanically precise apparatus for translating donors of consistent thickness in laser thermal printers, producing images with continuously adjustable laser thermal transferred density.

The use of larger printing spots hides spot placement errors and increases throughput in thermal media. However, the use of larger printing spots also lowers the system modulation transfer function (MTF) and makes it more difficult to balance or adjust the swath response to hide the intensity and placement errors, because interactions are extended across several neighboring writing spots instead of only to the nearest-neighbor spots. Nearest-neighbor interactions, which increase the output density provided by writing adjacent channels, occur when the optical writing spot is larger than the

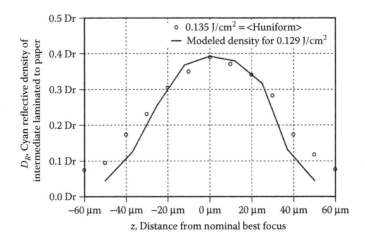

FIGURE 3.2 Experimental and measured depth of focus for a "singles" writing spot on a thin donor.

scan line spacing. Alternatively, it can be stated that the energy from a single laser is used less efficiently than the energy from two or more laser spots that operate in proximity in both space and time when the optical writing spot is more than one raster line wide. The increase in efficiency is significant in determining the overall throughput of the system. The price for this efficiency is the increase in the nearest-neighbor interaction among the channels and a decrease in contrast modulation around the rim of a half-tone dot on a print. Furthermore, use of the nearest-neighbor effect also requires the use of channel compensation and calibration methods[29,30] with a high fill factor, multispot print-head in order to strike a compromise among the densities exhibited by *single*, *pair*, and *triple* line patterns and an entire swath. Thus, the definition of the printing spot size is problematic, as the use of small writing spots increases MTF and makes nearest-neighbor effects more controllable but also reduces the effective depth of focus.

3.2.3 LASER/MEDIA INTERACTION—MULTISOURCE PRINT-HEADS

While there are viable laser thermal printing systems in which multiple laser sources work in parallel but in isolation, there are also many systems where the multiple laser sources are in sufficient proximity to incur laser source crosstalk or laser thermal media crosstalk (the nearest-neighbor effect) or both. Thermal interactions can occur between the temperature profiles produced in the donor by simultaneous exposure with multiple sources. These interactions are separate from thermally mediated influence of one source upon another source's emission, which might occur if multiple sources are mounted upon the same substrate or heat exchanger. In laser thermal printing, the thermal interactions of adjacent writing spots on the donor typically combine in a favorable way, effectively using energy that would be squandered by writing with a single spot. In particular, writing with multiple adjacent spots simultaneously enables some spots to exploit the skirts of their neighboring spots' exposure distributions.

There are several design architectures for multiwriting spot print-heads.[13,14] These can be grouped as those that have a line of adjacent writing spots or pixels spaced apart with a low fill factor or duty cycle (70% or less), and as those that have the adjacent writing spots immediately (or almost) adjacent for a high fill factor, respectively. However, a significant efficiency advantage can be attributed to the high fill factor, multisource laser thermal print-head over the low fill factor configurations. The heating of a strip of media by a neighbor is the predominant advantage exploited by the multiple sources so that the nearest-neighbor interaction requires only ~60%–80% of the exposure required by a series of single beam exposures to generate the same rise in dye layer temperature.

In some systems, a low fill factor print-head can be made to simulate a high fill factor print-head by tilting the print-head relative to the medium.[31–35] The tilting of a print-head relative to the medium is also an effective way to increase the printing resolution without having to fabricate the print-head addressing and pixel structures on a finer pitch. Figure 3.3 depicts a tilted print-head, showing that a high optical fill factor is provided by reducing the pixel pitch from the printing spot pitch (d) to the tilted printing spot pitch (Y). However, tilting the print-head introduces time delays between leading and lagging writing spots. Furthermore, the favorable nearest-neighbor thermal crosstalk effects at the media are somewhat reduced with a tilted print-head because some of that energy deposited by a leading spot escapes before

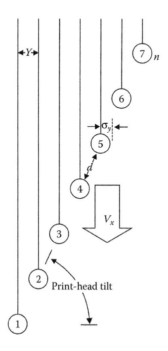

FIGURE 3.3 Print-head tilted to reduce spot pitch (Y), while maintaining a constant Gaussian beam radius.

the successive lagging spot source exposes an adjacent location. Nonetheless, a tilted low fill factor print-head can still provide considerable benefit even when accounting for the difference between leading and lagging thermal distributions. Thermal time dissipation conditions can occur where a significant portion of a preceding neighboring spot's heat from its exposure skirt remains available in the medium when the hottest part of the adjacent lagging beam arrives. Conversely, the second-neighbor interaction is predicted to be insignificant, as very little energy extends from one laser spot to another, two raster lines away.

In the case of the tilted print-head, lagging beams provide little benefit to the leading beams. In particular, the lagging spots are assumed to not contribute to dye transferred by the leading spots, relative to the experimental conditions of beam size, spacing, and thermal diffusion, because the thermal conditions present for a leading spot are unchanged by activating or deactivating the trailing beams. Because the leading laser in the swath of the tilted print-head of Figure 3.3 does not have the advantage of a thermal tail from any other beam, the leading laser transfers less dye if operated at the same power as the other lasers in the print-head, causing the artifact of a light line in the image. This swath edge artifact can be avoided by operating this leading laser as a *dummy laser*[33,35] at a power just below the threshold for transferring dye on its own. The second laser in the print-head enjoys nearly the same advantage of preheating by this preceding *dummy* beam as each subsequent beam does from its first-leading neighbor.

Alternately, interleaving or interlacing[32,36–38] enables the use of a low fill factor print-head with its writing spots spaced farther apart than the desired scan line spacing while avoiding the need to tilt the print-head. Interleaving can reduce interactions of nearest neighbors[25] while maintaining constant print-head scanning speed or step size in the slow scan direction. On subsequent passes, scan lines are written in the gaps between scan lines from previous swaths.

The principal disadvantages of interleaving are as follows: (1) scan lines are not written in their ordinal sequence, (2) part of the leading edge and trailing edge of the scanned area are never filled in, requiring extra scans to complete an image, and (3) production of a desired density requires as much area-averaged exposure as single-source printing because the nearest-neighbor interaction is insignificant as a result of the greater spacing of scan lines exposed during a single swath. This last point of insignificant nearest-neighbor interaction can be an advantage for interleaving because nearest-neighbor artifacts can be avoided,[39] such as difficulty balancing a print-head or accentuation of temperature profile shifts by varying scan line spacing. However, as interleaving lacks the nearest-neighbor interaction, no lasers are sacrificed as dummy lasers.

3.2.4 Laser/Media Interaction—Spot Size, Shape, and Profile

As many of the graphics applications require high-resolution printing in the 2400–3000 dpi range, the writing spot (or pixel) may need to be as small as 2.5–10 μm in size, although other applications have required less resolution (25–50 μm spots). In the array direction, the first-order writing spot size can be derived from the system dpi specification, although other factors, including the light profile, nearest-neighbor effects, and media effects (e.g., dot gain with inks) may alter the optical specification. In the cross-array (fast scan) direction, the motion of the drum and media relative to the writing spot and the resulting smear (convolution) of the light and heat across the media may motivate a cross-array printing spot specification that is different (narrower) than the array direction spot specification. In practice, the writing spots of laser light incident to the media are often round or square in cross section.

The image quality of the printed pixels can also be dependent upon the light profile within the writing spots. Many laser thermal printing systems have configurations in which at least one axis (typically the fast scan, which corresponds to cross-array or cross-emitter) presents a nominally Gaussian profile focused spot onto the media. In the orthogonal (slow scan or array) axis, the light profile presented to the media may also be nominally Gaussian. However, there are several systems, such as those that flood-illuminate the entire length of a modulator array, which present a nominally uniform light profile, at least on a per-pixel basis. In the exemplary instance that the writing spot is Gaussian in the fast scan direction and uniform in the slow scan direction, the motion of the Gaussian beam across the media during the pixel writing smears the fast scan energy (light and heat) into a more uniform profile. The image of a uniform per-pixel light profile at a modulator array will be at least somewhat rounded by diffraction and aberrations, with the net effect that the fast and slow scan per-pixel energy profiles will tend to converge. Systems with nominally uniform writing spot profiles can be expected to experience reduced nearest-neighbor interactions as compared with the nominally Gaussian writing spots.

The multimode lasing behavior intrinsic to many diode laser sources used in laser thermal printing can also affect both the depth of focus and the energy profile from the focused beam, resulting in degradation in the image quality of the printed pixels. In particular, the multiple modes within the converging light beam have a localized spatial coherence that causes that light to focus differently from the main beam, producing hot spots with different best focus positions and extents. If these hot spots persist long enough and are large enough to modify the local temperature profile, then these especially high-irradiance regions within the writing beam will thermally transfer dye more efficiently than that same amount of power spread uniformly over a larger area of donor. As a result, the image densities within the printed pixel can be nonuniform in a random way. This effect can be mitigated by various optical means, by the removal of higher order time variant modes[40] or by optics that homogenize or diffuse the mode structure.[41]

In Section 3.2.2, the point was made that the tightly focused laser beams should coincide with the media within a depth of focus defined by uniform image density criteria. One notable subtlety is that a focused Gaussian laser beam will interact with the media such that the nominal best focus waist position is shifted within the media. Conveniently, a Gaussian beam, normally incident on a dielectric slab, has the same radius at its narrowest waist as it does in the absence of that slab.[42] In Figure 3.4, the profile of the actual beam waist is represented by the thick, continuous curves propagating from left to right through a material with higher refractive index n_2 than n_1 of the surrounding air. The solid arcs are segments of the actual phase fronts for

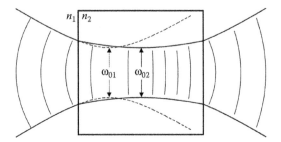

FIGURE 3.4 A Gaussian beam's waist has the same minimum radius ω_0 inside a medium (ω_{02}) as in air for normal incidence (ω_{01}).

that laser beam. The dashed curves are the profile that the beam waist would follow if the higher-index materials were removed.

The narrowest waist radius ω_0 is uniquely determined by the local waist radius ω, the local radius of curvature R of the phase front, and the wavelength of the beam λ, all determined at any single distance z from that narrowest waist along the beam's propagation direction[43,44] in a material of a single refractive index:

$$\omega_0^2 = \frac{\omega^2}{\left\{1+\left(\dfrac{\pi\omega^2}{\lambda R}\right)^2\right\}} \tag{3.1}$$

The location of the waist radius along the direction z of beam propagation obeys a similar equation[43,44]

$$z = \frac{R}{\left\{1+\left(\dfrac{\lambda R}{\pi\omega^2}\right)^2\right\}} \tag{3.2}$$

in which the radius of curvature and wavelength are the only refractive index-dependent parameters. The incident beam has an initial radius R_1, a beam waist ω_1 at the media interface on the air side, and a resultant radius R_2 and waist ω_2 on the side of the medium with the index n_2. Because the radii of curvature are perpendicular to the phase fronts and, therefore, can be considered *rays* of the beam, Snell's law relates their angles with the optical axis. Likewise, because the local waist just outside the medium ω_1 is identical to the local waist just inside the medium ω_2 by conservation of the beam's energy, then $\omega_1 = \omega_2$.

Therefore, the radius of curvature must enlarge by the medium's refractive index upon entering that medium. Thus,

$$\frac{R_1}{n_1} = \frac{R_2}{n_2} \tag{3.3}$$

The relationship between wavelengths of the same beam in two different media is

$$n_1\lambda_1 = n_2\lambda_2 \tag{3.4}$$

The reduction of wavelength in a higher-index medium cancels the enlargement of the radius of curvature. As a result, the beam waist maintains a constant narrowest radius ω_0, but the refracted beam waist plunges deeper into the higher-index medium than in air, by the ratio of the refractive indices.

$$z_2 = \frac{R_2}{\left\{1+\left(\dfrac{\lambda_2 R_2}{\pi\omega_2^2}\right)^2\right\}} = \frac{\dfrac{n_2}{n_1}R_1}{\left\{1+\left(\dfrac{\left[\dfrac{n_1}{n_2}\lambda_1\right]\left[\dfrac{n_2}{n_2}R_1\right]}{\pi\omega_2^2}\right)^2\right\}} = \frac{n_2}{n_1}z_1 \tag{3.5}$$

This offset in the position of the beam waist in the laser thermal media can be neglected to first order for donors that are much thinner than the distance from the focusing lens to the beam waist. However, the offset ultimately can affect the laser thermal print quality, in terms of the quantity of donor that is transferred and the resulting print density.

3.3   LASER SOURCE CHARACTERISTICS

Most of the laser thermal printer design architectures have employed laser diode arrays, although a few have employed discrete laser diodes, either fiber-coupled[17,18,45] or arranged into a multibeam print-head with secondary combining optics.[46] The majority of the new solutions in laser beam shaping for printing have emerged from the efforts to control the beam emitted from the laser diode arrays. Diode laser arrays[47] have been designed and fabricated in great variety, with arrays comprising various mode structures, including single-mode sources, single-by-multimode sources, and multimode sources. Various gain structures have also been used, including gain-guided structures, index-guided structures, and vertical cavity (VCSEL) structures, while the lasers span an emission wavelength range from the visible to the near IR. Laser thermal printers have been undertaken over a narrower wavelength range, from ~800 to 1016 nm, with the high-power IR laser diode arrays spanning a narrower range (~800–980 nm). These high-power laser diode arrays have proven to be generally robust, with lifetimes in excess of 10,000 h being reported.

The typical high-power laser diode array package (see Figure 3.5) has a compact structure, which provides working access to the emitted radiation, as well as quality electrical and thermal mechanical contact. These laser arrays can be configured in air- or water-cooled packages, as well as stacked to form a 2D array. The laser array itself, whose features are too small to discern in Figure 3.5, is a segmented source, consisting of a series of small, distinct light emitters. For state-of-the-art high-power

FIGURE 3.5   An exemplary laser diode array package, the BPC808-40C-622. (Courtesy of II-VI Laser Enterprise, Zurich, Switzerland.)

arrays, these emitters are multimode sources, which are periodically spaced apart over a substantial distance across the device.

In many of the designs, the lasers and laser arrays are operated in CW, as light sources that illuminate a separate light modulation device. The lasers themselves are thus significantly simplified because they lack the ability to directly modulate the emitted light. This is a particular advantage for laser diode arrays. Otherwise, the individual laser emitters, or groups thereof, must be addressed to provide drive signals on a per-channel basis. Providing individually addressable channels or emitters within a laser diode array is not trivial, not only because of the separation in the electrical pathways that is required, but also because optical crosstalk (phase locking) and thermal crosstalk between the emitters can degrade the quality of the modulation. The use of an external modulator or modulator array has several significant advantages, including the opportunity to optimize the modulation performance and to provide a large number of modulation channels (or pixels). Admittedly, an optical system that employs a modulator external to the laser source is inherently more complicated than a system with direct laser modulation.

High-power diode laser arrays are available at various rated output power levels, including 10, 40, and 80 W, with light emission in the near IR (790–980 nm). Historically, these devices were available from companies such as Opto Power Corporation (Tucson, AZ) and Spectra-Diode Labs (San Jose, CA). Presently, high-power IR laser diode arrays are available from numerous companies, including II-VI Laser Enterprise GmbH (Zurich, CH), Coherent Inc. (Santa Clara, CA), DILAS Diodenlaser GmbH (Mainz, DE), and Jenoptik (Jena, DE). The beam properties of the emitted light in the two directions (array and cross-array) are considerably different, relative to numerical aperture (NA), beam profile, beam coherence, and Lagrange (defined in Section 3.3.2).

3.3.1   Laser Coherence and Beam Shape Issues

As an example, an Opto Power OPC-A020 array comprises 19 multimode laser emitters, each 150 μm wide (w), which are spaced apart on a 650 μm pitch (p), for an overall array direction length of 11.85 mm. The large emitter-to-emitter pitch enables these lasers to provide very high output power levels from a very small area, while still minimizing thermal crosstalk effects between emitters. Typically, in the array direction, the light is emitted into a relatively small NA (~0.13 for the A020 laser), but with a very non-Gaussian angular beam profile. As these lasers have a fairly large emission bandwidth ($\Delta\lambda$) of ~3–4 nm for a laser, the output light is nearly incoherent temporally, with a short coherence length C_L:[48]

$$C_L = \frac{\lambda^2}{\Delta\lambda} = 0.2 \text{ mm} \qquad (3.6)$$

Of course, compared with an LED emitter with a 40 nm bandwidth and a coherence length of ~15 μm, these lasers are still relatively coherent. Approximating the array direction beam as an incoherent uniform source, the coherence width (or coherence interval) C_I of these lasers is small compared with the 150 μm array direction emitter width:[49]

$$C_I = \frac{2 \times 0.16 \times \lambda}{NA} \cong 2 \text{ μm} \qquad (3.7)$$

Given the further assumption that the lasers are largely free of filamentation effects, the light emitted in the array direction is spatially incoherent across each emitter. As a result, the overall array direction, near field light profile, across each of the emitters, as shown in Figure 3.6, is relatively flat topped with minimal rippling from intraemitter interference. In many lasers, there is a general roll-off to the array direction light emission profile, with the result that the array beam has both macro- and microspatial nonuniformities, although laser emitters with a generally uniform multimode emission have been reported.[50] In effect, in the array direction, the emitters approximate miniature incoherent or partially coherent extended sources, rather than the point light sources that most lasers approximate. Furthermore, as the emitters are not phase-coupled with one another, the light from the various emitters can be superimposed at the modulator without interference and the resulting rippling in the irradiance profile.

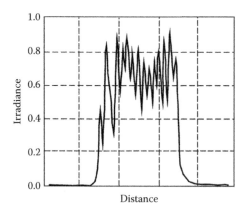

FIGURE 3.6 A measured near-field beam profile for a multimode emitter.

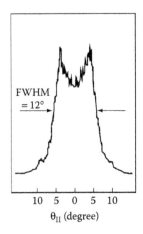

FWHM
= 12°

10   5   0   5   10
θ_{II} (degree)

FIGURE 3.7   A measured far-field energy distribution of a multimode emitter.

Likewise, Figure 3.7 shows a representative of far-field light distribution for the light output by an emitter in the multimode direction. While the angular extent is narrow (NA ~ 0.07–0.14) and the cutoff is sharp, the profile is not uniform but rather has sharp peaks and is often bimodal with pronounced dip in the center. The doubled-lobed structure is evidence of some near-field filamentation within the laser. The filamentation has a dominant spatial frequency that depends upon how hard the device is pumped, microscopic material parameters (linewidth enhancement factor and nonlinear index), and the stripe width. The dominant spatial frequency *Fourier transforms* into two lobes in the far field.[51]

By comparison, the light emitted in the cross-emitter direction is a nominally single-mode (TE_{00}) beam that behaves according to the principles of Gaussian beam propagation. Thus, both the near field and far field light profiles have Gaussian beam profiles, rather than the uniform and bimodal distributions seen for the multimode light. The light emitted in the cross-array direction is output over a much larger NA (~0.63), corresponding to a cross-array direction Gaussian beam $1/e^2$ emitting width of $H = 2 \times \lambda/(\pi \times NA) = 0.85$ μm. In the cross-array axis, the emitters typically have an epitaxially formed wave guiding structure that supports only one laser mode, effectively forcing diffraction-limited output. By definition, the light emitted in the cross-array direction (laser fast axis direction) of any given emitter is coherent over the beam width (typically approximated by the $1/e^2$ emitting width). Although each laser emitter outputs a coherent cross-array direction beam, there is no common phase relationship from emitter to emitter and, thus, the beams can be combined without cross-array phase-coupling and interference. The spatial coherence of the cross-array light is integrated over the entire array, as further modified by the cross-array *smile* and limitations from the optics.

3.3.2   LASER ARRAYS AS LIGHT SOURCES

There are numerous properties of the laser diode array, other than coherence, that must be well understood in order to design a laser thermal printer. In particular, it is of paramount importance that the light source be well matched to the application; otherwise, the system light efficiency will suffer. In this case, the optical spot at the media and the emission characteristic of the laser must correlate.

In classical optical terms, there are two concepts regarding the constancy or invariance of the propagation of light and the conservation of radiance[52] that apply. An optical source is characterized by both its physical and angular extents. The optical extent can be calculated two-dimensionally, as the etendué (product of spatial and angular emitting areas), or one-dimensionally, as the Lagrange (product of the spatial and angular emitting widths).[53] These quantities can be calculated for a source as a whole, or incrementally for any portion of the source, and then integrated over the entire source. In most optical systems, it is desirable that the etendué or Lagrange (the product of the physical and angular widths of the light) be conserved at any and every optical surface within the system to match the value calculated for the original source.

The conservation of etendué (or Lagrange) is closely related to the law of the conservation of radiance. Radiance is a radiometric term that quantifies the optical power density at a given surface, which is estimated as the optical power divided by the product of the emitting spatial area and the emitting angular area (in steradians).[52] Accordingly, per the law of the conservation of radiance, radiance is conserved or constant throughout an optical system, given that light absorption, scattering, and vignetting are ignored.

In most cases, it is sufficient to estimate the optical power and optical power density (irradiance = optical power per area) at key surfaces throughout a system, and otherwise simplify to tracking the etendué or the Lagrange during the design phase. Because etendué needs to be tracked only in the rare nonorthogonal system, Lagrange is more commonly used. Laser diode arrays are atypical light sources because they have dramatically different Lagrange values in the two meridians (array and cross-array), particularly when the laser array comprises at least one row of single-mode-by-multimode laser emitters. As discussed above, in the array direction, the typical laser array emitter is an incoherent or partially coherent source, and the emitter Lagrange can be estimated as $L = NA \times w/2$ (NA $\times$ half width). In the cross-scan direction, each emitter is nominally a coherent single-mode (TE_{00}) Gaussian beam light source, and the Lagrange can be estimated as $L = \omega \times NA = \lambda/\pi$.[54] In use, these values are calculated at the light source and are then tracked through the system, resulting in the desired spot size and light convergence at the target plane. The calculated values need not be absolute, but sufficiently accurate to be useful. As most light sources do not have hard edges to define the spatial or angular extents, but rather fall off according to some gradual profile, it is common practice to estimate these widths at the half maximum or ~10% intensity levels.

A typical laser diode array, comprising emitters that are multimode in the array direction and single mode in the cross-array direction, is depicted in Figure 3.8. Relative to a single emitter, the array direction corresponds to the junction direction beam ($\theta_{\parallel}$). In a single-mode laser, the emitted light in the junction direction is highly Gaussian with a modest divergence. However, in the present case of the multimode emitter, only the modest divergence is retained. The cross-array direction of the laser array also corresponds to the cross-junction direction of the laser diode structure.

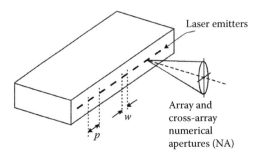

FIGURE 3.8 Emitting geometry of a linear laser diode array.

In the cross-junction direction, the emitted beam is generally Gaussian, with a relatively poor beam quality, and is emitted within a large angular divergence ($\theta_\perp$).

As a representative example, a laser array (the OPC-A020) has array direction emitting width (w) of 150 µm, and an array direction NA of ~0.13, and has an emitter Lagrange of $L = $ NA $\times$ $w/2$ ~ 9.75 µm. The array direction Lagrange for the entire laser array (19 emitters) is, therefore, ~187 µm, assuming that the array direction optics are designed to not see the nonemitting spaces ($p-w$) between the emitters. The latter goal, of collecting light from the laser array without seeing the spaces between the emitters, can be achieved through the use of an array direction-oriented lenslet array. Conversely, if the light from the laser array is collected with a single lens, rather than with the lenslet array, the array direction laser Lagrange would be fairly large ($L = 0.13 \times 11.85/2 = 0.77$ mm). This difference in the collected Lagrange can bear importance relative to the angular width (NA) or optical efficiency as the light encounters both the modulating devices and the printing media or target plane.

In principle, the Lagrange for the laser array in the cross-array direction is equivalent to the Lagrange of a single laser emitter. As the laser emitters nominally output single TE_{00} mode light as a classical Gaussian laser beam, the cross-array Lagrange is estimated as $L = \lambda/\pi$. Thus, an exemplary laser source emitting 830 nm light will have a cross-array Lagrange of ~0.26 µm. It should also be emphasized that the cross-array laser light is also typically emitted into a highly divergent beam (NA ~ 0.5–0.6), in contrast to the low divergence array direction beam (NA ~ 0.1).

Thus, it can be seen that the typical single-mode-by-multimode laser diode array has array and cross-array Lagrange values that differ by approximately 700 times. Although both the array and cross-array Lagranges are much smaller than the Lagrange of a modern LED emitter (L ~ 0.5 mm) or a xenon arc lamp (L ~ 3 mm), the difference between them is, nonetheless, substantial. The entire ensemble of differences in beam emission properties (mode structure, relative coherence, output NA, and brightness or Lagrange) help motivate the very different laser beam shaping optical designs that are employed in the array and cross-array directions of many of the laser thermal printing systems. In general, to design the cross-array optics with accuracy, the standard practices for Gaussian beam propagation must be followed. In many cases where the array direction light is multimode and generally incoherent, classical imaging optics and illumination optics design principles are appropriate. As might be expected, optical systems with these laser diode arrays typically employ numerous cylindrical lens elements so that the array and cross-array light beams can be shaped and directed independently. However, other specialty optics, including *fiber* lenses and lenslet arrays, are also frequently used in the immediate vicinity of the laser array[55] and throughout an entire optical system.

3.3.3 Other Laser Sources

In the preceding discussion, it has generally been assumed that the laser source is a laser diode array comprising a single row (a linear array) of single-mode-by-multimode laser emitters. Certainly, other laser sources are available and have been used in laser thermal printing systems.

As a simple extension to the presumed laser diode array structure, stacked laser diode arrays are also available, in which a multitude of rows of single-mode-by-multimode laser emitters are stacked in the cross-array direction. Stacked laser arrays are used in laser pumping applications, where the ensemble device is an effective source for high brightness IR laser light. The laser array to laser array spacing is relatively large (e.g., ~1.8 mm), which means that the cross-array Lagrange is significantly increased unless cross-array micro-optics, such as a lenslet array, reflective mirror array, or the like, is employed to *remove the spaces* between the arrays.[56]

As discussed further below, addressable single-mode laser diode arrays have also been used in laser thermal printer systems to write on the media directly. Lacking an external modulator, such systems can be configured in a straightforward manner, to image the laser array to the media plane at an appropriate magnification.[6] However, as such a system is susceptible to the failure of one or more laser emitters, alternative system configurations have been developed[7,57] to provide emitter redundancy, even with an addressable laser diode array. In this latter case, a laser diode array comprising a linear arrangement of single-mode array direction emitters was constructed, wherein the laser emitters were arranged in addressable subarrays.[58]

One exemplary IR (850 nm) laser diode array of this type consisted of 160 single-mode diode lasers, which are gathered into 10 groups or channels, separated by a 250 µm space between groups. Each channel or subarray is composed of 16 single-mode diode lasers spaced apart on a 50 µm pitch so that a subarray spans 750 µm. Each of the 10 channels is modulated as an ensemble by its own current driver. The individual single-mode laser emitters are spaced far enough apart that the emitted beams are not coherent (phase locked) one to another. The beams can be combined within the optical system without incurring significant interference effects.

With this type of laser array, the Lagrange per emitter is nominally identical in the array and cross-array directions, at $L = \lambda/\pi = 0.85 \ \mu m/\pi = 0.27 \ \mu m$. However, in the array direction, there is a Lagrange per subarray (16 emitters) of 4.3 µm, as well as an overall potential array direction Lagrange for the entire 160 emitters (spaces removed) of ~43.3 µm. In practice, the array and cross-array beams may not have the same precise emission characteristics (such as NA). In particular, the cross-array light may be index guided and thus more tightly confined, whereas the array direction light may be only gain guided and may emit from a larger active region at a smaller NA. It can again be anticipated that it might be important to use array direction optics (such as a lenslet array) that omits the spaces, rather than optics (such as a single combining lens) that sees the whole laser array. As previously, the array and cross-array direction laser beams possess significantly different properties, thus imparting comparable differences to the laser beam shaping optics employed in the two directions.

3.4 ARRAY AND CROSS-ARRAY OPTICS

In the cross-array direction, the typical design intent is to collect the light from the laser array, transmit the beam through the array direction optics with minimal crosstalk effects, and focus the beam onto the media plane. Typically, the printing application defines the desired spot size relative to the pixels (or *dpi*) required. The incident beam's NA is typically determined by the depth of focus required by the media handling system. In the case where an external light modulator is used, there may be further significant system constraints imposed on the cross-array beam size and NA as part of optimizing the light to the modulation device.

The cross-array optics, which are only shown in the most basic way in Figure 3.9, typically include a *fiber lens* and one or more cross-array lenses. The *fiber* or *rod* lenses are typically microlenses with very short focal lengths (~100–200 μm), which both allow spherical aberration to be minimized, enhanced by special corrections designed into the lenses. For example, the fiber lens or fast axis collimator (FAC) can be a gradient index cylindrical microlens from Doric Lenses Inc., an aspheric microlens from LIMO GmbH, or a hyperbolic cylindrical microlens from Blue Sky Research, which provide further aberration control.[59] These lenses are typically employed to collimate the cross-array light beams, although they can be used to reduce the laser divergence to a lesser extent (to less than full collimation). Use of a rod lens also allows the high NA cross-array light beams to be quickly controlled before the propagating light beams become awkwardly large. Moreover, the cross-array direction optics can also be packed in close to the laser array, helping to reduce the overall size of the system. Concepts for self-registering microlenses have also been developed to enable light collection from a laser array.[60]

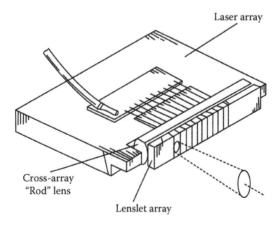

FIGURE 3.9 Laser diode array package with integrated cross-array and array optics. (From U.S. Patent 5,212,707.[55])

The array direction lenslet array or slow axis collimator (SAC) is a monolithic array of cylindrical lenses, molded from fused silica or Ohara S-TIH53 glass. The lenslet array is most commonly anamorphic and operates only in the array direction light, but arrays with integrated cross-array direction power are also available (e.g., the LIMO GmbH FAC-SAC collimation modules).

3.4.1 LASER SMILE ERROR

Most simply, light transfer from the laser array to a modulator array or the media plane can be accomplished by imaging the emitting surface to the target plane with the two planes conjugate to each other. In the direct printing systems, the laser emitters are imaged directly onto the media plane,[6] and thus the cross-array laser characteristics directly affect image quality. In other systems,[9,11,21] in which the emitted beams are intermingled, the cross-array laser characteristics can have a less direct impact on image quality but are still significant with regard to system light efficiency, modulation contrast, and depth of focus.

Generally, the cross-array laser beam shaping optics comprise an arrangement of cylindrical lenses, except that the problem of laser array *smile* (see Figure 3.10) is an added complication. In the prior discussions of the cross-array properties of the laser diode arrays, the laser emitters were considered to be identical in emissive properties, having a nominal emitter Lagrange of λ/π and located in a perfect linear arrangement. However, as can be seen in Figure 3.10 (which is a near field projected image), the laser emitters are typically not located in a perfect linear arrangement, but are offset from the ideal, usually with a slowly varying pattern error spanning a few microns. The net effect of this diode laser fabrication error is that not all laser emitters on the same substrate emit beams from the same plane. Most simply, smile is a lack of straightness in the array over its length.

Certainly, an uncorrected smile error can manifest itself as a printing artifact or inefficiency. In the printing systems where the laser emitters are mapped directly[6] or indirectly[9,11,21] to the media to form printing spots, smile can create directly viewable artifacts. Most directly, the inaccuracy of the diode laser light source positions is projected onto the medium as an arch in the sequence of writing spots, which appears as a smile. Alternatively, a smile error creates banding when a tilted printhead is used, as the distance or pitch between emitters (writing spots) is no longer constant. By comparison, in many of the printing systems involving flood illumination of a modulator array, the emitted beams are intermingled,[9,11] as an uncorrected smile error can cause the cross-array Lagrange to be enlarged significantly. In effect, several microns of smile error across the laser array can increase the cross-array Lagrange by 10 times or more from the nominal Gaussian λ/π. This increase can

FIGURE 3.10 A line of printing spots exhibiting laser smile error.

significantly impact coupling through a modulator (modulation contrast, efficiency) and the spot formation at the media (depth of focus, spit size).

Laser array manufacturers have improved their ability to control or reduce smile error, and laser bar technology based on hard-soldered CuW submounts can deliver *out-of-factory* smile error of 1 μm or less, and most manufacturers offer low-smile packages. For example, devices with smile values of only 0.3–0.4 μm over a 5 mm bar length are reasonably available. Laser array smile develops in the fabrication process as the mechanical stress across the array changes. Most typically, smile is exhibited as an arched pattern of emitters, although other more complicated pattern errors ("s" or "w") can occur. Assuming that the supply of laser arrays have smile errors that follow a simple "s"- or "c"-shaped arc, the smile error can be compensated for by bending the fiber lens to a matching arc.[59,61] The resulting line of collimated laser beams travels in parallel to the optical axis but with slight offsets. Cylindrical power crosstalk or skew ray effects are negligible because the bent fiber lens has a few microns displacement over a 10–15 mm laser array length.

3.4.2 LASER SMILE CORRECTION

Despite these gradual improvements in laser array technology, the system impact of smile error is sufficiently damaging that numerous means[5,61–64] for smile correction have been developed. Smile correction at or near the fiber lens is necessary for the optical systems that image intermingled beams,[9,11,21] as the beam deviations incurred by the emitter offsets are overlapped and no longer separable once the beams have been intermingled.

There is greater freedom to develop smile correctors for optical systems that image the laser emitters, either directly[45] or indirectly,[7,57] to the media. Exemplary smile correctors have also been developed that can correct most, if not all, of the smile in an array, independent of the shape of the smile error across the array. One such smile corrector,[62] shown in Figure 3.11, comprises a series of glass plates inserted into the optical path in collimated space. By properly tilting each of the plates, the position of that beam at the pupil is shifted to correct the smile at the medium's plane.

Other exemplary smile correctors allow individual cross-array lenses[62,63] or mirrors[64] to be adjusted on a per-emitter basis, adjusting the various beams into coplanar or common optical planes. These latter methods are limited to the practical

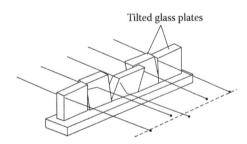

Tilted glass plates

FIGURE 3.11 Laser smile correction using an array of tilted plates. (From U.S. Patent 5,854,651.[62])

dimensions of the beam correcting offset mechanisms, and thus work well with laser arrays that have laser subarrays[58] or large area multimode emitters positioned at large pitch distances. Alternatively, optical aberrations, either natural or induced, have been used[5] to broaden the apparent cross-array emitter size, thereby desensitizing the system to smile error but at the cost that the cross-array Lagrange has been increased.

Fast axis collimators, as manufactured, can suffer from imperfections at the submicron level. Taken together, misalignment errors, bar smile, and facet curvature introduce optical errors including wavefront distortion, wavefront curvature, and pointing error. However, these errors can be compensated with refractive phase plates before the wavefront forms a caustic. In the simplest form, when only pointing (smile) correction is required, the phase plate can be made in the form of either a microprism array or a continuously varying wedge surface. For example, PowerPhotonic Ltd (UK) offers a range of smile correction phase plates[65] that correct for imperfections in fast axis collimation using continuously varying wedge surfaces that correct for the postcollimation pointing of light from each emitter. These phase plates correct for a parabolic-type smile error, expressed as P-V pointing error of 0.5–3.0 mrad, and are available in increments of 0.5 mrad. Also, these correction phase plates (see Figure 3.12) are available integrated into the SAC, providing a single element to be aligned during the laser bar microlensing process.

3.4.3 Operating Dependence of Laser Smile

Notably, laser smile also depends upon external conditions, including device mounting and temperature. In the latter case, smile correlates with laser current and smile can increase by multiples of 100 nm, going from threshold to a nominal current. Lasers can be scanned and binned depending upon the type of the smile (C, S, or W) and the P-V smile value. Each emitter can then be assigned a single average phase compensator.

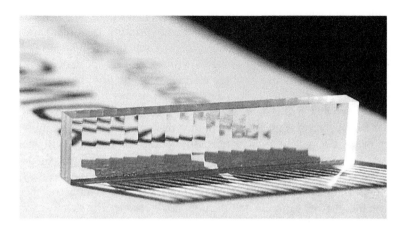

FIGURE 3.12 Phase plate smile corrector. (Courtesy of PowerPhotonic Ltd., Fife, UK.)

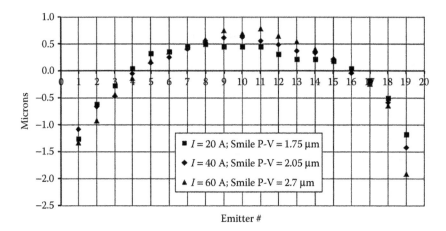

FIGURE 3.13 Measured fast axis laser smile versus emitter position and laser current. (Courtesy of Kodak Canada ULC, Burnaby, BC.)

In addition, in printing applications, laser diode bars often operate at different currents depending upon the sensitivity of the printing plate that is imaged. It is important to note that the smile of the laser bar has a temperature dependence. As shown in Figure 3.13, in the case of conductively cooled bars, laser smile can correlate with the laser current. Thus, increasing the laser current will increase the output power, but it will also increase the spot size in the fast axis direction through increased bar smile.

3.4.4 OPTICAL ALIGNMENT ISSUES

Typically, laser thermal printing systems have rather different alignment tolerances for the constituent optics in the array and cross-array directions. This generality can break down, depending upon the system configuration.

In cases where the laser array is directly addressed and directly imaged to the media with largely spherical optics, the spot-to-spot pitch will be largely determined by the laser array fabrication process, and the array printing (slow scan) and cross-array printing (fast scan) directions will generally see similar optical tolerances. The fast scan motion of the drum or media will tend to blur the spot in the cross-array direction, potentially easing those tolerances. If the print-head is integrated as a closely packed array of spots, then there will be swath-to-swath positioning tolerances, driven by the need to minimize a visual artifact known as banding.

Should the laser array employ laser emitters that are single mode by multimode, the laser beam shaping optics will typically be anamorphic, and will include a variety of cylindrical optical elements. Relative to the laser arrays, the single-mode direction typically corresponds to the cross-array direction, while the multimode emitter orientation corresponds to the array direction. The asymmetries naturally align, such that the array direction of the laser array corresponds to the array direction of the printer. It can be understood that these systems[9,11] are generally very sensitive to crosstalk of the array beam into the cross-array direction. For example, a slight θ_Z (about the optical axis) tilt of an array direction cylinder lens will transfer optical power into the cross-array direction, which could cause dramatic changes in the beam propagation (spot size in critical planes), as well as a rotation effect. Likewise, a slight θ_Z tilt of a cross-array direction cylinder lens will likely have a more dramatic effect on the cross-array beam propagation than on the array direction light propagation.

Not surprisingly, although the rotational alignment tolerances for the cylindrical lens elements can be as much as ~10 arc/min, tolerances of ±2 arc/min are not unheard of. By comparison, the spatial mechanical alignment tolerances for the cylinder lens elements can be a relatively relaxed ±50 μm, as compared with the submicron positional tolerances seen in other laser applications.

3.5 DIRECT LASER TO MEDIA SYSTEMS

As a design architecture for a multichannel laser thermal printer, the system configurations in which a series of discrete laser sources correspond directly to a series of laser printing spots provide potentially the most compact designs with the lowest cost structures. In many cases, the laser emitters are directly addressable with image data, and the laser emitters are mapped to the media. Alternatively, the laser beams can be routed to the media indirectly, by means of optical fibers. The laser sources can also be equipped with individual external modulation devices before being mapped to the media. In any of these cases, the resulting multichannel laser system is vulnerable to the degradation or failure of individual laser sources, although some of the systems offer the potential to field-replace failed lasers. Although this discussion is targeted at linear printing systems, primarily using one-dimensional (1D) laser arrays, system concepts have also been developed for writing with 2D arrays of laser diode sources.

3.5.1 LASER EMITTERS MAPPED TO THE MEDIA

Aside from *contact printing,* wherein an addressed array of light sources would print image pixels without any intervening optics, the simplest and most obvious architecture for providing a laser thermal printer is to image an array of directly addressable laser diode emitters to the media plane. In one exemplary system[6] developed by Eastman Kodak Company (shown in Figure 3.14), a linear array of laser diodes is monolithically integrated on a single semiconductor substrate, but is individually addressable and independently controllable from each other. The simplified optical system uses spherical optics to image the laser array directly onto the media, such that each laser diode corresponds to a pixel in a line on the print. Heat is generated on a per-pixel basis, resulting in dye evaporation and transfer from the donor to the receiver. A similar system[66] is described for high-resolution electrostatic printing, in which an array of surface emitting lasers is imaged in magnified fashion onto the media. Both of these systems are susceptible to a variable response from laser emitter to laser emitter across the laser array. However, while emitter variability can be potentially corrected for *via* a calibration process, any outright emitter failures would provide likely uncorrectable line artifacts.

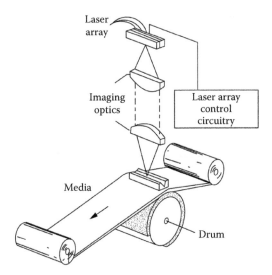

FIGURE 3.14 Printer configuration with an addressed laser array imaged directly to the media. (From U.S. Patent 4,804,975.[6])

Many of the optical systems discussed below are motivated by the perceived need to desensitize the printing system to laser emitter variability or failure. However, even within this class of simplified systems that use directly imaged laser arrays, source failure was taken seriously. In one case,[19] the multitude of laser emitters are driven as a group, and the light from the entire array is directed onto the media to form a single writing spot. The emitter redundancy is provided at the cost of a low printing throughput (speed and resolution). More recently, it has been proposed[67,68] that the diode laser arrays are sufficiently robust that laser emitter to media-mapped printing systems can be undertaken without the need for redundancy.

An exemplary optical system,[67] developed by Scitex Digital Printing Inc. for an image setter (shown in Figure 3.15), collects light from an addressable laser array,

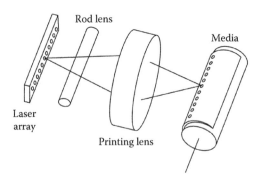

FIGURE 3.15 Directly addressed laser array printer with anamorphic optics. (From U.S. Patent 5,986,819.[67])

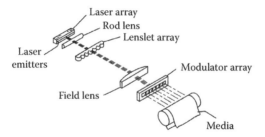

FIGURE 3.16 Laser printing system with the laser emitters optically mapped to provide adjacent nonoverlapping illumination of a modulator array. (From U.S. Patent 6,356,380.[68])

consisting of a multitude of single-mode-by-multimode emitters. The system provides an anamorphic optical system with a cylindrical rod lens and a spherical printing lens, such that a row of nominally circular 20 μm printing spots is provided. As in many later systems, the printing lens is double-telecentric (telecentric in both object and image planes) so that the printing spots have a common depth of focus across the drum. In the cross-array direction, where the narrow, single-mode emitter is to be mapped to the target of a 20 μm width, the working distance between the emitter and the rod lens is controlled to provide the desired width at the media plane as a defocus spot width. Depending upon the cross-array NA needed at the media plane, this system potentially requires tight control on the drum and media placement. Otherwise, motion of the media through a defocused beam may cause undesired spot size and shape variation.

As another alternate system,[68] shown in Figure 3.16, the light from the laser diode emitters of a nonaddressable laser diode array source is mapped to an external spatial light modulator array. This system, which was developed by Barco Graphics, employs array and cross-array micro-optics, including a rod lens and a lenslet array. In the array direction, the laser emitters are mapped (either far-field projections or near-field images) to the modulator array such that each emitter illuminates one of a series of adjacent regions of the modulator array. In the cross-array direction, the rod lens works with other optical elements to image the beam directly onto the modulator array. The modulator array is, in turn, imaged onto the media by a printing lens (not shown). This system is nominally a critical (or Nelsonian) illumination system[69] in which the illumination source is imaged onto the object, and it is thus sensitive to the detailed behavior of the source. While outright failure of the laser diode emitters across a laser diode array may be a diminishing concern, output variability, either emitter to emitter, or as a light emission profile (near field or far field) across the emitters, could detrimentally impact either of these systems.

3.5.2 MULTIPLE DISCRETE LASERS MAPPED VIA FREE-SPACE OPTICS

Although many of the laser thermal printer designs have employed high-power laser arrays, either directly addressed or indirectly modulated, some viable systems have been provided that employ a series of discrete laser diode sources. The Polaroid Helios™ systems[46,70] are an example of a design approach in which a series of

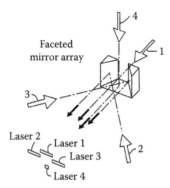

FIGURE 3.17 Printer configuration with a laser array synthesized from free-space lasers. (From U.S. Patent 5,161,064.[51])

free-space propagating beams are combined to form a print-head that directs the printing spots onto a media mounted on an external drum. As with many of the other laser thermal media, the Helios media, which comprises a light-sensitive carbon particle layer sandwiched between donor and receiver polymer sheets, is a threshold media. To accommodate the image quality needs of the initial Helios medical imaging market, the 8 × 10 in. monochrome prints were printed with pixels having grayscale (256 levels) modulation. The Helios system provided the grayscale modulation by writing each pixel with a *halftoning* process in which each pixel, consisting of a series of subpixels, is written by a combination of four independently modulated laser diodes.

The basic Helios system,[46,70] as shown in a simplified form in Figure 3.17, comprised four laser beams that were combined *via* a multifaceted mirror to subsequently follow parallel optical paths through the remaining optics (see Figure 3.18) that form the printing spots on the media. The mirror array provides two offset opposing facets that redirect two of the laser beams (Lasers 2 and 3) into parallel paths. The third beam (Laser 1) passes through a gap between the two opposing facets, while the fourth beam (Laser 4) deflects off a tilted mirror located in the gap between the two opposing facets. This system provided a unique configuration of four printing spots, with each printing subpixel structure within the overall 90 × 90 μm pixels. As shown, two elongated spots lie offset along a common axis, a third elongated spot is horizontally centered to the first two spots, but vertically offset, and the fourth spot has a smaller size and is positioned similarly to the third, but with the opposite vertical offset. The fourth and smallest beam is designed to have about one-seventh of the energy of the other three beams so that accurate tone scale reproduction could be provided for the highest density regions of the print. Although Figure 3.17 does not show this detail, the third beam is not only vertically offset from the first two, but it also has some horizontal overlap with these beams as well. The combination of offset and overlap reduces print artifacts in the written pixel that would originate with irradiance variations from diffractive interactions at the mirror facet edges.

The four laser diode lasers are high-power (500 mw) IR (820 nm) single-emitter lasers, with single-mode (~1 μm wide, $1/e^2$ NA ~ 0.5)-by-multimode

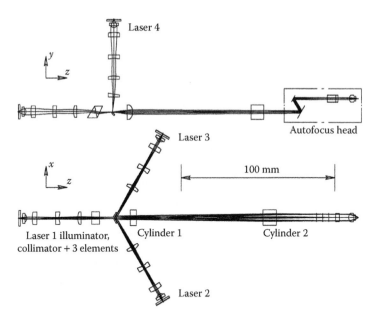

FIGURE 3.18   The Helios optical system from Polaroid, with four beams combined to form a printing spot with subpixel addressing.[70] (Courtesy of Polaroid.)

(~100 μm, NA ~ 0.07) emission structures. The laser beam shaping optical systems (see Figure 3.18), which are largely identical for the four lasers, begin with a fast (NA ~ 0.55) molded glass aspheric spherical collimating lens. Ultimately, the desired elongated printing spots provided at the media plane are ~34 μm long × ~3 μm wide, with the small spot being only ~5 × 3 μm in size. The focusing objective, provided to form the four spots onto the media, is a similar lens to the collimator and had a 0.47 NA. Notably, the combined system requirements of subpixel grayscale printing define this system to be fast (high NA) at the media plane and able to image the small printing spots. As a consequence, the depth of focus required at the media plane is short. The system is also required to present the printing beams telecentrically to the media plane, such that the writing beams are parallel to each other as well as perpendicular to the media. The telecentricity requirement is more severe in the horizontal direction (multimode axis) than in the vertical direction, because the scanning motion of the writing spots across the media relaxes the tolerances in the vertical (or slow scan) direction.

In this system, the printing spots are formed by imaging the laser emitters onto the media plane. Thus, the optical system must provide a differential laser beam shaping, such that the cross-array beam is ultimately magnified by about three times, while the emitters are ultimately demagnified by about three times in the multimode emission direction. It is also desired that the NAs in the two meridians be nearly equal at the media plane. As a result, an ~8 to 9:1 anamorphism is required somewhere within the beam shaping optics.

As shown in Figure 3.18, prior to the faceted mirror array, each of the four laser beams encountered its own beam shaping illumination optical system, comprising the collimator and a three-element beam expander. These illumination systems provide magnified intermediate real images of the laser emitters in both the multimode and single-mode directions. As the beam expanders receive collimated beams and output beams that focus to about 10 times magnified images of the emitters, the beam expanders are nearly afocal (similar to the classic element Galilean beam expander). The illumination systems are arranged radially about the mirror array, with the magnified real images of the laser emitters nominally imaged onto the mirror array facets. The mirror facets act as field stops to control the array direction sizes of the final emitter images, thereby desensitizing the system to illuminator magnification variations. The four beams are redirected by the mirror array such that four nominally parallel beams propagate downstream toward the focusing lens.

Before reaching the focusing lens, the four beams travel through a common anamorphic collimator comprising two crossed cylinder lenses, which is used to equalize the printing NAs to the media plane. A short focal length, positive cylinder lens is located close to the mirror array to collimate the high NA light in the cross-emitter (single mode) direction. A long focal length, positive cylinder lens is located closer to the focusing lens, to *collimate* the light in the slow NA (multimode emission) direction. The final collimated beams are presented to the focusing lens, which can be equipped with an autofocus mechanism. Taken as a whole, the Helios system anamorphically applies Gaussian beam propagation design to the cross-emitter direction, and imaging optical design in the array direction. The array direction system uses critical illumination,[69] with the source profiles ultimately imaged onto the media, which makes the printing sensitive to the array direction near field emitter light profiles. Polaroid addressed this issue by developing laser diodes with more uniform laser near field profiles than were commercially available.[50]

While the illumination systems for lasers 1 to 4 are nominally identical, the illumination systems for lasers 2 and 3 are the only ones that are actually identical, and the illumination systems for lasers 1 and 4 are provided with prismatic wedges. These wedges provide telecentricity correction, as well as coma correction, needed by the vertically offset laser beams (lasers 1 and 4) that traverse the short focal length first cylindrical collimator in off-axis positions. In the case of laser 1, a pair of prisms (BK-7 and SF-1) is used, while a single prism and a mirror facet angle adjustment are used for laser 4. As laser 4 is provided with an identical 100-μm-wide emitter to that of lasers 1–3, and the printed spot is to be reduced in size (5 vs. 34 μm), the laser 4 beam must be vignetted somewhere in the optical path. This can be accomplished by masking the laser 4 mirror facet, or by reducing its size.

The Helios system underwent several generations of change, first to use higher power lasers and then with improved designs that enabled more printer features and greater throughput on larger media. In one generation,[50] the printer comprised eight laser diodes, arrayed about a multifaceted mirror, to form two adjacent writing pixels, each with the four beam subpixel writing spots. As the multifaced mirror of this later system is cut from a monolithic surface, it is much easier to fabricate than its predecessor (see Figure 3.17). In this system, the anamorphic optics (a cylindrical microlens for the cross-emitter fast axis,[71] and a secondary cylinder lens to correct astigmatism) were designed into the illuminator, located prior to the mirror array. The complete illumination systems, including spherical lenses, provided magnified (48× and 6.6×, respectively, for the multimode and cross-emitter directions) real images of the laser emitters onto the mirror facets. As before, the mirror facets both redirect and truncate the beams to the desired sizes before the eight laser beams are focused onto the media by a two-element objective lens system. The Helios print-head technology has been used in other applications, including image setting, and as a high-power pumping source for fiber lasers.[72,73]

3.5.3 ARRAY PRINTING WITH DISCRETE LASER OPTICAL SYSTEMS

The Polaroid Helios system is an example of a laser printer in which assemblies of discrete lasers have been used to assemble laser thermal printing heads. The Helios system used a complex optical path, partly because of the subpixel structure but also because of the common free-space optics used to focus the light to the media. As an alternative, Presstek LLC (Hudson, NH) developed a series of light engines in which each of the directly addressed discrete lasers had its own optical path to collect and focus light onto the media. An extended laser printing array is formed from an ensemble of lasers and associated optics assembled across the length of the printing drum. This Presstek print-head has been successfully used in numerous graphics applications, such as CTP, computer-to-proof, and direct on-press imaging, with the Heidelberg Quickmaster DI printing press being a notable example.

This approach has the virtue that the laser beam shaping optical systems are inherently simple. In one configuration,[17] as shown in Figure 3.19, the light from an IR laser diode is collected by a collimating lens and then focused onto the input end of an optical fiber by a coupling lens to form a fiber pigtailed laser unit. The light traverses the length of the optical fiber, and the fiber output light is collected and focused onto the media by a similar two-element lens output optical system. Standard optical communications connectors, such as an SMA connector, can be used to mate the optical fiber to the output optical system. Preferably, the laser emits light with a small divergence (NA < 0.3), and the finer NA is likewise limited, so as to maximize the depth of focus of the focused light at the media. As is typical in the laser thermal systems, the laser emitter outputs light with both a low NA and a large NA (>0.3). To compensate, the fiber input optical system could also be equipped with a divergence reduction cylinder lens (not shown), which is located prior to the collimator, and which works on the fast axis light. Depending upon the design, this

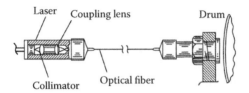

FIGURE 3.19 Printer configuration with a synthesized laser array having fiber-coupled lasers. (From U.S. Patent 5,351,617.[17])

printer can produce printed spots between ~12.5 μm and 50 μm in size. Alternatively, to avoid the light loss incurred by coupling to the optical fiber, as well as any optical noise from fiber bending or reflection-induced mode hopping in the laser, the optical system can be configured to have the input optics focus the light directly onto the media, rather than into an optical fiber.

As shown in Figure 3.20, a full printer is achieved by arraying a series of laser optical systems across the length of the drum. Each laser optical system is responsible for printing its image pixels within its respective portion of the drum, as the drum is rotated and the print-head is moved laterally. The system places a premium on the optomechanical alignment tolerances between the series of adjacent laser optical systems being maintained over the length of the drum. Similarly, it can be difficult to control focus across the synthesized laser array, as the focal distances are not defined with one mechanical reference (as with the fiber pigtailed print-head discussed in the following section) but depend upon the fabrication and calibration of the individual laser optical systems.

Further improvements have been made to this approach, including the use of an annular baffle within the fiber output optics[40] to limit ghost reflections and depth of focus loss caused by high-NA light emerging from the optical fiber. In addition, a controlled angle diffuser can be inserted prior to the baffle to smooth out hot spots in the multimode light that emerge from the optical fiber so that more uniform image pixels are printed, thereby improving the image quality.[41] Smoothing out of the sharp multimode light peaks can also improve the depth of focus to the media plane, although the beam Lagrange is presumably increased. As previously noted, this system can be

FIGURE 3.20 A digital offset printing press using Presstek's laser thermal imaging head technology. (Courtesy of Presstek LLC, Hudson, NH.)

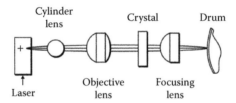

FIGURE 3.21   Printer configuration with a synthesized laser array using IR laser-pumped EO-crystals. (From U.S. Patent 5,822,345.[75])

sensitive to focus variations across the printing head, which can be caused by both variations in the laser optical systems and misalignment of the printing head to the drum. To address this variability of the spot focus from one laser optical system to another, provision has been made to individually shim each of the respective output optical systems[74] in order to adjust the focusing distance out of each assembly. With this approach, the focal position for each channel can be tuned to reach the needed working distance, although a small variation in the printing spot size may result, as the conjugate distances (fiber to focusing lens to the drum) are being changed.

A novel design alternative to this system, as shown in Figure 3.21, has the directly addressed IR laser acting as a pump laser to an external laser crystal.[75] The 808 nm pump laser light encounters a cylindrical microlens provided for divergence reduction and a spherical focusing lens, which together focus the pump light onto the end face of the laser crystal. The laser crystal, which may be a Nd:YAG crystal, for example, produces a low-NA single-mode TEM_{00} 1064 nm output beam, which is focused on the media by a focusing objective lens. The reduction in the beam divergence, compared with the original pump source, more than compensates for the power lost in wavelength conversion, thereby providing an overall increase in brightness of the light available for printing. Provisions are also made to enhance the response time of the laser crystal, for optimizing the pulse width to control the shape of the printing spots, and to optimize the thermal and mechanical mounting of the crystal in order to minimize printing spot size variation.

Presstek developed another print-head technology,[45] branded under the ProFire trade name, which uses a series of IR laser diode sources, each containing a laser driver board and a laser diode array that provides four uniquely addressable laser beams. Each laser emitter is optically coupled to a corresponding optical fiber, having a 60 μm core and a 0.12 NA. The light output from each group of four optical fibers is then imaged onto the media by a single lens assembly to provide four writing channels. This system provides 21 μm laser spots to the media, enabling 2540 dpi plate printing.

3.5.4   FIBER ARRAY PRINT-HEAD

Compared to many of the prior systems, having laser emitters mapped directly to the media plane, an alternative, highly integrated optical fiber print-head was developed by Eastman Kodak Company for the KODAK APPROVAL Digital Color Proofing System. This print-head provides a compact optical head with replaceable individual laser diodes. These systems use laser thermal technology to image cyan, magenta, yellow, and black dyes onto an intermediate sheet that the customer laminates to paper stock. Color images are printed on a press using cyan, magenta, yellow, and black inks. This process may be replicated on a press by imprinting more or less ink to create the tone scale. The high writing resolution of laser thermal technology does an excellent job of simulating the halftone printing process.

The first fiber array print-head printer[18,32,37] was developed using a series of butt-coupled fiber pigtailed diode lasers individually spliced to corresponding print-head single-mode optical fibers (5 μm core), where the print-head optical fibers were brought together in a pattern of adjacent V-grooves. The fiber exit faces were reimaged onto the media by a print lens to provide an array of printing spots.

In order to reduce the spot pitch (increase the optical fill factor), the optical fibers are progressively etched down from the initial 125 μm diameter cladding, to a mere 18 μm diameter. The reduced optical fibers were then assembled into the V-grooves, which were etched[76,77] into crystalline silicon, and held in place with ultraviolet (UV) curing cement. The V-groove base started at the input end with 250 μm pitch V-grooves and, through a series of progressively smaller V-groove structures, reached the output end with 20 μm pitch V-grooves. The optical fill factor of the print-head could be further increased by tilting the print-head or by interleaving the scan lines, or by doing both simultaneously. Unfortunately, the fusion coupling of the fiber pigtailed laser diodes to the print-head fibers both reduced the system light efficiency to ~10% and caused channel output power instability.

The APPROVAL Digital Color Proofing System print-head,[33,78] which has been built-in 30-channel and 64-channel versions, uses a similar architecture, except the single-mode lasers and single-mode optical fibers were replaced with multimode lasers and multimode optical fibers, and dummy channels were added to improve printing uniformity.[33,35] Figure 3.22 shows a sketch of the optical path of a fiber V-groove printing system. Each diode laser source comprises an array of high-power 830 nm laser emitters that output single-mode-by-multimode light. This light is coupled by a cylindrical lens into the individual multimode optical fibers (NA ~ 0.24) to yield fiber pigtailed lasers with ~400 mW output per fiber. Although the fiber supports 0.24 NA light, ~90% of the coupled optical power is contained within a generally Gaussian beam with NA ~ 0.12. The pigtailed optical fibers are coupled into the print-head optical fibers (also NA ~ 0.24) by means of standard industry ST connectors, which have a high positioning accuracy. To further enhance the

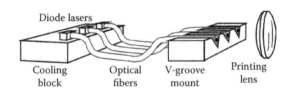

Diode lasers

Cooling   Optical   V-groove   Printing
block   fibers   mount   lens

FIGURE 3.22 Printer configuration with a laser array synthesized from discrete fiber-coupled laser diodes and a V-groove fiber mount. (From U.S. Patent 4,911,526.[18])

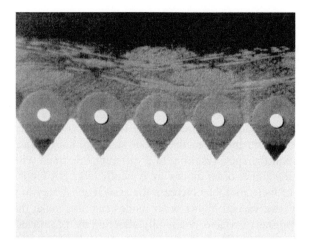

FIGURE 3.23 Portion of the end face of a V-groove optical fiber array.

coupling reliability, the pigtailed fibers can have a smaller core (50 μm) than the print-head fibers (52 μm).

The multimode print-head optical fibers are glued into V-grooves etched in silicon[76,77] at 130 μm spacing between centers. The end face of the V-grooves, as shown in Figure 3.23, is polished to provide a smooth and coplanar output surface from which the printing light beams exit the fiber array. A printing lens, working at a demagnification of 2.2:1, images the array of fibers onto the drum, where the donor is mounted over the intermediate. In one version of the system, the printing lens had an acceptance NA at the fiber of 0.12, with the result that the higher-order, large NA light overfilled the print lens, and is thereby clipped. As it operates at a modest magnification, the print lens[79] supports relatively large fields and NAs at both the object [fiber array (~2 mm field and ~0.12 NA)] and image planes (~1 mm field and ~0.264 NA). The integrated print-head, comprising the assembled V-groove fiber array and the print lens, writes swaths of image data along a nominal helix spiral pattern as the head is moved laterally across the drum. This print-head could also be tilted at an acute angle to increase the apparent print-head resolution by compensating for the gaps between the optical fiber cores.[34] The fiber-to-fiber placement error can be measured, as can the spot-to-spot spacing, in both the along-array and cross-array directions. The array is tilted to achieve the desired spot-to-spot spacing. Adjusting the angle of the array compensates for variations in lens magnification. The data for each channel are digitally delayed to align pixels to a line normal to the fast scan direction on the proof. An autofocus mechanism can be used to ensure optimal focus of the writing spots on the print media.

As might be expected, this system is sensitive to *fiber* noise. For example, movement of the fibers as the print-head traverses the drum can launch light into higher-order modes. Likewise, energy can be scattered into higher-order modes as the result of fiber-to-fiber irregularities, crimps, sharp bends in the path, and so on. Laser mode hopping, which is of direct or indirect lasing instability, can cause similar effects. As higher-order modes typically comprise high NA light, the relatively large fiber NA (0.24) used in this system means that light launched into those modes is not automatically attenuated. At the cost of a reduced light efficiency, the original 0.12 NA print lens used vignetting to remove these unstable higher-order modes. Alternately, the system can be configured with a faster printing lens [NA (~0.48) to the media plane] that allowed most of the higher-order-mode light (and power) through to the media plane.

Noise from higher mode structure, whether from fiber movement or mode hopping, can be problematic and interactive. Mode structure noise rotates and changes the structure within the spot, resulting in a different line trace within the image. Furthermore, as the higher-order modes are typically large NA modes, they provide a smaller depth of focus and thus increase the system sensitivity to focus position. Finally, noisy or time-variant higher-order modes can also make the light profiles within the printing spots variable, potentially affecting the final print uniformity.

Laser mode hopping can be minimized with careful coupling of the optical fiber to the diode laser. Optical reflections returning to the diode laser cause mode hopping, changing the output power and wavelength.[80] Multiple diode laser channels reduce the sensitivity to individual laser noise, as each channel may be considered an uncorrelated random noise source. The multiple diode laser elements that constitute each laser source further reduce the individual channel noise. Images created using a single laser can channel exhibit artifacts as a result of laser startup transients caused by changes in diode laser temperature, optical power, wavelength, or mechanical position. Additional channels break up the appearance of these artifacts. Modulation of the laser with image data is another method of breaking up the artifacts in the prints and making them less noticeable.

3.5.5 FIBER ARRAY PRINT-HEAD WITH ADDRESSABLE LASER ARRAYS

Although the fiber pigtailed multichannel print-head described in previous Section 3.5.4 is technically straightforward, the total cost of the fiber pigtailed laser print-head becomes excessively expensive as the number of channels increases. The electronics cost to support the print-head also grows rapidly as the number of channels increases. About 30 fiber pigtailed channels is a reasonable compromise because the ratio of the print-head cost to total system cost is less than 20%.

One approach to reduce the cost of the fiber array head is to use an individually addressable laser diode array (IALDA) coupled to a fiber array. As an example, CreoScitex Corporation developed a system[81] in which each independently addressable laser emitter is coupled to an optical fiber. The print-head routes an input bundle of optical fibers to an output optical bundle via a connector board, and further provides two-fiber alignment V-groove assemblies, one at the IALDA side and one at the imaging side, as illustrated in Figure 3.24. Despite having a provision for spare emitters, printers that use direct imaging of IALDAs found only modest implementation because of the limited redundancy and relatively high cost of service.

Notably, however, for laser array to fiber array coupling, this print-head used a customized version of LIMO monolithic fiber-coupling lens, having a common fast axis (FA) lens on one side and slow axis (SA) lenslet array on the other. As shown

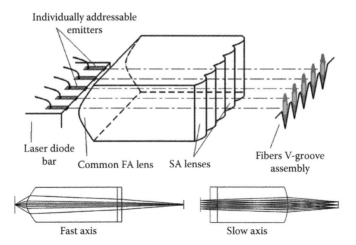

FIGURE 3.24 Anamorphic monolithic array and cross-array lens, perspective and cross-sectional views.

in the cross-sectional views of Figure 3.24, this lens is highly anamorphic, with the image plane of the cross-array or FA lens and the focal plane of the array direction or SA lens coinciding at the fiber position. The exit NA is intended to be equal in both directions, and the spot size in the SA direction is then meant to be approximately 10% smaller than the fiber core diameter.

3.6 MODULATED SUBARRAY LASER PRINTING

Although the system of Figure 3.22, with multiple individually fiber-coupled lasers, and the system of Figure 3.24, with an addressable laser array with fiber-coupled emitters, both provided laser to pixel addressing, as they both lack redundancy, they thus require laser substitution to correct a failed emitter. As an alternative approach to reducing the print-head cost, a print-head architecture was developed in which the laser array is provided with multiple addressable groups of single-mode diode lasers on the same substrate.[58,82]

The associated optical system[7,82] provides a sophisticated design that employs multiple microlenslet arrays and spot and pupil reimaging to overlap the beams from each group of these single-mode diode lasers into one spot to gain optical power and redundancy of emitters for reliability. This print-head, which was developed by Eastman Kodak Company, has been used in dry image setting systems by both Matsushita and Dainippon Screen.

3.6.1 OPTICAL CONFIGURATION OF A MONOLITHIC MULTICHANNEL PRINT-HEAD

The key attribute of the monolithic multichannel print-head is that it uses a directly addressable laser diode array, provided with laser subarrays, which allows the system to provide source redundancy without resorting to an external modulator array.

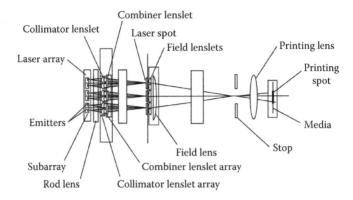

FIGURE 3.25 Array direction optics of a laser printer with addressed laser emitter subarrays. (From U.S. Patent 5,619,245.[57])

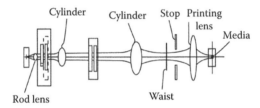

FIGURE 3.26 Cross-array direction optics of a laser printer with addressed laser emitter subarrays. (From U.S. Patent 5,619,245.[57])

The optical layout,[7,14,57] shown in Figure 3.25 for the along-array direction and in Figure 3.26 for the cross-array direction, employs an array direction intermediate imaging concept in which the respective beams from each of the addressed subarrays are collected into a printing beam, and the ensemble of printing beams are imaged to the media.

The laser diode array[58,82] that powers this system comprises 160 single-mode, 835 nm diode lasers gathered into 10 groups or subarrays, which are separated by 250 μm spacings. Each channel, which is composed of 16 single-mode diode lasers spaced at 50 μm intervals, is 750 μm wide and driven by its own current driver. The total length of all of the single-mode diode lasers in the array is ~10 mm. As the fill factor of the laser array is very low (~6%), the challenge of optical system design is to increase the fill factor to nearly 100% on the medium without losing much optical power, while introducing a reasonable optical system complexity. Another requirement is that all 16 single-mode lasers overlap at the printing spot of their channel. Because the 16 single-mode lasers within a channel are mutually incoherent, each channel behaves optically like a multimode laser.

In addition to optical power gain for each channel, overlapping of the 16 single-mode diode lasers increases reliability. If a single-mode diode laser degrades in power or ceases to emit, the others can compensate the power loss by slightly increasing their emitted power. The multichannel print-head system reliability and robustness are improved significantly as a result.

In the array direction, the print-head[7,57] uses a series of refractive lenslet arrays and field (or combiner) lenses to collect and gather the channels of beams to the media. The array direction optical system employs a classical optical design approach of object-to-image conjugation, intertwined with pupil-to-pupil conjugation, to obtain the image without loss of brightness. In greater detail, each of the 16 laser emitters in a given laser subarray is collimated by a collimator lenslet. Only two of the 16 emitters in each channel are depicted in Figure 3.25 for clarity. A second lenslet array (the combiner lenslet array) is provided with one combiner lenslet per laser subarray. The combiner lenslets focus the respective collimated beams from a given laser subarray in overlapping fashion, such that 10 separated beams (channels) are available downstream as imaged laser spots, which are presented nominally telecentrically to an intermediate image plane. The combination of a combiner lenslet with a focal length of 50 mm, and a collimator lenslet, with a focal length of 200 µm, is to magnify a ~4-µm-wide emitter by ~250 times to an ~1-mm-wide image (laser spot). Note that the 16 collimated beams are directed into the input aperture of the respective combiner lenslet such that, in total, the ensemble of 16 beams fills the aperture of the combiner lenslet and forms a combined Gaussian beam. As a result, although a given emitter has a miniscule NA (~0.1/250) at the imaged laser spots, the composite NA from the ensemble of 16 laser beams is a relatively large ~0.1.

The print media can be colocated at the image plane occupied by the imaged laser spots. However, both the spot size and the spot pitch may be incompatible with the printing specifications. Most simply, a printing lens could be provided to reimage the laser spots directly to the media plane to provide an array of addressed printing spots. A field lens near the intermediate image plane diverts the beams so that they pass through the aperture stop of the printing lens. If only this field lens is used, then a line of separated images of the beam combiner lenslets would be projected onto that stop, and a slight misplacement of that stop would severely vignette at least one of the outermost laser channels but not affect the central channels. This problem is remedied[57] with the addition of another lenslet array, consisting of field lenslets. Each of these lenslets effectively reimages a corresponding combiner lenslet to the plane occupied by the aperture stop. In this context, the field lens causes each of the combiner lenslet's images to be superimposed onto all of the others at the stop of the printing lens, equalizing the vignetting for all of the channels. As a result, the printing lens images the field lenslet array onto the medium and, as each field lenslet is nominally filled by the light of a given 1 mm laser spot, the optical fill factor at the media plane is high (~100%). The print lens reduces the spot size at a magnification of one-fortieth such that a row of 25 µm printing spots is presented telecentrically to the media.

The cross-array optics, shown in Figure 3.26, can be configured in numerous ways but, nominally, a variety of cylinder lenses is used to shape the beams. Gaussian beam waists are not formed to be coincident with the laser spot images, but to fall near the aperture stop of the printing lens, and at the media plane. The light from the single-mode diode lasers is first collected by the rod lens. The rod lens reduces the NA of each beam from the diode lasers in cross-array direction. The optical system can form either a round spot or an elliptical spot, according to the relative magnifications in the array and cross-array directions. A two-cylinder lens system in the cross-array direction produces an elliptical spot with a 2:1 aspect ratio on the media, whereas a three-cylinder system produces a round spot.

After the fiber lens and combining lenslet array are aligned with the monolithic diode laser array inside the diode laser array enclosure, optical power vs. current is measured for each channel of an array both at the combining plane, 50 mm from the diode lasers, and at the focal plane intended to be coincident with the surface of the written medium.[83] The typical efficiency of the optical system from diode laser to image-recording medium is approximately 70% for the print-head. Theoretical calculation suggests that the system optical efficiency should be about 80%–85%, implying that about 10% additional loss is observed. There are several unaccounted losses, such as excessive loss in the combining lenslet array, optical misalignment, and contamination of optical surfaces by dirt. This system can also be outfitted with a smile corrector,[62] positioned between the combiner lenslet array and the laser spot images, to provide a dramatic increase in the effective depth of focus at the media plane.

3.7   LASER ARRAY AND MODULATOR ARRAY SYSTEMS

In recent decades, laser thermal print-heads with a large number of channels have been developed for various applications, including CTP plate setters for the graphic arts market.[22] In general, the more channels in a print-head, the less laser power needed from each channel to maintain the same printer productivity, and thus the speed of the print-head relative to the medium can be slower. In the case of media mounted on a rotating drum, this means a lower r/min of the drum while exposing the media, and less time to accelerate and decelerate the drum to that lower r/min. In a flatbed plate setter, providing more channels in the print-head allows a lower print-head velocity while exposing the media, as well as less acceleration time, to reverse the print-head direction to that lower print-head velocity.

The preferred system architecture to achieve a large number of printing channels uses an integrated print-head wherein laser light illuminates a spatial light modulator array, which is subsequently imaged to the media plane. For example, this type of print-head can typically deliver 12–25 W of optical power to the media, compared to a total of about 10 W from the fiber optically coupled lasers and 6 W from the monolithic diode laser array. This design architecture also delivers the large number of high-power writing channels (e.g., 256 channels) more cost effectively than the system architectures discussed previously. As an additional advantage, the writing channels stitch seamlessly because they are derived from a continuous line of laser beam illumination. The design and performance of such systems is very dependent upon the properties of the spatial light modulator array.

This design space was first extensively developed by Xerox Corporation,[84–86] which developed a complete solution including laser array sources, a viable modulator array technology, and the basic optical system configurations to transfer light from the laser array, through the modulator array, and to the media plane. Subsequent to these developments by Xerox, modulator array/laser array printing systems have been developed by others, and used in products, including the Creo Trendsetter thermal plate setter[9] and the KODAK NEWSETTER TH180 Platesetter System (developed by Kodak Polychrome Graphics™) for newspaper printing.[10] These later systems were enabled by numerous technology improvements, in various areas, including for laser diode arrays, micro-optics, and spatial light modulator arrays.

3.7.1 THE XEROX TIR MODULATOR ARRAY SYSTEM

An early printing system that combined a laser or laser diode array with a spatial light modulator array having a large number of channels (>5000) was developed by Xerox for electrostatic printing applications. This system is enabled by the total internal reflectance (TIR) modulator,[87,88] shown in Figure 3.27, which comprised an adjacent row of pixels formed as individual patterns of electrodes on the top surface of an electro-optic substrate. The electrode patterns are formed as interlaced fingers to provide a structure of alternating polarity electrical fringe fields when voltage is applied. The electrical fields penetrate the electro-optical substrate, which is typically either lithium niobate ($LiNbO_3$) or lithium tantalate ($LiTaO_3$), to produce localized changes in the indices of refraction. Phase differences are imparted to the transiting light beam, which, in turn, result in diffraction patterns when the light is directed to a Fourier plane within the printing lens. The initial versions[84,85] of the TIR modulator did not employ electrodes patterned directly on the electro-optic substrate, but rather a special multichannel silicon driver chip provided with electrodes that were proximity-coupled by *contact* with the electro-optic substrate. Although this approach provided an easy means for realizing a large number of pixels, the presence of an air gap between the electrodes and the substrate caused the required drive voltages to increase.

When proper spatial filtering is applied to discriminate between the light patterns of the modulated and unmodulated light, this Schlieren-type optical system provides the means to furnish an addressable array of pixels when the spatial light modulator array is imaged to the media plane. As the initial Xerox systems[85,86] used coherent laser sources (HeNe lasers) and imaging of the light diffracted around the stop, the best results were obtained with Gaussian apodized stops (rather than square profile stops) as side lobe interactions were reduced. Although the TIR modulator is a transmissive device, which is generally advantageous, optimal operation of the modulator requires the light to attain grazing incidence in the region underneath the electrodes, which can be a significant impediment to the optomechanical system design.

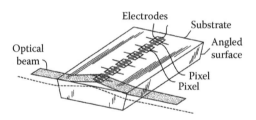

FIGURE 3.27 The linear TIR spatial light modulator.

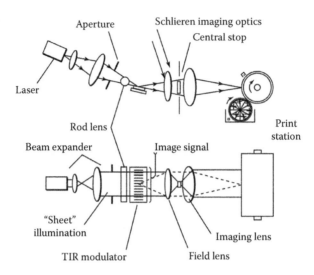

FIGURE 3.28 The Xerox printing system with TIR modulator array. (From US 4,591,260.[89])

As shown in Figure 3.27, the input and exit faces can be cut at an angle,[85] to enable an in-line optical system configuration.

The Xerox system,[85,86,89] shown in Figure 3.28, introduces many of the basic elements required to optimize this type of printing system, including anamorphic laser beam shaping optics, *sheet* or *line* illumination to the modulator array, and imaging optics to couple light onto the print media. The unspecified laser source outputs a single beam, with a collimated meridian (aligned to the array direction of the modulator) and a divergent direction. The illumination system comprised an array direction beam expander, used to present collimated light to the modulator, and a three-element anamorphic cross-array optical system, used to focus the light to the modulator array. In the cases where a Gaussian beam laser source is used,[86] the illumination system could be equipped with an apodizer, to uniformize the spatial light profile to within a few percent, but at the cost of a 50% light loss.

The printing lens comprised a field lens portion and an imaging lens portion. The central stop blocks the zero-order array direction diffracted light so that the higher-order diffracted light can be imaged to the media. As such, this system emphasizes modulation contrast over optical throughput.

As an alternative,[8] Xerox also developed a system concept that uses a laser diode array as the light source where light from the multitude of laser emitters is used to flood-illuminate the modulator array. Laser coherence (or the lack thereof) contributes significantly to the actual performance of this type of system. In particular, the laser light must be sufficiently incoherent to avoid significant interference fringes. Accordingly, Xerox described a laser diode array comprising closely packed single-mode laser emitters, wherein the emitters are located in two parallel rows, with the emitters spaced at the same pitch, but 90° out of phase from one row to the other. As the resulting laser emitters are far enough apart to avoid phase locking, they lack a common phase and can be combined without interference. The beams were combined in the far field, without any uniformizing optics, to produce a generally Gaussian illumination distribution at the modulator array.

3.7.2 Spatial Light Modulators

This laser thermal printer design architecture, which combines a laser source (often a laser array) and a spatial light modulator array, would seem readily adapted to a wide variety of modulator technologies. In actuality, the applied power densities, geometry, and modulation speed requirements often limit the device technology choices, thereby eliminating LCD arrays and many micromechanical shutters from consideration. Moreover, the general linear arrangement of the system (including the laser diode array and the swath printing motion of the print-head relative to the media) favors a linear modulator array. Despite these limitations, several modulator array technologies have potential and have been applied in printing systems. Candidate device technologies include polarization modulators (such as PLZT),[90,91] AOMs, digital mirror array modulators (DMDs),[92,93] micromechanical grating modulators,[94–96] and electro-optic grating modulators.[84,87,88,97,98] Compared with many other applications that employ modulator arrays, the modulation contrast required to print on the typical graphics art media is quite low and is threshold rather than grayscale-based, such that a moderate (10–25:1) dynamic range is often sufficient.

The specifications on the spatial light modulator array can be considered more explicitly. The spatial light modulator array typically needs 200 or more independently addressable channels, and must be able to sustain a laser beam power density of about 1 kW/cm². This laser beam power density is rather high, and depends upon modulator pixel size, fill factor, and optical efficiency. The modulator must also provide a minimum contrast ratio of ~10:1 (the ratio of channel *on* irradiance to the channel *off* irradiance), and work well at semiconductor, diode laser bar source wavelengths, typically between 800 and 980 nm. In addition, modulated beam rise and fall times should be less than ~2 μsec.

From the optical designer's perspective, the optimal modulator array would be a transmissive device, used at normal incidence, which either absorbs the incident light that is modulated to the *off* state or reflects it back toward the light source. However, most of the viable modulators[87,88,94–98] are Schlieren-type devices, which impart phase modulation to the incident light, and require angular or Fourier plane filtering downstream. The quality of the modulation depends upon the angular filter design. The design of the angular filter in turn impacts the design of the printing lens, as access to an internal stop plane may be required, and also creates an inherent tradeoff between modulation contrast and system efficiency. In addition, the degree of partial coherence of the emitted laser light can affect the modulation performance of the Schlieren-type modulators.[99] In lower power optical systems and applications, other modulator technologies, including absorptive devices,[100] may also be viable.

Potentially, optimization of the modulator performance can also limit both the array and cross-array optical designs. For example, in the array direction, the length of the modulator array and the allowed NA for optimal response (modulation contrast, minimal crosstalk, etc.) may determine the array direction Lagrange supported by the system. Furthermore, to enhance the uniformity of the response across the modulator array, the laser beam shaping optics may be required to present the modulator with spatially uniform and telecentrically oriented incident light. Similarly, in the cross-array direction, the modulator response characteristics and the pixel extent may impose system constraints, including the amount of laser smile tolerated.

3.7.3 An Array Print-Head with a Fly's Eye Integrator

In a printing system that employs a spatial light modulator array, the resulting print quality is dependent upon the uniformity of response of both the modulator array (pixel to pixel) and the illumination to the modulator array. Although a variety of methods for improving illumination uniformity could be considered, including those that use diffusers and integrating cavities, use of either light pipes (integrating bars) or fly's eye integrators is often appropriate in applications where it is necessary to substantially preserve the brightness while providing uniform illumination. For example, in the photolithographic printing of integrated circuits,[101] a multitude of designs have been used where the light from excimer lasers, YAG lasers, or arc lamps has been made uniform by fly's eye or light pipe integrators. These design approaches are extendable into laser thermal printing, particularly in the instance that a spatial light modulator array needs uniform illumination.

One exemplary laser thermal printing system[10,11,102] that combines a laser diode array light source, a spatial light modulator array, and light uniformization means is shown in Figure 3.29. The print-head, developed by Eastman Kodak Company, combines classical imaging optical techniques (object-to-image conjugation with pupil-to-pupil conjugation) and traditional light integration illumination optics (the fly's eye integrator). In particular, the fly's eye integrator, which is a portion of the array direction optical system of optics and micro-optics, contributes to flood illuminating a spatial light modulator array with uniform light. In the cross-array direction, the light from each emitter is focused to form a beam waist at the modulator, which is confined within a narrow width corresponding to the defined active height of the modulating pixels. In addition, the array and cross-array optics illuminate the modulator array with a long, narrow line of light of uniform radiance, while largely preserving the brightness of the laser diode array source (less transmission and other losses),

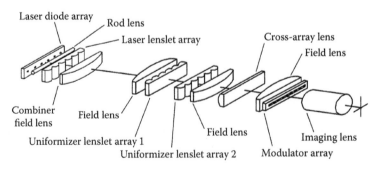

FIGURE 3.29 Laser printer using fly's eye optical integration to illuminate a modulator array. (From U.S. Patent 5,923,475.[11])

and providing redundancy relative to the emitters. The illuminated modulator is imaged telecentrically to the media plane by the print lens to create a line of closely packed writing spots. Depending upon the type of modulator array used, such as polarization type or a Schlieren type, the filtering means would be positioned in the vicinity of the printing lens, as appropriate.

As with many of the other laser thermal printing systems, the system of Figure 3.29 uses largely anamorphic (cylindrical) laser beam shaping optics, with separate optical systems designed for the array and cross-array directions. The operation of the array direction illumination optics (the laser lenslet array, the combiner field lens, several field lenses, and two uniformizer lenslets), as illustrated in Figures 3.29 and 3.30, is to collect light from each laser emitter, and magnify and redirect it such that the entire length of the modulator array is illuminated. In general, the fly's eye integrator, which includes the two uniformizer lenslet arrays and the immediately adjacent field lenses, is designed to uniformly illuminate the modulator by dividing the light from each emitter into N_2 multiple beams. These N_2 beams are overlap imaged over the full length of the modulator array. For clarification, the image conjugate relationships in the system are shown as follows: planes a_0, a_1, and a_2 are conjugate to each other, as are planes b_0 and b_1. Plane a_0 corresponds to the front surface of the laser array, while plane b_0 corresponds to the back focal plane of the laser lenslet array. As a first stage of light integration, the laser lenslet array is used in combination with the combiner field lens to image the beams from each emitter in overlapping fashion to an intermediate illumination plane a_1. The laser lenslet array has N_1 lens elements, with each lens element corresponding to a given laser emitter. Although the light from the various emitters has been overlapped at the a_1 plane, the emitters have not been subsampled and mixed angularly or spatially. As a result, any systematic problems in the light profile across the emitters, such as the edge roll-off shown in Figure 3.6, are not removed, although such effects are averaged. Put another way, the light integration through the first stage is incomplete in that all points within the illuminated area do not see light from all points on each of the sources (emitters). However, if the near-field laser uniformity is improved[50] consistently, the uniformization optics could be proven unnecessary.

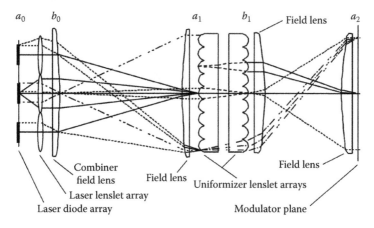

FIGURE 3.30 Light transfer within the fly's eye integrator optical system.

Thus, a second integration stage is used, which consists of the two uniformizer lenslet arrays and the adjacent field lenses. The preuniformizer field lens is used to create telecentric illumination at the intermediate plane a_1. The light profile at plane a_1, which is the magnified, overlapped, and averaged image of the emitters, is parsed into N_2 beams, where N_2 is the number of lenslets in the first uniformizer lenslet array. The corresponding N_2 lenslet elements in the second lenslet array work together with the postuniformizer field lens of Figure 3.29 to image the lenslets of the first uniformizer array in a magnified and overlapping fashion onto the a_2 plane (the modulator plane). The modulator plane field lens of Figure 3.29 functions in the same manner as does the preuniformizer field lens to provide telecentricity in the illumination. Thus, the intermediate a_1 illumination plane is subsampled by the N_2 uniformizer lenslet pairs to create a more uniform radiance distribution of light at the a_2 plane. The more N_2 lenslet pairs that are used in the fly's eye integrator, the better the averaging. In general, the goal is to reduce the residual nonuniformity to just a few percent.

The laser beam shaping of this type of system can be further understood with reference to a specific laser diode array, such as the Opto Power OPC-A020 laser that was discussed previously. Again, the A020 array has 19 multimode laser emitters, each 150 µm wide, which are spaced apart on a 650 µm pitch, for an overall array length of 11.85 mm. In the array direction, the 830 nm light is emitted into a relatively small NA (~0.13), with a non-Gaussian angular beam profile and a relatively flat topped, but noisy, spatial profile (see Figure 3.6). The light emitted in the cross-array direction is output as a quickly divergent Gaussian single-mode beam (NA ~ 0.63).

In a first-order design, using the OPC-A020 laser, the refractive laser lenslet array (2.47 mm EFL) is designed to *collimate* light from the full field of each of the emitters. The laser lenslet array reduces the array direction Lagrange by effectively removing the spaces between emitters (~0.187 vs. 0.77 mm, otherwise). The focal length of the first field lens was chosen to overlap the N_1 beams at the a_1 intermediate image plane with the illuminated width bound by the constraints of the uniformizer lenslet array manufacture. Unless these constraints (size limitations on lenslet width, size limitations on the overall size of the array, or limitations on the sag height of the power surfaces) effectively limit the system, the design will be set up on the basis of convenience and conservation of brightness.

The fly's eye assembly further directed the light to a modulator array, which, in this first-order design example, comprises 256 pixels, each 63.5 µm wide, for an overall device length of 16.25 mm. The design NA at the modulator plane is 0.023, which would be acceptable for most candidate modulator technologies. The uniformizer lenslet arrays, which were identical, each comprised six cylindrical lenslets, each 1 mm wide. The combiner field lens had a nominal focal length of 99 mm, such that the 6 mm overall width of the first uniformizer lenslet array is filled with light. The lenslet elements of the uniformizer lenslet arrays had 8.0 mm focal lengths to ensure that the output faces of the lenslets at the b_1 plane were filled with light. The 130 mm postuniformizer field lens provided the appropriate magnification to illuminate the full length of the modulator array. The print lens demagnified the modulator array at 1/6 times to provide a 2.7-mm-wide line of printing spots with an array direction NA of ~0.14.

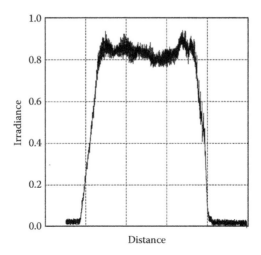

FIGURE 3.31 A measured irradiance profile of array direction modulator plane illumination.

Figure 3.31 illustrates the array direction illumination quality observed at the modulator plane with a prototype system based on the above design. The system yielded ~±6% uniformity within the nominally uniform area created by the fly's eye integrator, although better results are achievable. System light efficiency in the main beam at the modulator plane is ~69%, with minimal light lost to the side lobes.

The cross-array optics of this exemplary system included a rod lens that is mounted to the diode laser assembly and a set of cylindrical lenses to provide a Gaussian beam waist at the modulator array, such that the light fits within the pixel height. Smile correction is used to effectively reduce the laser smile from ~10 to ~2 µm residual. The illustrated system shows the printing optics as spherical, but anamorphic printing optics can be used if necessary.

During the system design process, detailed analysis and optimization using lens design software (such as ZEMAX or Code V) or illumination design software (such as Light Tools) or both, can be used to control lens aberrations and verify system performance. In addition, the first-order design can also be modified to help system efficiency. For example, an underfill factor can be applied during the layout of the uniformizer lenslet arrays so that the second array is slightly underfilled to compensate for broadening of the beams at the b_1 plane induced by both lens aberration and aperture diffraction. There is an opportunity, particularly with refractive lenslets, to experience light loss from scatter and diffraction at the seams where adjacent lenslets meet. Likewise, an overfill factor can be allowed at the modulator plane to allow for edge broadening at either end of the modulator. A similar system[11] can also be designed in which the far-field light profile, rather than the near-field light profile, is input into the fly's eye integrator for light uniformization to then provide illumination to the modulator array. The difference can be regarded as providing a Koehler illumination input, rather than a critical illumination type input.[69]

3.7.4   AN ARRAY PRINT-HEAD WITH AN INTEGRATING BAR HOMOGENIZER

A variety of other laser thermal printing systems have been designed with other approaches to light homogenization, as well as other important features. As one example, a laser thermal printing system developed by Kodak Polychrome Graphics provided array direction light homogenization with an integrating bar[10,21] instead of a fly's eye integrator. Integrating bars or light pipes, which have also been widely used in projection systems, operate by a process of internal reflections to overlap and homogenize the light traversing their length. These bars are generally solid dielectric rectangular structures with the input and output faces at the opposing ends. Light propagates along their length and totally internally reflects (TIR) when it encounters the glass-to-air interface on the sides. The degree of uniformization is largely dependent upon the length of the integrating bar and the NA of the input beam. If the input light fills the input face, then light uniformization can occur without any significant loss of source brightness. In the case of the laser printer, the system is configured as shown in Figure 3.32 to provide numerous reflections in the array direction, such that the array direction light is made uniform. The exit face is then array direction imaged to the modulator plane. Conversely, the cross-array direction light underfills the integrating bar and propagates at a low NA, such that it sees the integrating bar as a thick window, and not as a uniformizer. As in many of the other systems, this printer can also use an array direction laser lenslet array (not shown in Figure 3.32) to reduce the effective Lagrange.

As another example, an IR (850 nm) laser thermal printing system[9] is provided with two mirrors that deflect and redirect the sloping portion of the emitter light profile from one side of the array beam to the other, thus adding a compensatory way to the roll-off on the other side. This design approach, developed by Daniel Gelbart of Creo Incorporated (shown in Figure 3.33), has a modified critical-type illumination system[69] that potentially corrects the macrononuniformities (roll-off) in the array direction, but may have little benefit in smoothing out the micrononuniformities, and will also only work well when the light profile is generally symmetrical. This approach also reduces the system brightness because of the increased angular

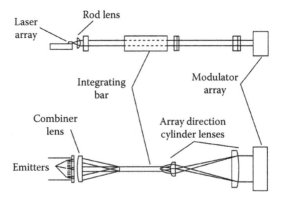

FIGURE 3.32   Laser printer illumination of a modulator array, by means of an integrating bar. (From U.S. Patent 6,137,631.[21])

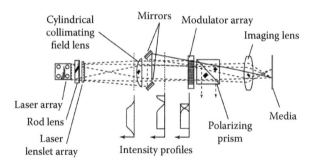

FIGURE 3.33 Laser printer using two mirrors to improve illumination uniformity to a modulator array. (From U.S. Patent 5,517,359.[9])

spread of the illumination to the modulator, but the optical system is likely tolerant in the array direction to such an increase. In the cross-array direction (not shown in Figure 3.33), the rod lens focuses the light to the modulator plane. The printing lens then relays the beam focus to the media plane. As shown, the modulator is a polarization-type device (a PLZT modulator), and the system is equipped with a polarization prism to distinguish between the modulated and unmodulated light.

This system[9] also uses the potential simplification of combining the laser lenslet array and combiner field lens (see Figure 3.33) into a single element.[103] To begin with, the laser lenslet array is located at a working distance greater than the lens element focal length so that the emitters are imaged and magnified at the modulator plane. In addition, the pitch of the laser lenslet array is slightly smaller than the emitter pitch (the scale of Figure 3.33 obscures this detail) on the laser array (776.3 vs. 787.5 μm, for example), such that the lenslets are used as off-axis imagers, relative to the optical axis of the system. This off-axis imaging causes the magnified images to shift inward, such that the magnified images overlap at the modulator plane. Similar results can also be achieved by fabricating the lenslets on the same pitch as the laser emitters but with the optical axis of the lenslets shifted appropriately. The lenslet arrays used in this type of laser thermal printer, and several of the others previously described, can be designed and fabricated with diffractive (or binary) optical methods. The lenslet arrays can also have more complex designs, including aspherical power.[104]

Certainly, the use of light pipes (straight or tapered, made of glass bars or mirrors) in light valve illumination systems can produce highly uniform illumination, at least with respect to low-spatial frequency variations across the field. However, as shown in Figure 3.34, high-spatial frequency nonuniform features, which are caused by residual interference effects, can occur. These interference effects are present even in systems with multiple quasi-monochromatic and mutually incoherent sources. The light pipe forms at its exit multiple virtual images, each one presenting a different reflection of the source at its entrance. These virtual sources are mutually coherent and will form an interference pattern even if the main source has very low coherence length. Figure 3.34 presents results of irradiance profile measurement at the light

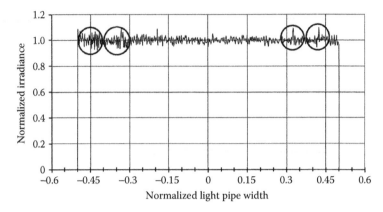

FIGURE 3.34   A measured irradiance profile at the light pipe exit, when illuminated with an array of mutually incoherent emitters. (Courtesy of Kodak Canada ULC, Burnaby, BC.)

pipe exit illuminated with an array of mutually incoherent emitters. The sharp spikes and dips can be clearly seen on the top of the otherwise exceptionally flat profile.

Although most laser thermal print media are too insensitive to see such high-frequency spatial nonuniformities, these effects can cause problems for other processes. As a corrective measure,[105] a coherent nonuniform beam can be transformed into a noncoherent uniform beam using a light pipe whose aspect ratio is chosen so that the path lengths of the rays from the apparent sources to the exit of the light pipe are sufficiently different and greater than the coherence length of the laser light. In addition, a multiregion retardation plate is placed at the entrance of the light pipe to reduce the coherence length of the laser light. The minimum required light pipe aspect ratio (length divided by width) depends upon the input beam divergence angle θ:

$$R_{min} = \cot \theta \tag{3.8}$$

For a chosen aspect ratio R (preferably $\geq$ 1.5–2x R_{min}) and laser coherence length C_L, the minimum width of the light pipe can be calculated from the Equation (3.1):

$$W_{min} = C_L \left(R + \sqrt{1 + R^2} \right) > 2R\, C_L \tag{3.9}$$

Figure 3.35 illustrates a similar solution,[106] in which a random phase mask is positioned between the fast axis collimated laser diode bar and the light pipe entrance aperture. The phase mask provides areas with different thicknesses, and rays from the same emitter will pass through different portions of the phase mask and will acquire small phase retardation differences, preferably about half wave. As phase is then scrambled, coherence or interference effects are reduced. Alternate methods to reduce the spatial coherence of laser arrays while using diffusers in combination with light pipes are also known.[107]

In addition to providing redundancy and increasing available output power, increasing the number of laser emitters can also help to reduce the apparent level of

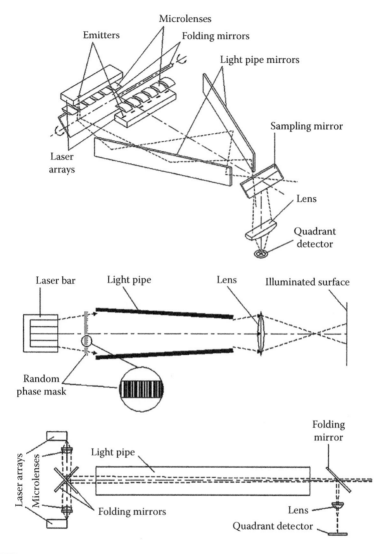

FIGURE 3.35 Dual laser array combining: perspective and cross-sectional views. (From U.S. Patent 7,209,624.[110])

interference effects by averaging light from more independent sources. Notably, off-the-shelf laser diode bars typically have 10–19 emitters for 30% fill factor bars and 40+ emitters for 50% fill factor bars. Unfortunately, increasing the fill factor much further reduces the individual emitter brightness, thus reducing the gains.

Alternatively, light from multiple laser devices can be combined down a common illumination optical path. For example, wavelength multiplexing with dichroic combiners can be employed, but the laser bandwidth increase (e.g., wavelengths λ_1, λ_2, λ_3... as 810 nm, 825 nm, and 850 nm, respectively) can make the downstream imaging optics much more difficult. Polarization multiplexing, in which a half-wave plate rotates the polarization of light from one laser device and a polarization combiner redirects that light onto the optical axis of the other laser, can also be used. However, this method of combining lasers is not suitable for systems that use polarization-dependent modulators such as the Xerox TIR modulator of Figure 3.27.

Therefore, it is more common in these systems to combine light from multiple lasers in angle space.[108–110] Light can be combined in either the array or cross-array directions, but the latter can require that smile be well controlled or corrected, so there is sufficient Lagrange available. In either case, laser arrays can be placed in parallel along the long axis, and the emitted light from at least one laser device can be deflected into the common optical path.[108,109] As shown in Figure 3.35, light from two collimated laser bars[110] is directed by means of two folding mirrors to the entrance aperture of a tapered light pipe. Array direction light from each of the two arrays is mixed, and the net Lagrange is nearly the sum of the beam products of the two bars. A minimum beam product value is achieved in case of straight light pipe, whereas, in the fast axis direction, the beam product is essentially unchanged.

As each laser diode bar independently provides an incident line of illumination light to the spatial light modulator, care is required to ensure that these multiple lines coincide at the center of the modulator. Thus, as shown in Figure 3.35, a sampling mirror directs a small part of the power to a quadrant detector. The signal from the detector is then used by a servo system to adjust the two folding mirrors to precisely overlap the two illumination lines.

3.7.5 ALTERNATE ARRAY PRINT-HEAD DESIGNS

Modulator array-type printers[12,111] have also been explicitly developed for use with the grating light valve (GLV) modulator.[93,94] In one exemplary system,[12] developed by Agfa Corporation, the illumination beam is focused onto the modulator array, and the zero-order diffracted light is imaged to the media plane. An illumination system,[112] shown in Figure 3.31, and designed in conjunction with this system, provides an array direction (slow axis) cylindrical lenslet array pair, comprising a collimating lens and an imaging lens for each emitter. This system magnifies each emitter to create nonoverlapped magnified images of the multimode near-field laser profiles, such that these images nearly touch, thereby providing a high fill factor line of illumination. This slow axis light propagates forward and encounters the cylindrical combiner collimator lens, which provides far-field images of each emitter in overlapping fashion as illumination to the modulator array.

In the fast axis (cross-emitter) direction, the rod lens collimates the beam and the array direction optical elements are regarded as windows. The fast axis narrowing lens then makes the beam slowly convergent. The cross-array beams are subsequently intercepted before the modulator array by a third cross-array cylinder lens, such that a collimated cross-array far-field *image* is presented to the modulator array.

This system is somewhat similar to the fly's eye-based illumination system[11,102] presented previously, although that system reimaged and homogenized the near-field light profiles, rather than the far-field light profiles. However, the system of Figure 3.36 does not actually provide the means to correct a systematic repeating variation in the far-field spatial light profile from emitter to emitter, if such a

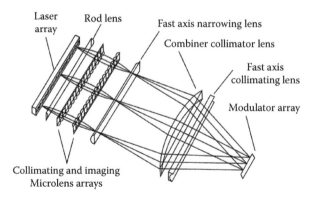

FIGURE 3.36   Laser printer illumination with array direction imaging of the emitters for an improved fill factor. (From U.S. Patent 6,433,934.[112])

pattern should occur (see Figure 3.7). Effectively, this system provides Koehler-like illumination in the array and cross-array directions to the modulator.

In general, the modulator-array-type systems will benefit by the use of smile correction optics,[61–64] so that the cross-array beams can be combined with minimal Lagrange growth. Alternately, a cross-array optical design was proposed[44] that does not require additional optics or mechanics, in which aberrations are deliberately introduced into the cross-array optics as a means of smile correction. However, as the cross-array beam is broadened, the source radiance is effectively reduced.

The applicability of the modulator-type laser thermal printing system can potentially be further extended by combining the light from multiple diode laser arrays,[5,56,108] in order to hit higher power levels at the media. In addition, there are other system concepts in which a modulator array is illuminated by a laser but without consideration for source redundancy. In one instance,[20] a single high-power laser beam, such as from an argon laser, is split into a multitude of beams by a pair of beam splitters to illuminate a pair of AOM-type modulator arrays, such that each modulator pixel receives an individual beam. While this system is susceptible to the failure of the laser source, the beam splitting arrangement does provide the means to illuminate each modulator pixel identically. In another exemplary system,[113] an AOM array can be flood-illuminated by one or two laser sources, which may be either single-mode or multimode laser sources. Although the two-laser case (which combines beams by means of a polarization beam splitter) may provide a minimum of source redundancy, the design lacks provisions to uniformly illuminate the modulator array.

3.8   THERMAL EFFECTS IN OPTICS

These laser printing systems have been discussed relative to enabling processing of printing plates by thermal ablation. In higher energy systems, such as IR laser systems used for laser cutting and laser welding, and pulsed systems, energy densities can be high enough to cause significant thermal problems in the optics themselves.

Although the energy levels are not high enough to cause laser-induced damage (e.g., compaction or cracking), both thermal defocus and thermally induced stress birefringence[114] can cause significant problems at these light levels used in laser thermal printing.

3.8.1 THERMAL DEFOCUS

In general, optical design approaches for reducing laser-induced thermal defocus have emerged from the laser materials processing field, where lasers are used to cut, drill, and weld metals and other materials. In laser systems, thermal defocus generally occurs with dn/dT, the change of refractive index with temperature, although thermal soak temperature increases and thermal gradients can cause different effects. These sensitivities can be described by two terms, each of which depend upon dn/dT: the thermal glass constant (γ), which represents the thermal power change that is due to an optical material, and the thermo-optic constant (G), which accounts for the thermal response of an optical material to radial gradients.[115] In one example, using thermal glass constant (γ), an IR laser system[116] is described in which light from a 20 kW fiber laser is to be imaged to a material, and three simultaneous equations are solved to achromatize and athermalize the lens so as to maintain focus over a wide temperature range.

Although the laser printing systems typically operate at appreciably lower power levels than those in 1–20 kW laser thermal processing systems, a mere 80 W of CW optical flux can cause significant thermal defocus if the laser wavelength is even only moderately close to absorption peaks of one or more of the optical materials used (e.g., absorption coefficient $\alpha \approx 0.001$ mm^{-1}) in the lens. Depending upon the system design, and depth of focus or Rayleigh range thereof, the resulting focus shifts may or may not be significant. Although use of autofocus may resolve any problems that occur, it still can be valuable to design the optics to reduce these effects. In addition to applying a design approach[116] that reduces thermal defocus, careful glass selection can be key. Such efforts can emphasize using glasses with low-absorption (e.g., fused silica) or near-zero thermal gradient sensitivity (e.g., calcium fluoride-variant glasses such as N-PK51A). It can also be valuable to select glasses for one or more elements in a multielement lens that have a negative gradient dn/dT value, while the other elements have a positive dn/dT value, so that dn/dT values compensate.[116] Notably, calcium fluoride and other fluoride glasses (e.g., BaF_2 and LiF_2) satisfy the first requirement by having negative dn/dT values. However, it should be noted that many of these materials, including crystalline materials such as CaF_2, are susceptible to thermal shock and fracturing, so care is required in their use.

3.9 OPPORTUNITIES AND CONCLUSIONS

Since the late 1980s, environmental concerns have demanded new printing technology that is environmentally conscious and dry, free from any wet chemical processing. Laser thermal printing meets this requirement while it provides the same image quality as silver halide in terms of resolution and density for reflective and transparent images as well as printing plates. Laser thermal's requirement for more laser energy than silver halide can be satisfied by several multichannel print-head architectures.

The laser thermal print-heads that were developed to meet these needs span a range of viable optical design architectures, including the direct laser to media systems, optical fiber-based systems, the modulated subarray laser approach, and the spatial light modulator array-based systems. A range of enabling technologies, including micro-optics, high-power laser diode arrays, spatial light modulator arrays, and new optical media, have helped to make these systems possible. The associated laser beam shaping designs, while generally evolutionary, have employed the lasers, micro-optics, and other components, in complex and elegant designs that combine classical imaging and Gaussian beam optics into unitary systems. Furthermore, many of these systems, and the designs that enabled them, have been proven viable in the market place.

Certainly, as new components emerge, there will be further opportunities to design higher performing and lower cost systems. For example, optically pumped fiber lasers could be a useful high-power laser light source for these types of systems. Moreover, there are market opportunities beyond laser thermal printing, such as color laser image projection[117–119] and the manufacture of organic LED devices,[120] which can potentially benefit the design approaches and solutions developed to support laser thermal printing.

REFERENCES

1. Leger, J. and Golstos, W., Geometrical transformation of linear diode-laser arrays for longitudinal pumping of solid state lasers, *IEEE J. Quant. Electron.*, 28(4), 1088–1100, 1992.
2. Goring, R., Schreiber, P., and Poβner, T., Microoptical beam transformation system for high-power laser diode bars with efficient brightness conservation, *SPIE Proc.*, 3008, 202–210, 1997.
3. Fan, T., Sanchez-Rubio, A., Walpole, J., Williamson, R., McIngailis, I., Leger, J., and Goltsos, W., Multiple-laser pump optical system, U.S. Patent 5,081,637, 1992.
4. Head, D. and Baer, T., Apparatus for coupling a multiple emitter laser diode to a multi-mode optical fiber, U.S. Patent 5,436,990, 1995.
5. Ben Oren, I., Ronen, J., Steinblatt, S., and Komem, A., Optical system for illuminating a spatial light modulator, U.S. Patent 5,900,981, 1999.
6. Yip, K., Thermal dye transfer apparatus using semiconductor laser diode arrays, U.S. Patent 4,804,975, 1989.
7. Kessler, D. and Endriz, J., Optical means for using diode laser arrays in laser multibeam printers and recorders, U.S. Patent 5,745,153, 1998.
8. Thornton, R., Incoherent, optically uncoupled laser arrays for electro-optic line modulators and line printers, U.S. Patent 4,786,918, 1988.
9. Gelbart, D., Apparatus for imaging light from a laser diode onto a multi-channel linear light valve, U.S. Patent 5,517,359, 1996.
10. Sarraf, S., Light modulator with a laser or laser array for exposing image data, U.S. Patent 5,521,748, 1996.
11. Kurtz, A. and Kessler, D., Laser printer using a fly's eye integrator, U.S. Patent 5,923,475, 1999.
12. Reznichenko, Y. and Kelley, H., Optical imaging head having a multiple writing beam source, U.S. Patent 6,229,650, 2001.

13. Gelbart, D., *High power multi-channel writing heads, in IS&T 10th International Congress on Advances in Non-Impact Printing Technologies,* pp. 337–339, 1994.
14. Kessler, D., Optical design issues of multi-spots laser thermal printing, *Proc. IS&T/OSA Annu. Conf.,* pp. 215–219, IS&T, Springfield, VA, 1996.
15. Kurtz, A., Optical systems for laser thermal printing, *SPIE Proc.,* 5525, 11–30, 2004.
16. Marshall, G.F., ed., *Optical Scanning,* Marcel Dekker, New York, 1991.
17. Williams, R., Pensavecchia, F., Kline, J., and Lewis, T., Method for laser-discharge imaging a printing plate, U.S. Patent 5,351,617, 1994.
18. Hsu, K., Owens, J., and Sarraf, S., Fiber optic array, U.S. Patent 4,911,526, 1990.
19. Etzel, M., Image recorder with linear laser diode array, U.S. Patent 4,978,974, 1990.
20. Tanuira, H., Optical beam splitting method and an optical beam splitting modulation method, U.S. Patent 4,960,320, 1990.
21. Moulin, M., Illumination system and method for spatial modulators, U.S. Patent 6,137,631, 2000.
22. Baek, S.H., Haas, D., Kay, D., Kessler, D., and Sanger, K., Multi-channel laser thermal printhead technology, in *Handbook of Optical and Laser Scanning,* Marshall, G. F., ed., Marcel Dekker, New York, pp. 711–767, 2004, Chap. 14.
23. DeBoer, C., Sarraf, S., Weber, S., Jadrich, B., Haas, D., Kresok, J., and Burberry, M., Instant transparencies by laser dye transfer, *IS&T 46th Annu. Conf.,* pp. 201–203, IS&T, Springfield, VA, 1993.
24. Sarraf, S., DeBoer, C., Haas, D., Jadrich, B., Connelly, R., and Kresock, J., *Laser thermal printing, in IS&T 9th International Conference of Advances In Non-Impact Printing Technologies,* pp. 358–361, IS&T, Springfield, VA, 1993.
25. Lamberts, R., Sine-wave response techniques in photographic printing, *J. Opt. Soc. Amer.,* 51, 982–987, 1961.
26. Thomas, W. Jr., ed., *SPSE Handbook of Photographic Science and Engineering,* Wiley, New York, p. 424, 1973.
27. Levene, M., Scott, R., and Siryj, B., Material transfer recording, *Appl. Opt.,* 9, 2260–2265, 1970.
28. Sirohi, R. and Kothiyal, M., *Optical Components, Systems, and Measurement Techniques,* Marcel Dekker, New York, p. 95, 1991.
29. Sanger, K., Mackin, T., and Schultz, M., Method and apparatus for the calibration of a multichannel printer, U.S. Patent 5,291,221, 1994.
30. Sanger, K., Mackin, T., and Schultz, M., Method of calibrating a multichannel printer, U.S. Patent 5,323,179, 1994.
31. Kitamura, T., Beam recording apparatus effecting the recording by a plurality of beams, U.S. Patent 4,393,387, 1983.
32. Haas, D. and Owens, J., Single-mode fiber printheads and scanline interleaving for high-resolution laser printing, *SPIE Proc.,* 1079, 420–426, 1989.
33. Baek, S.H. and Mackin, T., Thermal printer, U.S. Patent 5,164,742, 1992.
34. Guy, W. and Mackin, T., High resolution thermal printers including a print head with heat producing elements disposed at an acute angle, U.S. Patent 5,258,776, 1993.
35. Baek, S.H. and Mackin, T., Thermal printer capable of using dummy lines to prevent banding, U.S. Patent 5,278,578, 1994.
36. Tsao, S., Apparatus for arranging scanning heads for interlacing, U.S. Patent 4,232,324, 1980.
37. Haas, D., Method of scanning, U.S. Patent 4,900,130, 1990.
38. Starkweather, G. and Dalton, J., Error reducing raster scan method, U.S. Patent 5,079,563, 1992.
39. Haas, D., Mackin, T., Sanger, K., and Sarraf, S., Interleaving thermal printing with discontiguous dye-transfer tracks on an individual multiple-source print-head pass, U.S. Patent 5,808,655, 1998.

40. Sousa, J., Williams, R., Ruda, M., and Foster, J., Apparatus for laser-discharge imaging and focusing elements for use therewith, U.S. Patent 5,764,274, 1998.

41. Sousa, J., Method and apparatus for laser imaging with multi-mode devices and optical diffusers, U.S. Patent 6,210,864, 2001.

42. Yariv, A., *Solutions Manual for Optical Electronics*, Holt, Rinehart, and Winston, New York, pp. 27–29, 1991.

43. Yariv, A., *Quantum Electronics*, Wiley, New York, p. 129, 1967.

44. Kogelnik, H., Propagation of laser beams, in *Applied Optics and Optical Engineering*, Vol. VII, Shannon, R. and Wyant, J., eds., Academic Press, New York, pp. 155–190, 1979, Equations 85 and 86.

45. Wolfenden, B., Capitalizing on digital technology, Photonics Spectra, November 2001.

46. Clark, P. and Londono, C., Radiation source for a printer, U.S. Patent 5,161,064, 1992.

47. Diode Laser Arrays, in *Cambridge Studies in Modern Optics*, Botez, D. and Scifres, D., eds., Cambridge University Press, New York, 1994.

48. Reynolds, G., DeVelis, J., Parrent, G. Jr., and Thompson, B., *The New Physical Optics Notebook: Tutorials in Fourier* Optics, SPIE Optical Engineering Press, Washington, DC, 1989, Chap. 11.

49. Born, M. and Wolf, E., *Principles of Optics*, 6th ed., Pergamon Press, Oxford, 1980, Sec. 10.4.

50. Goodman, D., Roblee, J., Plummer, W., and Clark, P., Multi-laser Print Head, *SPIE Proc.*, 3430, 6–26, 1998.

51. Marciante, J. and Agrawal, G., Lateral spatial effects of feedback in gain-guided and broad-area semiconductor lasers, *IEEE J Quant. Electron.*, 32(9), 1630–1635, 1996.

52. Nicodemus, F., *Radiometry, Applied Optics and Optical Engineering, Vol.* IV, in Kingslake, R., ed., Academic Press, New York, 1965, Chap. 8.

53. Welford, W., *Aberrations of Optical Systems*, Adam Hilger, London, 1989.

54. O'Shea, D., *Elements of Modern Optical Design*, Wiley-Interscience, New York, 1985.

55. Heidel, J., Zediker, M., Throgmorton, K., and Harting, W., Array of diffraction limited lasers and method of aligning Same, U.S. Patent 5,212,707, 1993.

56. Lang, R., Laser diode arrays with optimized brightness conservation, U.S. Patent 6,240,116, 2001.

57. Kessler, D. and Simpson, J., Jr., Multi-beam optical system using lenslet arrays in laser multi-beam printers and recorders, U.S. Patent 5,619,245, 1997.

58. Endriz, J., Diode laser source with concurrently driven light emitting segments, U.S. Patent 5,594,752, 1997; and U.S. Patent 5,802,092, 1998.

59. Holdsworth, A. and Baker, H., Assessment of micro-lenses for diode bar collimation. *SPIE Proc.*, 3000, 209–214, 1997.

60. Bielak, R., Self registering microlens for laser diodes, U.S. Patent 5,420,722, 1995.

61. Blanding, D., Bent smile corrector, U.S. Patent 6,166,759, 2000.

62. Kessler, D. and Blanding, D., Optically straightening deviations from straightness of laser emitter arrays, U.S. Patent 5,854,651, 1998.

63. Gelbart, D., Microlensing for multiple emitter laser diodes, U.S. Patent 5,861,992, 1999.

64. Harrigan, M., Optical compensation for laser emitter array non-linearity, U.S. Patent 5,629,791, 1997.

65. McBride, R., Trela, N., Wendland, J., and Baker, H., Extending the locking range of VHG-stabilized diode laser bars using wavefront compensator phaseplates, *SPIE Proc.*, 8039, 80390F, 2011.

66. Askinazi, M. and Burton, H., Laser array printing, U.S. Patent 5,461,413, 1995.

67. Steinblatt, S., Plotting head with individually addressable laser diode array, U.S. 5,986,819, 1999.

68. Whitney, T., Apparatus for imaging light from multifaceted laser diodes onto a multi-channel spatial light modulator, U.S. Patent 6,356,380, 2002.

69. Goodman, D., Basic optical instruments, in *Geometrical and Instrumental Optics*, Malacara, D., ed., Academic Press, New York, 1988, Chap. 4.

70. Clark, P. and Londono, C., Printing medical images with high-power diode lasers, *OSA Proceedings of the International Optical Design Conference*, pp. 377–382, OSA, Washington, DC, 1994.

71. Baxter, K. and Goodman, D., Laser assembly with integral beam-shaping lens, U.S. Patent 5,793,792, 1998.

72. Goodman, D., Gordon, W., Jollay, R., Roblee, J., Gavrilovic, P., Kuksenkov, D., Goyal, A., and Zu, Q., High brightness multi-laser source, *SPIE Proc.*, 3626A, 1999.

73. Singh, R., Chin, A., Zu, Q., Dabkowski, F., Jollay, R., Bull, D., Fanelli, J., Goodman, D., Roblee, J., and Plummer, W., Description and applications of high brightness multi-laser-diode system, *SPIE Proc.*, 3945A, 2000.

74. Sousa, J., Method of calibrating distances between imaging devices and a rotating drum, U.S Patent 6,091,434, 2000.

75. Sousa, J., Foster, J., and Mueller, W., Diode-pumped laser system and method, U.S. Patent 5,822,345, 1998.

76. Miller, C., *Optical Fiber Splices and Connectors*, Marcel Dekker, New York, pp. 266–267, 1986.

77. Kaukeinen, J. and Tyo, E., Method of making a fiber optic array, U.S. Patent 4,880,494, 1989.

78. Baek, S.H. and DeBoer, C., Scan laser thermal printer, U.S. Patent 5,168,288, 1992.

79. DeJager, D. and Baek, S. H., Thermal printer system with a high aperture micro relay lens system, U.S. Patent 5,258,777, 1993.

80. Sarraf, S., Apparatus and method for eliminating feedback noise in laser thermal printing, U.S. Patent 5,420,611, 1995.

81. Gelbart, D., Pilossof, N., and Weiss, A., Individually addressable laser diode arrays based imaging systems with increased redundancy, U.S. Pat. Pub. 20030210861, 2003.

82. Endriz, J., Diode laser source with concurrently driven light emitting segments, U.S. Patent 5,802,092, 1998.

83. Wyatt, S., Integration and characterization of multiple spot laser array printheads, *SPIE Proc.*, 3000, 169–177, 1997.

84. Turner, W. and Sprague, R., Integrated total internal reflection (TIR) spatial light modulator for laser printing, *SPIE Proc.*, 299, 76–81, 1981.

85. Sprague, R., Turner, W., and Flores, L., Linear total internal reflection spatial light modulator for laser printing, *SPIE Proc.*, 299, 68–75, 1981.

86. Sprague, R., Turner, W., Hecht, D., and Johnson, R., Laser printing with the linear TIR spatial light modulator, *SPIE Proc.*, 396, 44–49, 1983.

87. Sprague, R. and Johnson, R., TIR Electro-optic modulator with individually addressed electrodes, U.S. Patent 4,281,904, 1981.

88. Sprague, R., Thick film line modulator, U.S. Patent 4,376,568, 1983.

89. Yip, K., Imaging system utilizing an electro-optic device, U.S. Patent 4,591,260, 1986.

90. Mir, J., Linear light valve arrays having transversely driven electro-optic gates and method of making such arrays, U.S. Patent 4,707,081, 1987.

91. Kitano, H., Saito, I., Shingaki, K., Matsubura, K., and Masuda, T., Electro-optical light shutter device and printer apparatus using same, U.S. Patent 4,887,104, 1989.

92. Hornbeck, L., Active yoke hidden hinge digital mirror device, U.S. Patent 5,535,047, 1996.

93. Hornbeck, L., Multi-level deformable mirror device with torsion hinges placed in a layer different from the torsion beam layer, U.S. Patent 5,600,383, 1997.

94. Bloom, D., Sandejas, F., and Solgaard, O., Method and apparatus for modulating a light beam, U.S. Patent 5,311,360, 1994.

95. Bloom, D., Corbin, D., Banyai, W., and Straker, B., Flat diffraction grating light valve, 5,841,579, 1998.
96. Kowarz, M., Spatial light modulator with conformal grating device, U.S. Patent 6,307,663, 2001.
97. Nutt, A., Ramanujan, S., and Revelli, J., Optical modulator for printing, U.S. Patent 6,211,997, 2001.
98. Ramanujan, S., Kurtz, A., and Nutt, A., Grating modulator array, U.S. Patent 6,084,626, 2000.
99. Ramanujan, S. and Kurtz, A., Laser printer utilizing a spatial light modulator, U.S. Patent 6,169,565, 2001.
100. Nochebuena, R. and Paoli, T., Electroabsorptive asymmetrical fabry-perot modulator array for line printers, U.S. Patent 5,414,553, 1995.
101. Konno, K. and Kanagawa, M., Light illuminating device, U.S. Patent 4,497,015, 1985.
102. Kurtz, A., Design of a laser printer using a laser array and beam homogenizer, *SPIE Proc.*, 4095, 147–153, 2000.
103. Ota, T., Multi-beam laser light source and multi-beam semiconductor laser array, U.S. Patent 5,465,265, 1995.
104. Nishi, N., Jitsuno, T., Tsubakimoto, K., Murakami, M., Nakatsuka, M., Nishihara, K., and Nakai, S., Aspherical multi lens array for uniform target irradiation, *SPIE Proc.*, 1870, 105–111, 1993.
105. Fan, B., Tibbetts, R., Wilczynski J., and Witman, D., Laser beam homogenizer, U.S. Patent 4,744,615, 1988.
106. Reynolds, M., Random phase mask for light pipe homogenizer, U.S. Pat. Pub 20080239498, 2008.
107. MacKinnon, N., MacAulay, C., and Stange, U., Apparatus and methods relating to wavelength conditioning of illumination, U.S. Patent 6,781,691, 2004.
108. Simpson, J. Jr., Multiple laser array sources combined for use in a laser printer, U.S. Patent 6,064,528, 2000.
109. Wang, T., Shinkoda, I., Goldstein, K., and Reynolds, M., Method and apparatus for illuminating a spatial light modulator with light from multiple laser diode arrays, U.S. Patent 6,853,490, 2005.
110. Reynolds, M., Dets O., Shiue S., Shinkoda I., and Speier I. Apparatus and method for illumination of light valves, U.S. Patent 7,209,624, 2007.
111. Ide, A., Yamasa, H., and Yamamoto, Y., Electro-static printer having an array of optical modulating grating valves, U.S. Patent 6,025,859, 2000.
112. Reznichenko, Y., Davydenko, V., and Lissotchenko, V., Illumination system for use in imaging systems, U.S. Patent 6,433,934, 2002.
113. Gross, A., Laser marking apparatus including an acoustic modulator, U.S. Patent 5,309,178, 1994.
114. Kurtz, A. and Bietry, J. Lens design with reduced sensitivity to thermally induced stress birefringence, *Appl. Opt.*, 52(18), 4311–4322, 2013.
115. Rogers, P. and Roberts, M., Thermal compensation techniques, Chap. 39, in *Handbook of Optics*, McGraw-Hill, New York, 1994.
116. Scaggs, M. and Haas, G., Thermal lensing compensation objective for high power lasers, *SPIE Proc.*, 7913, 2011.
117. Kurtz, A., Kruschwitz, B., and Ramanujan, S., Laser projection display system, U.S. 6,577,429, 2003.
118. Kowarz, M., Brazas, J., and Phalen, J., Electromechanical grating display system with spatially separated light beams, U.S. Patent 6,411,425, 2002.
119. Amm, D. and Corrigan, R., Optical performance of the grating light valve technology, *SPIE Proc.*, 3634, 71–78, 1999.
120. Kay, D., Tutt, L., and Bedzyk, M., Using a multichannel linear laser light beam in making OLED devices by thermal transfer, U.S. Patent 6,582,875, 2003.

4 Practical UV Excimer Laser Image System Illuminators

Jeffrey P. Sercel and Michael von Dadelszen

CONTENTS

4.1 INTRODUCTION

The illumination section of an excimer laser system is central to its performance, harnessing the full potential of this powerful production tool's strengths and capacity. This optical device serves to shape the beam, control its divergence, control the beam profile through the use of different homogenizers, and geometrically increase the efficiency of a process.[7] The techniques outlined herein are not merely theoretical; in fact, they have been proven and are continually being proven and optimized on a daily basis in production facilities around the world.

4.2 CHARACTERISTICS OF EXCIMER LASERS

Excimer lasers are high average power ultraviolet (UV) laser sources with many significant characteristics that make them ideal for high-resolution materials processing. Excimer lasers operate at a variety of user-selectable UV wavelengths from 157 to 351 nm. This allows processes to be optimized based on absorption. For example, submicron layers of materials can be removed with each laser pulse. This characteristic alone makes excimer lasers remarkably different from other laser types.

Short UV excimer laser wavelengths can be projected onto material with very high resolution. Even with the use of simple lenses to shape and direct the beam, micron resolution is easily achieved.

Excimer laser beams can cover relatively large areas of material with effective processing results. Thus, near-field imaging can be used to project a mask image onto a workpiece so that complex features can be patterned (Figure 4.1).

Excimer laser beams are not perfectly uniform in terms of intensity over the area of the beam. Therefore, only a portion of the area of the raw laser beam is usable for high-uniformity materials processing. In some cases, only the most uniform section or *filet* of the beam will be selected for use, and the nonuniform section of the laser beam will be discarded. Moreover, as an excimer laser ages, its beam quality changes. This can affect process control and therefore must be addressed.

It stands to reason that high duty-cycle processing is much more cost effective if optical techniques are employed in order to use a larger fraction of the available laser power. Furthermore, due to the premium price associated with UV photons, high beam utilization—known as the beam utilization factor (BUF)—is often a key economic factor that can qualify or disqualify an otherwise technically feasible application. Beam efficiency enhancers (or beam shaping optics and beam homogenizers) can be employed to shape the beam to expose the open area of the mask and simultaneously render the laser energy uniform.

To efficiently use a higher percentage of the available UV excimer laser beam energy, optical techniques are employed to ensure uniform irradiation over large areas in effectively three dimensions. Imaging optics is used to control feature accuracy over larger fields of view (2D). Beam shaping or beam homogenizer techniques are used for exposure control (3D). The latter comprise the illumination section.

We will review excimer laser beam delivery techniques and examine actual industrial case histories of UV excimer laser processing and the optics required to effectively use these lasers in cost-effective manufacturing situations.

As a final note, a figure of merit for an effective UV industrial system is the number of optics required to perform a given materials processing function. Designs that minimize the number of optics have shown to be the most effective (and most cost effective) for a wide range of common production applications.

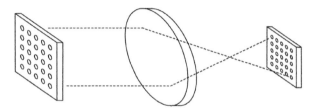

FIGURE 4.1 Near-field imaging.

4.2.1 Photo-Ablation with Excimer Lasers

The method of materials removal with excimer lasers is unique and is a direct function of the laser's characteristic form and energy type. Known as laser (photo) ablation, this occurs when small volumes of materials absorb high peak power laser energy. Each laser pulse etches a fine submicron layer of material and the ejecting material carries the heat away with it. Depth is obtained by repeatedly pulsing the laser and depth control is achieved through overall dosage control. Etch rate per laser pulse is usually plotted against on-target laser fluence to characterize a material (Figure 4.2).

Generally, most materials exhibit strong UV absorption characteristics, such that most of the UV energy is absorbed in low volumetric, submicron depths. This suggests that excimer laser beams can be used to process materials at relatively low-focus intensities. Excimer lasers tend to be available in short pulse width and high pulse energies, and it is clear that effective materials processing occurs at relatively large areas of focus (i.e., a 500 mJ UV beam at 1 J/cm² target fluence theoretically can expose an area of up to 7×7 mm²).

Conversely, compared with UV absorptivity, most materials tend to absorb visible, near IR, and IR laser energy less readily, due to either volume absorption or surface reflection. In practice, these longer wavelength lasers overcome their lower relative optical absorption in materials with very high focal irradiance of the workpiece. In fact, in most applications, these longer wavelength lasers usually operate at the far field (i.e., a small spot at the focal point of a lens). If the absorption depth of these longer wavelength lasers is large, then higher irradiance is required to vaporize deep volumes. High irradiance is also used to break (damage and reduce) the surface reflectivity of materials, thus allowing energy to couple to workpiece materials mainly through thermal processes.

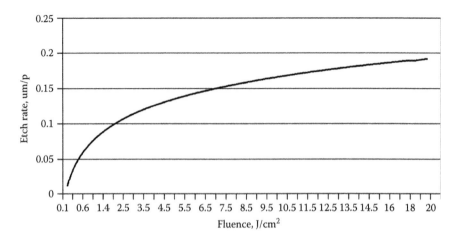

FIGURE 4.2 Etch rate versus fluence.

4.2.2 OVERVIEW OF EXCIMER LASER APPLICATIONS

Excimer lasers have been used for numerous industrial applications for more than two decades. Their unique features include a large area, short pulse width UV beam that is capable of causing material interactions at relatively low energy densities.

Processes that favor excimer lasers (e.g., micromachining) usually require high-accuracy material removal of both 2D feature geometry (image quality), as well as 3D geometry (process depth control).

Numerous other applications are well suited for large area, high power UV excimer laser beams. The smooth, uniform excimer beams are efficiently used (generally at low fluence and over large process areas) for applications such as semiconductor processing, lithography, eye surgery, display glass annealing, substrate cleaning, surface alloying, selective materials removal, planarization and polymer restructuring, surface treatment, pulsed laser deposition, and photochemical reactions. They are also used for dry cleaning objects such as wafers or flat panel displays, as well as for micromachining, marking, and annealing.

4.2.3 EXCIMER OPTICAL SYSTEM COMPONENTS

An excimer laser's optical system can be subdivided into the following three main sections:

1. The light source or laser
2. The illumination system
3. The imaging section

4.2.4 THE LIGHT SOURCE: UV EXCIMER LASER BEAMS

Excimer laser beam profiles vary with laser types. X-ray preionized lasers, for example, tend to be characterized by square, flat-topped beam shapes that are also extremely homogenous, but these lasers have not survived the test of time in production applications. As a result, most industrially suitable excimer lasers feature beams that are characterized by imperfect beam shapes. Consequently, they suffer from the effects of unevenness that are depicted in the exaggerated beam profiles illustrated in Figure 4.3.

Excimer lasers are gas discharge lasers and, as a result of degradation or changes in multiple components (and therefore the discharge uniformity) over time, the characteristics of the beam change. For example, as the gas fill ages, changes in gas chemistry occur that alter the electrical discharge properties of the laser gas. This causes the width of the discharge to increase, resulting in beam growth in the short axis. The beam profile in the short axis gradually changes from a near Gaussian to a flat-topped beam profile. Along with beam growth, the divergence and incoherence of the laser beam increase, resulting in reduced peak pulse energy in the far field. Increased high voltage across the discharge also widens the beam. The typical excimer laser stabilizes its output based on total pulse energy. Therefore, as beam area increases, power density drops. This is a concern for applications that require

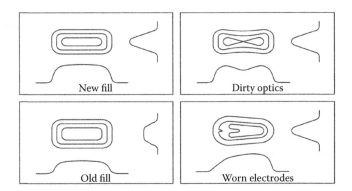

FIGURE 4.3   Beam profile examples.

tight control of power density on target. Due to these beam profile characteristics, the illumination section of the laser optical delivery system is required to shape, homogenize, and redistribute the laser energy. Finally, as the optics in a laser become contaminated or wear out, the beam profile also changes.

4.2.5   ISSUES IMPACTING BEAM QUALITY

1. Uniform, stable beam shape is critical to proper delivery of laser energy to the target.
2. Proper resonator optics alignment is critical to good beam quality.
3. Beam profile changes with HV changes or during halogen injections.
4. Regular resonator optics cleaning is essential to maintaining good beam quality. Inside surfaces of resonator optics become contaminated by dust formed inside the laser vessel during the discharge process.
5. Beam nonuniformities become exaggerated as the gas fill ages.
6. A trapezoidal or split beam can occur when preionization pins or electrodes are worn.
7. The use of a beam profilometer can aid in diagnosing beam quality problems.

4.3   NEAR-FIELD IMAGING: THE CONCEPT OF MASK ILLUMINATION

Excimer laser systems employ near-field imaging beam delivery techniques. Near-field imaging is simply the use of an imaging lens to image an object at a plane other than that of the focal point of the lens. Near-field imaging contrasts with far-field imaging, where lasers are used at the focal point of the lens. In near-field imaging, a mask is imaged at its conjugate plane. Near-field imaging provides control of spot size and hole geometry, and is capable of producing complex, large area features. It involves the use of a mask to project a pattern of laser light onto a part (see Figure 4.1). The features contained in a pattern are then projected onto the target material at a magnification determined by the relative positioning and the focal lengths of the optical elements. This technique is the basis for excimer laser micromachining in many materials processing systems.

Imaging systems that provide demagnification are most commonly used because compression of the beam is required to achieve fluence levels that result in ablation of the target. Simple lens systems can provide resolutions down to 10 μm, while more complex systems can achieve submicron resolutions with higher cost optics.

In excimer laser micromachining, *imaging systems* are used to control the power and spatial distribution of UV light on target. In an imaging system, the power distribution of the light at the object (mask) plane is relayed to the image (work) plane. The simplest laser imaging system consists of a mask and an imaging lens. More complex systems may include beam conditioning optics and automatic mask aligners. Shaping, manipulating, and directing the beam *via* laser processing optics, illuminators, and imaging optics enhance excimer flexibility and applicability.

Typically, in excimer laser micromachining, only a fraction of the UV beam is used to ablate or expose small features into a material. Very precise features ranging in size from one to hundreds of microns in size are produced. These features can be positioned and aligned accurately to submicron tolerances. The excimer beam, in this case, is used at low efficiency by selecting a uniform central portion of the beam— the *filet*—and throwing away or not utilizing the remainder of the available beam energy. While this may seem to be a low-efficiency process, it is widely used and the illumination section of the beam delivery system incorporates few (or no) optics.

4.3.1 BUF Concept

One engineering challenge presented by the use of mask imaging involves the concept of BUF. BUF becomes a significant factor in an application such as via (and even smaller microvia) drilling, wherein multiple holes of specific sizes and depths must be drilled in printed circuit boards (PCBs) in order to establish interconnections between layers of circuitry (*vias*) and to connect external components to specific layers of circuitry. A given PCB can require—based upon its dimensions, circuit density, and number of layers—hundreds or even thousands of vias and microvias. Furthermore, microvia drilling applications are exacting in terms of via hole size and quality, as well as drilling speed. These parameters must be optimized. For example, current requirements call for microvia diameters in certain designs in the range of 75–150 μm, while higher density via designs require hole sizes on the order of 25 to 100 μm.

The simplest mode of imaging (e.g., drilling one via at a time) will yield BUF of less than 0.5%, resulting in very low overall via drilling rates. The laser beam must be utilized more efficiently to offer economic payback. If 100% of the available beam was used to drill vias, thousands of vias per second could be drilled. The author has pioneered many techniques that result in high BUF and has significantly increased the cost effectiveness of excimer laser processing systems. A trade-off for productivity versus flexibility is usually the case. Generally, the more *tooled up* an illumination system is, the less flexible it will be (Figure 4.4).

The raw beam usually suffers from nonuniformity, which normally crops a significant percentage of available beam energy. The amount of the beam that can actually be used for processing—referred to as the *usable fraction of the beam* (UFB) is

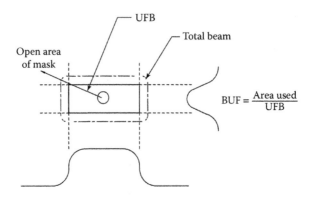

FIGURE 4.4   BUF diagram.

a fraction or filet of the entire raw beam. However, beam homogenizers can convert, capture, reconcentrate, and reconfigure much of the nonhomogeneous beam output to make it usable and can thus increase the UFB to over 95%. In addition, when the beam uniformity is improved over the full mask area, the vias contained within the mask will drill uniformly (Figure 4.5).

Proper mask illumination is a key factor in imaging beam delivery design. The illuminator section is independent (although interrelated) and is thus integral to the overall imaging system performance. The most basic role played by the illuminator section of the beam delivery system is matching the entrance numerical aperture (NA) of the imaging lens. See Figure 4.6. For instance, the laser mask area may be larger than the optimum f-number (diameter) of the imaging lens, which would normally overfill the imaging lens and cause aberrations and optical loss. In simplest terms, the first basic illuminator is a field lens that focuses the light into an optimum image lens entrance aperture.

FIGURE 4.5   Beam homogenizer profile.

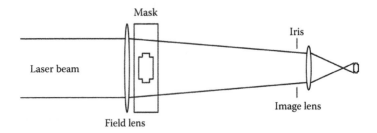

FIGURE 4.6   Field lens.

Beyond this, more complicated illuminators will simultaneously homogenize or shape the laser beam to provide a uniform laser intensity profile onto the mask.

Figure 4.7 shows an illuminator–homogenizer for a reflective Schwarzchild imaging objective. Here, the laser beam overlaps (becomes homogenous) at the mask plane, and is matched to the objective entrance aperture to avoid the central obscuration common to these kinds of image lenses.

Homogenizers tend to add considerable beam decollimation, making the illumination design of a homogenized beam a complex task that results in a fixed NA to which the imaging lens must be optimized. Illumination systems and imaging systems must be considered together in order to optimize the optical system's performance.

In high BUF applications, the illuminator section does all of the above and can also simultaneously optimize the laser intensity profile at the mask to make the most efficient use of laser photons. In this instance, the goal is to maximize the exposure of only the open area of the mask, and thus avoid wasting photons on blank mask areas. Figure 4.8 gives an illustration of using spot separation to improve BUF.

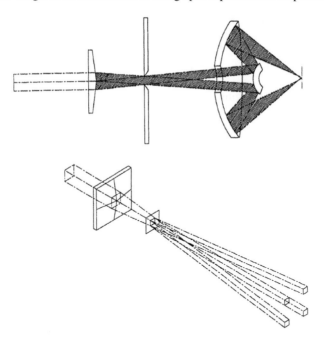

FIGURE 4.7 Homogenizer illuminator.

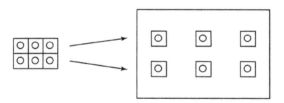

FIGURE 4.8 Geometrical spot separation and mask illumination.

4.3.2 MASK IMAGING

The simplest laser imaging system consists only of a mask and an imaging lens. Here, the raw laser beam is directed onto a mask and the mask is then imaged onto the part (Figure 4.9). This is commonly used for small area masks where the light impinged on the focusing lens is inside the optimum *f*-number of the lens.

As shown in (Figure 4.6), the first improvement to the basic mask imaging techniques is the addition of an illuminator section field lens. This lens directs the beam through the center of the imaging lens to reduce problems (such as spherical aberration and barrel distortion), and demonstrates the importance of illumination optics on imaging optics performance. Here, we see that the image lens performance is enhanced by the proper use of an illumination field lens.

4.3.2.1 Conventional Imaging Lenses

We define conventional lenses as imaging lenses of singlet, doublet, or multiple elements where the beam crosses over at the lens focal point and diverges as it comes to focus at its image plane (Figure 4.9). Beyond the focal plane, the beam tends to spread away from the beam centerline. One consequence is that for conventional imaging systems, as the lens to image distance varies, not only does the image go in and out of focus, but the size of the image also varies.

4.3.2.2 Telecentric Imaging Lenses

A telecentric imaging lens transfers the mask pattern to the part with near normal incidence over the field of view (Figure 4.10). As the lens-to-image distance varies, the image will still move in and out of focus, but the size of the image itself will remain almost constant, especially when compared to the conventional imaging lens.

The need for telecentric imaging systems is illustrated by considering the example of ink jet printer orifice drilling. If one is drilling multiple orifices per step and repeats this process, then the direction of the ink flow will be dictated by the orifice-drilling angle. Deviations from the ideal telecentric condition where the holes are

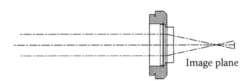

FIGURE 4.9   Conventional imaging lens.

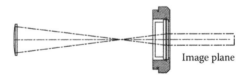

FIGURE 4.10   Telecentric imaging lens function.

drilled through thicker materials at an angle will quickly show up as inaccuracy in the ink patterns.

Thus, we can see that the illumination section of the beam delivery system has a powerful effect on the imaging system, by controlling image ray directionality.

4.3.3 Illumination Section Components

We have seen examples of how a laser beam is imaged onto a part, and how the illumination section can affect the performance of the imaging section. In summary, the following types of components can make up an imaging type laser beam delivery:

1. Variable attenuator
2. Beam shaping telescopes
3. Field lens or telescope field lens
4. Mask beam scanner
5. Coordinated motion mask
6. Beam homogenizer
7. Spot array generator (mask efficiency enhancer) (see Figure 4.11)

4.3.3.1 Variable Attenuator

A variable beam attenuator (Figure 4.12) is a common item included in the illumination section of the optical system of an excimer laser. Single plate walk-off type and dual plate zero walk-off type attenuators are common.

FIGURE 4.11 Prism telescope.

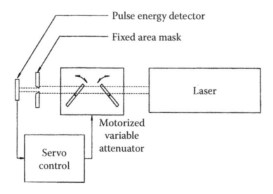

FIGURE 4.12 Variable attenuator configuration.

The angle-tuned attenuator operates when a laser beam is transmitted through an attenuator plate that is coated with a special angle sensitive dielectric coating. The attenuator plate is positioned through a variable angle to the beam, which then changes the optical transmission of the plate. A single plate will cause some beam walk-off. A dual plate system will compensate for this walk-off.

In most cases, the single-plate attenuator offers the best economy and does not affect the laser process because the laser beam walk-off is very small compared to the size of the laser beam. This is especially true when it is used with a beam homogenizer, such as a fly eye lens array, because the mask beam profile is less sensitive to the input beam position and shape.

4.3.3.2   Beam Shaping Telescopes

The next step up is the addition of a beam shaping telescope. This telescope can be a single axis beam expander (using cylindrical lenses) or a spherical beam expander using spherical lenses. In some cases, the beam size is reduced in order to intensify the laser energy on the mask (see Figure 4.13).

Beam expanders reduce the beam divergence and beam compressors will increase the laser beam divergence. This fact needs to be considered when designing the optical system.

A further type of beam shaping telescope is a dual anamorph. This can consist of two independent sets of cylindrical lens telescopes, but is more efficiently made up of two cylindrical lenses with an additional spherical lens that serve to replace two cylinder lenses. These dual anamorph telescopes control the beam size in two independent directions.

Beam shaping is the simplest technique for matching the laser beam to image mask features. Spherical and cylindrical telescopes may be employed to expand, compress, or change the aspect ratio of an excimer beam. By using beam shaping, the size and shape of the beam can be made to match the opening in the object mask, thereby increasing beam utilization efficiency. Beam compression, before

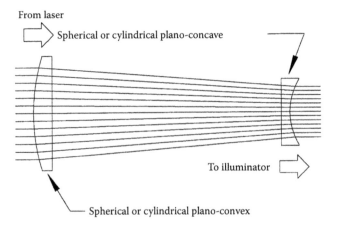

FIGURE 4.13   Beam compression telescope.

the mask, is used to increase the fluence through the mask so that on-target power is achieved with a lower demagnification imaging system.

In addition, it is commonplace to introduce off collimation to a telescope in order to expand or compress the laser beam to match a mask, and also to act as a field lens to match the spot size of the beam to the optimum f-stop of the objective lens.

One other example of a telescope is the prism telescope (see Figure 4.11). Here, two matched sets or roof prisms act as a geometrical spot separator. By changing the distance between the prisms, the collimation of the beam remains constant, but the two spots can be adjusted in separation to allow optimum mask illumination of separated features. This type of telescope has been employed in industrial applications such as marking and hole drilling where two features are separated beyond the beam size. Here, the laser power required to process the part was substantially reduced by expanding the laser beam to cover a basically blank mask. The net result was a laser installation that was very efficient, and allowed the laser power to be turned down where consumables were reduced by a factor of five. This shows the effect of a simple illuminator device on the economic performance of an excimer application.

4.3.4 CONVOLUTION OF EXPOSURE INTENSITY

The use of a moving (dragging) mask is a common technique, either by moving the part during a mask exposure, or by dynamically altering the mask while the laser is pulsing and the part is fixed.

Mask motion during laser exposure with an excimer laser can reshape an optical surface such as the eye, or produce arrays of retro reflectors, molds, and many other 3D features. For instance, the simplified diagram shown in Figure 4.14 illustrates how a diamond-shaped mask imaged onto a part that is moving produces a micro-machined feature with a V-shaped cross section. Ramping the part stage speed, or

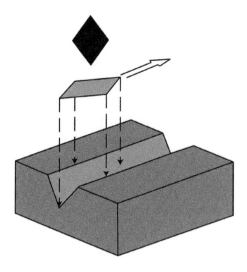

FIGURE 4.14   Mask dragging.

varying the laser pulse rate at constant stage speed will also create shaped features such as cylindrical surfaces or smooth ramps. These techniques require properly integrated motion control of numerous axes in simultaneous motion, but mostly all require a proper mask illumination technique.

Laser dosage control and exposure uniformity are dictated by the process parameters. For processes with very critical exposure levels, care must be taken to ensure that the laser beam is uniform and the scanning pattern is controlled such that the laser energy and power total dosage per unit area are even.

4.3.5 SCANNED MASK IMAGING: ISCAN ILLUMINATOR

If the illumination intensity profile of the laser beam on the mask is varied, then the resulting micromachined feature will track the intensity profile and exhibit a shaped bottom profile. This technique is shown in the scanning mask imaging technique discussed in Figure 4.15.

ISCAN™ is the trade name for J. P. Sercel Associates' scanned mask image micromachining, whereby a laser beam is scanned over a mask in order to process a larger area.[10] One application of the ISCAN is to effectively homogenize the laser beam by convolution of intensity in the scanning mirror direction. Here, the convolution of multiple Gaussian beam profiles can be made to average into a top hat profile.

One other application for ISCAN is to effectively enlarge the laser beam. In this case, when a mask area is larger than the laser beam area for a given energy density requirement, the laser beam is simply scanned over the fixed mask. When conducted properly, the laser beam profile is convoluted over the mask, resulting in a uniform exposure of the target area. The beam is usually scanned over a field lens, which is large enough to cover the entire mask. This field lens then directs the masked laser light into

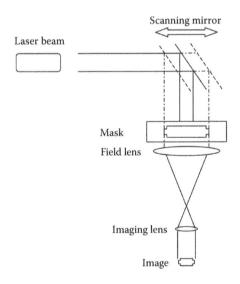

FIGURE 4.15 Imaged mask scanning ISCAN.

the optimum *f*-number of the imaging lens for best image performance. The scanned mask technique allows very large areas of material to be processed with a single mask.

4.3.6 COORDINATED OPPOSING MOTION

The coordinated opposing motion (COMO) technique involves moving both the part and the mask in a coordinated motion with respect to each other. Here, a relatively small aperture lens is used to cover a relatively larger area of imaging on the part. The beam remains fixed with respect to the mask and image lens. For instance, if a *four times* reduction ratio is used, the mask will be four times larger than the target image and the part will move four times slower than the mask and in equal and opposite directions. In some cases, a noninverting lens can be employed to eliminate the opposing motion requirement.[1]

COMO is now commonly referred to as the *step and scan* technique. New generation lithography stepper systems use this concept to reduce the optical complexity of the imaging lens, which is exactly why it was invoked in the late 1980s with excimer micromachining. System accuracy is limited by the precision of the stage and the mask-to-part overlay alignment technique used. The mask stage in a COMO system is typically two to four times larger than the part site to be laser processed, depending on the optical magnification from the mask to the part. Each individual chip site is represented as a single mask site on the mask stage that is addressed under computer control and step-and-repeat. Mask-to-part overlay is accomplished with CCD cameras and reticules to ensure proper overlay accuracy. When telecentric lenses are coupled with premask illumination optics to further improve BUF, the result is a high throughput system that has been shown to produce high-quality parts quickly, reliably, and cost effectively.

The COMO configuration (see Figure 4.16) is used in a micromachining system designed and built by J. P. Sercel Associates. This system is capable of drilling from

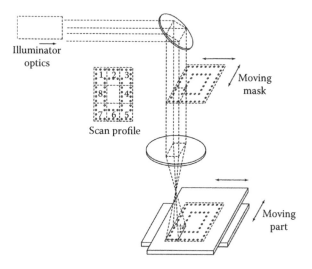

FIGURE 4.16 Coordinated opposing motion technique.

1000 to 10,000 vias/min, or generating complex features in substrates and thin films at production rates. Here, the mask is illuminated to maximize BUF. The mask is then scanned in the COMO mode, while the part is moved into its coordinated position. High-volume processing is achieved through high BUF, using mask-filling efficiency generators that are selected under computer control.[9]

A telecentric imaging lens transfers the mask pattern to the part with near normal incidence over the field of view. Processing can occur with either a continuous scanning motion or step and fire modes, depending upon the mask illumination optics selected.

4.4   BEAM HOMOGENIZERS

Because of the well-earned reputation of laser beams to change shape over time, many beam homogenizing devices have been developed in an effort to ensure a stable illumination source for imaging systems. Whether simple or complex, all beam homogenizers work on the same principle. The beam is divided into subsections that are superimposed on top of one another, so that a loss of power density in one area of the beam does not affect the uniformity of the mask exposure plane. Generally, simple homogenizers split the beam into a few pieces and are capable of averaging out symmetrical beam profile changes. More complex homogenizers cut the beam into more pieces in an attempt to average out asymmetric beam shape problems.[8]

4.4.1   BEAM FLANKS AND EDGE EFFECTS: PROCESSING AT THE HOMOGENIZED PLANE

When a laser beam is homogenized, one possible technique for processing the laser beam is to use the homogenized plane as the position for materials processing. One technique uses a contact mask. Feature sizes down to 25 μm have been achieved this way. An additional technique is to laminate an expendable contact mask to the workpiece as an integral component of the final product.

Surface effectuation such as annealing, surface texturing, surface remelting, doping, selective material removal, and surface cleaning are examples of applications where the laser beam is used to cover a large area, usually without the need for patterning, or with contact-mask patterning.

Frequently, the flanks of the homogenized beam are too gradual to be used for processing at the homogenized plane. There are many processes that cannot tolerate a gradual beam flank intensity transition and, in this case, the design of the beam homogenizer is more critical. This is because when using the homogenized beam at its primary plane, there are no opportunities to clip off the flanks.

Figure 4.17 shows a reasonable homogenizer beam profile from a nonimaging fly's eye array with a rectangular beam shape. As can be seen from the figure, the beam flanks exhibit some slope. For applications where this kind of intensity change is not acceptable, the homogenized beam would need to be reimaged through a mask

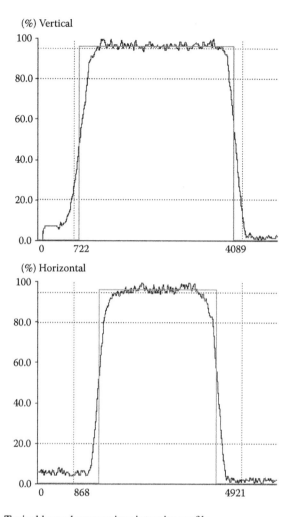

FIGURE 4.17   Typical beam homogenizer intensity profile.

where the flanks can be clipped off. Otherwise, a more complicated (and expensive) imaging homogenizer would be needed. In general, homogenizers are characterized by their signal to noise ratio or line edge response, that is, the amount of energy that resides in the homogenous zone, compared to the amount of energy in the flanks or nonuniform section.

4.4.2   IMAGING THE HOMOGENIZED PLANE

When a mask is illuminated by a beam homogenizer and then imaged onto a work-piece, the result is called imaging the homogenized plane. Here, the intensity flank of the homogenized beam profile is less critical. This is because the flanks can be clipped out by the imaging mask.

The following sections discuss various types of beam homogenizers and their uses. In industrial applications, both the total beam delivery efficiency and the economics of purchasing and replacing optics for a production floor application are important. In general, the less optics that need to be purchased and replaced, the better. This is balanced by system performance and the total efficiency of the system; the net result is economic payback.

We will now look at a few different types of homogenizer types, although there are numerous choices available.

4.4.2.1   Beam Homogenizers by Type: Biprism

Biprism homogenizers use roof prisms to divide the beam into converging segments. These prisms have two or more facets, and each axis may be independently controlled, or both axes may be homogenized by a single prism.

These homogenizers have the advantage of requiring few elements (as few as one) with the corresponding low relative cost and optical losses.

Good results have been obtained with two, three, and four or more faceted prisms. Each axis of the beam will have facets in orthogonal directions.

The disadvantages of prism type homogenizers are summarized as follows:

1. The distance to the homogenized plane depends strongly upon beam size and divergence entering the homogenizer.
2. The size of the homogenized beam depends upon the beam size entering the homogenizer for biprism types. For higher facet types, the beam size must be matched exactly to the facet sizes.
3. The level of achievable homogeneity is usually lower than that with other homogenizers due to the number of beamlet overlaps.
4. It is difficult to manufacture roof prisms with angles small enough to keep the beamlet divergences acceptably low for imaging applications. In practice, this limits the reimaging magnifications to low values.

Referring to Figure 4.18, the relevant dimensions for the biprism homogenizer are as follows:

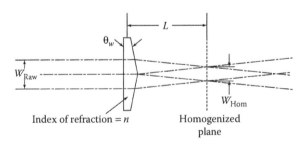

FIGURE 4.18   Biprism homogenizer.

$$W_{\text{Hom}} = W_{\text{Raw}}/2 \tag{4.1}$$

$$L = \frac{W_{\text{Raw}}}{4\,\tan(a\sin(n\sin(\theta_W)) - \theta_W))} = \frac{W_{\text{Raw}}}{4(n-1)\theta_W} \quad \text{for small } \theta_W \tag{4.2}$$

4.4.2.2 Off-Axis Cylinder Lens

The off-axis cylinder lens divides the beam into three segments in each dimension by passing the beam through two half-section cylinder lenses, separated by some distance. The portion between the two lens sections allows the beam to pass through unattenuated. The portions impinging on the lens sections are refracted and super-imposed on the unattenuated portion.

Referring to Figure 4.19, the nominal distance to the homogenized plane is simply given by

$$L = F(D + W/2) \tag{4.3}$$

However, the optimum configuration may not involve complete overlap of the refracted beamlets. Therefore, the distance, L, could be significantly shorter than nominal. This homogenizer can also be placed asymmetrically, such that more of one side of the incoming beam is refracted through the cylinder lens than the other side. This can be useful for asymmetric beams.

The advantages of this homogenizer are as follows:

1. The relative fractions of the beam, which are unattenuated and are refracted, are set by the separation between the lens half sections.
2. The amount of overlap between beamlets is set by the distance from the lens set to the chosen homogenized plane. This allows an optimum level of overlap (homogeneity) to be selected.

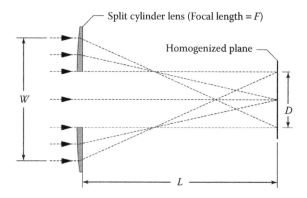

FIGURE 4.19 Schematic diagram of off-axis cylinder lens homogenizer.

3. The beam can be divided into nine (3 × 3) beamlets. By selecting the optimum overlap in each dimension, this enables a high degree of homogeneity to be achieved.
4. This homogenizer has the lowest optical losses of all the homogenizers, and is relatively inexpensive.

The main disadvantage of this homogenizer is that as the laser beam shape changes with time, the optimum settings (lens section separation, distance to homogenized plane) will also change.

This homogenizer works best with beams having good symmetry in both dimensions.

4.4.3 Fly's Eye Beam Homogenizer

The *fly's eye* homogenizer derives its name from the resemblance of its primary component to the arrays of fixed lenses seen on insect (fly) compound eyes. These homogenizers are also referred to as lenslet arrays and imaging or nonimaging beam integrators. The two main types of fly's eye homogenizer (imaging and nonimaging) are illustrated in Figures 4.20 and 4.21. Both types use either arrays of crossed cylinder lenses or arrays of square spherical lenses to divide the beam into beamlets. These beamlets are then passed through a spherical lens to be overlapped at the homogenization plane. The fundamental difference between the two types

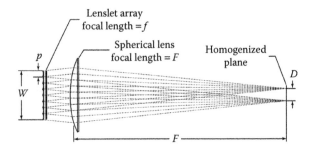

FIGURE 4.20   Schematic illustration of nonimaging fly's eye homogenizer.

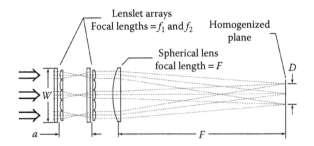

FIGURE 4.21   Schematic illustration of imaging fly's eye homogenizer.

of fly's eye homogenizers is that the imaging version uses two lens arrays and a spherical lens, whereas the nonimaging version uses a single lens array and a spherical lens.

The number of elements in the lenslet array is variable, with smaller numbers being less expensive and larger numbers providing better homogenization. These homogenizers typically use from 5×5 to 7×7 lenslet arrays, even though values up to more than 10× are commonly employed.

Due to the difficulty and expense involved in manufacturing square-shaped spherical lenses with small dimensions and high pointing accuracy, crossed cylinder lenses are more commonly used to create the same effect. As long as the working distances are sufficiently long, the axial separation of the principal planes of the two lens arrays will have a negligible impact on the final result.

In practice, the two most important factors used in deciding whether to use an imaging or a nonimaging fly's eye homogenizer are the type of application and the Fresnel number. For large area mask illumination with large Fresnel numbers, experience shows that nonimaging homogenizers work exceptionally well. For applications requiring highly uniform beam profiles, where the homogenized plane is the working plane (planarization, deposition, etc.) and where the Fresnel number is small, an imaging homogenizer yields superior results.

Between these two extremes, the best options are less certain, and some experimentation may be required to identify the optimum homogenizer configuration. In this chapter, we aim to give some general, practical guidelines to aid in the selection of the optimum fly's eye homogenizer, as applied to excimer lasers and their applications. The following sections describe three types of fly's eye homogenizer in more detail. Note that for ease of discussion, the ensuing sections do not include the placement of a field lens at the mask (homogenized) plane. Such a lens can reduce the input beam size (and thus the NA requirements) for the objective lens by a significant value (e.g., 25%–50%). The optimum field lens focal length is highly case-by-case dependent, especially if telecentric operation of the imaging lens is required, but is usually close to the mask to imaging lens distance. If a field lens is implemented, then great care must be taken to avoid placing any optical components further down the beam line in potential focal *hot spots*.

4.4.4 NONIMAGING FLY'S EYE HOMOGENIZERS

It has been reported that (from a theoretical perspective at least) the nonimaging fly's eye homogenizer is prone to relatively poor performance,[1] with a tendency to exhibit both diffraction effects or diffuse edges. Diffraction effects can lead to interference patterns at the homogenized plane. Diffuse edges at the homogenized plane can result in significantly reduced efficiency in an imaged mask application, or in poor dosage uniformity for applications where the homogenized plane is the process plane.

However, our practical experience shows that nonimaging homogenizers are a preferred solution for many applications.

The paper by Henning et al.[1] gives an excellent derivation of the parameter spaces that pertain to excimer laser applications, and we shall refer to these results during this discussion.

4.5 BASIC HOMOGENIZER DESIGN CONSIDERATIONS

Figure 4.20 illustrates the basic geometry of the nonimaging fly's eye homogenizer as noted earlier, the lenslet array can be either a single 2D array of individual spherical lenses of square dimensions, or two crossed 1D arrays of cylindrical lenses. The 2D spherical lenslet array has the advantage of using fewer optical elements and surfaces in the beam line. This is always an advantage in excimer laser (UV) applications, especially at 193 or 157 nm.

On the other hand, these arrays are typically more difficult and more expensive to manufacture than crossed cylindrical lens arrays. This chapter is equally applicable to either configuration, and the ultimate decision is left to the system designer.

As discussed later in this chapter, the beam entering the homogenizer section has usually been through some type of preconditioning. Due to the highly astigmatic nature of most excimer laser beams, this preconditioning is typically in the form of a dual anamorphic beam expander.

Referring to Figure 4.20, the dimension of the beam at the homogenized plane is given by

$$D = \left| \frac{pF}{f} \right| \tag{4.4}$$

Note that this value is independent of separation between the lenslet array and the spherical lens. Note also that f can be positive or negative (convex or concave lenslets). The most compact geometries are derived when concave lenslets are used, since the spherical lens may then be placed very close to the lenslet array without concern for optical damage that may be caused by the focusing effects of convex lenslets.

The realistic values that can be selected for the various parameters in the above equation are derived from the requirements for the specific application. Of particular importance are

1. The desired homogenized plane dimensions (e.g., mask/object dimensions for an imaging beam delivery system).
2. The required dimensions at the image plane (work surface) for an imaging beam delivery system.
3. The maximum NA for the beam delivery system.

Consider the following example. The desired dimensions of the homogenized plane are defined by the nominal laser pulse energy and the desired energy density at the homogenized plane. For instance, reticule type masks are damaged by energy densities >0.5 J/cm². A laser pulse of, say, 250 mJ would therefore require the area of the homogenized plane to be

$$A \geq \frac{0.25 \text{ J}}{0.5 \text{ J/cm}^2} = 0.5 \text{ cm}^2 \tag{4.5}$$

For a square mask, this corresponds to a value for D (see Figure 4.20) of

$$D \geq \sqrt{A} = 0.71\,\text{cm} \tag{4.6}$$

For a near-field mask imaging beam delivery system, the desired beam dimensions at the image plane are defined by the pulse energy to the work surface and the required energy density. Using the preceding example, assume that no energy is lost in the beam delivery process, and that the full 250 mJ is delivered to the image plane. Furthermore, assume that an energy density of 2 J/cm² is required at the image plane. A quick calculation shows that this requires a demagnification value for the imaging system of $M < 2$.

Allowing for actual losses in the beam delivery system, a reasonable value for the imaging system demagnification would be $M = 3$.[*]

It is beyond the scope of this section to provide a detailed discussion of the effects of the NA and objective lens focal length of the beam delivery mask imaging system, as well as the inclusion of a field lens at the mask plane, on image resolution and depth of focus. Large NA objective lenses are expensive, frequently prohibitively so, and while they provide high resolution, they also have very small depths of focus.

For this example, assume an objective lens with a focal length of 150 mm, and a maximum NA of 0.1. This corresponds to a maximum beam dimension (diagonal) through the objective lens of 3 cm, which corresponds to a maximum rectangular beam size through the lens of $\approx 2 \times 2$ cm.

As shown in Figure 4.20, the beamlets exiting the homogenized (mask) plane are divergent. The maximum divergence half angle is derived from the following relationships (as given in the LIMO Application Production Notes[2]):

$$\tan\theta = \frac{1}{2}\left(\frac{W}{F} + \frac{p}{f}\right) \qquad \text{(Convex lenslet array)} \tag{4.7}$$

$$\tan\theta \leq \frac{1}{2}\left(\frac{W + D - 2p}{F}\right) \qquad \text{(Concave lenslet array, } D > p) \tag{4.8}$$

$$\tan\theta \approx \frac{1}{2}\left(\frac{W - D}{F}\right) \qquad \text{(Concave lenslet array, } D < p) \tag{4.9}$$

where:

$$W \approx np \tag{4.10}$$

---

[*] Demagnification is a reduction of mask onto part and is commonly expressed in terms of whole numbers rather than fractions for ease of use (e.g., a mask demagnification of a factor of 1/2 would be expressed as a magnification of 2).

For a 150 mm objective lens, with a mask-imaging demagnification value of 3, the mask to lens distance is thus

$$s_1 = 15 \text{ cm } (1 + M) = 60 \text{ cm} \tag{4.11}$$

The objective lens to image plane distance is

$$s_2 = 15 \text{ cm}\left(1 + \frac{1}{M}\right) = 20 \text{ cm} \tag{4.12}$$

Transitioning from a beam dimension of 0.7 cm at the mask to a dimension of 2 cm at the objective lens 60 cm away requires a maximum divergence of

$$\tan \theta \leq \frac{1}{2}\left(\frac{(2-0.7)}{60}\right) \approx 0.01 \tag{4.13}$$

Assume a concave lenslet array. Furthermore, assume a lenslet dimension (p) of 0.5 cm, and a 5 × 5 dimensional array ($W \approx 2.5$ cm). In this case $D > p$, and we use Equation 4.5, which immediately yields the result:

$$F \geq 110 \text{ cm} \quad \text{(homogenizer condenser lens)} \tag{4.14}$$

We will use a value for F of 150 cm.

For the required value of D (≈ 0.71 cm), Equation 4.4 gives

$$f \approx -105 \text{ cm} \tag{4.15}$$

In summary, we have defined an imaging beam delivery system with the following approximate parameters (excluding a field lens):

Lenslet array	5 × 5 lenslets, 0.5 cm facet width, ~–105 cm lenslet focal length, 2.5 × 2.5 cm total
Spherical condenser lens	~150 cm focal length, >3.5 cm clear aperture
Homogenized (mask) plane	150 cm from spherical lens, 0.71 × 0.71 cm, ~0.5 J/cm² energy density
Imaging system demagnification	3×
Objective lens	150 cm focal length, >3 cm clear aperture, 0.1 NA, 120 cm from mask plane
Image plane	20 cm from objective lens, ~2 J/cm² energy density

While every beam delivery system is different, depending upon laser pulse energy, the specified energy densities at the mask and image planes, and the desired resolution, and depth of focus, the preceding values are fairly typical of a standard large area, mask-imaging beam delivery system. The corresponding nonimaging fly's eye homogenizer is matched to the system. This example illustrates how important it is to design the homogenizer for the specific application.

Having defined a practical homogenizer configuration, we now consider the quality aspects of the homogenized beam.

4.6 QUALITY OF THE HOMOGENIZED BEAM

There are four issues to consider when evaluating the quality of the beam at the homogenized plane: large-scale homogeneity, diffraction effects, coherence length effects, and astigmatism. There are also some mitigating factors, which are discussed later in this section.

4.6.1 LARGE-SCALE HOMOGENEITY (INTEGRATION OF OVERALL LASER BEAM PROFILE)

The large-scale homogeneity of the homogenized plane is simply a function of the uniformity of the incoming beam and the number of beamlets into which it is divided. The best way to evaluate the expected large-scale homogeneity is to profile the beam after the preconditioning optics, divide the profile into the number of beamlets, and superimpose them. The resulting superimposed profile will give a good prediction as to the level of homogeneity that can be expected without diffraction and coherence effects and so on.

If the profile and divergence characteristics of the raw beam are known, then it is reasonable to use ray tracing software to model the preconditioning optics and derive an anticipated large-scale homogeneity.

This is the step where the number of facets in the lens array should be selected. If the following sections indicate there may be issues with the selected geometry (e.g., with diffraction effects, etc.), then it may be necessary to change the number of facets.

4.6.1.1 Diffraction Effects

Henning et al.[1] have shown that the dominant, observable diffraction effects in a nonimaging fly's eye homogenizer will be due to Fresnel diffraction. Using the results from the paper by Henning et al.[1] (see Equation 4.3.1 in Reference 1), but ignoring lens wavefront distortions, the effects of pure diffraction were numerically calculated, and are shown in Figure 4.22. Note that this derivation maps the diffraction from one facet only. It also assumes perfectly collimated light and ignores any line width effects, as well as assuming fully coherent light.

Fresnel diffraction is purely a function of Fresnel number. Higher Fresnel numbers give sharper edges and smaller percentage diffraction variations. It is evident from this figure that practical nonimaging fly's eye homogenizers should have Fresnel numbers >10, with values ≥100 being preferable.

For the current geometry (see Figure 4.20), the 1D Fresnel number is given by the following relation:

$$N_F \approx \frac{pD}{4\lambda F} \qquad (4.16)$$

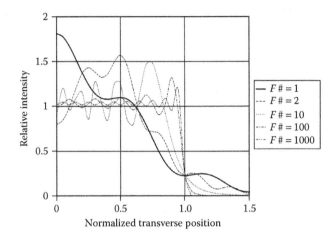

FIGURE 4.22 The normalized intensity distributions at the homogenized plane as a function of distance from the optical axis (normalized to the half width) for various Fresnel numbers, using perfectly collimated light and assuming fully coherent light.

where λ is the laser wavelength, and other terms are defined in Figure 4.20.

Consider, again, our example. In this case (for 248 nm light),

$$N_F \approx \frac{0.005 \times 0.007}{4 \times 248 \times 10^{-9} \times 0.15} \approx 235 \qquad (4.17)$$

This is well into the range where pure diffraction effects are minimized.

4.6.1.2 Coherence Length Effects

The next important phenomenon to impact beam quality at the homogenized plane is the finite coherence length of excimer lasers. Excimer lasers exhibit very short coherent distances, typically 0.01 to 0.1 mm. These short coherence lengths act to *dampen* the diffraction phenomena described in the previous section. As described in the paper by Henning et al.[1] in a rigorous analysis, the coherence length term should be convoluted with the diffraction equation to derive the fundamental results. Because we wish to develop practical guidelines for homogenizer design, we will consider the coherence length effect independently, and evaluate its impact on the homogenized plane beam uniformity.

Referring to Equation 4.2 and Henning et al.[1] the partial coherence intensity distribution is derived from the convolution of the Fresnel number dependent coherent distribution with a coherence length-dependent term. The coherent length term defines an effective 1/e damping range over which the coherent diffractive structure is effectively washed out. A shorter coherence length results in a longer damping range. The damping range, C_{damp} (normalized to the facet lens half height), is as follows:

$$C_{\text{damp}} \approx \frac{1}{\sqrt{2}\pi N_F \sigma_p} \qquad (4.18)$$

where:

$$\sigma_p = \frac{2 \times L_c}{p} \qquad (4.19)$$

L_c is the coherence length. Coherence damping has the benefit of helping smooth out the diffraction effects,[1] but also has the disadvantage of stretching out the flanks of the homogenized plane. As discussed in the pending section on mitigating factors, there are other phenomena that can also smooth out the diffraction, so the best results are obtained when the deterioration of the homogenized plane edge quality due to coherence length effects is not excessive.

Considering our example again, and assuming a coherence length of 0.05 mm, Equation 4.16 immediately yields the following:

$$C_{\text{damp}} \approx 0.05 \text{ (normalized)} \approx 0.12 \text{ cm (actual)} \qquad (4.20)$$

This is a significant distance, and will definitely have a smoothing effect on the Fresnel diffraction, as well as softening the flanks of the homogenized plane.

Note that oscillator–amplifier and short-pulse excimer lasers, where the oscillator is a frequency-multiplied dye laser, for example, can have significantly longer coherence lengths than those discussed here, and Fresnel diffraction effects will be more intense.

4.6.1.3   Astigmatism

Most raw beams in industrial excimer lasers are highly astigmatic, with varied dimensional, profile, and divergence properties between the two main dimensions, as well as varying coherence lengths. This is less true for multipass oscillator–amplifier and unstable resonator configurations, as well as for some X-ray preionized excimer lasers, but it must still be taken into account.

The astigmatic beam intensity profile is not the main concern in homogenizer design given that this is the primary reason for using a homogenizer in the first place. The astigmatic divergence and dimensions are, however, very important.

Apart from specialized resonator optics design, there are two methods that are known to minimize detrimental effects due to beam astigmatism.

4.6.1.4   Dual Anamorphic Beam Expander

If the two dimensions are independently expanded and collimated, then it is possible to closely match them to the homogenizer. This is usually achieved via two transverse cylindrical beam expanders. Not only can the beam axes be independently magnified, but judicious selection of the speeds of the two telescopes can also result in more closely matched divergences.

4.6.1.5   45° Rotated Homogenizer

In order to minimize the number of required lenslets, the homogenizer is usually in the same orientation as the expanded laser beam. In this case, apart from corner diffraction effects, the two beam dimensions are independently homogenized. Even with the best anamorphic beam expander matching this can still result in severe homogenized astigmatism. This effect can be greatly reduced, however, if the homogenizer is rotated 45°. The two axes of the homogenizer thereby *share* the two laser beam axes and astigmatism effects can be greatly reduced. The technique has been shown to steepen the homogenized beam flanks and create higher flank uniformity for each homogenized beam axis.

The disadvantages of this approach are twofold. First, it does not meet the obvious requirement for a larger lenslet array, and second, the resultant homogenized beam is rotated, which may require an even higher number of lenslets to achieve adequate large-scale homogenization. However, in some large area beam dosage applications, the rotated homogenized image is a better fit to masks that are diamond shaped. The diamond-shaped masks have been shown to improve the area coverage over overlapping laser pulses where otherwise square beams will produce high-contrast edges from row to row of laser scans.

It is always best to match the beam expander as closely as possible to the raw beam properties, and only rotate the homogenizer with respect to the beam if absolutely necessary.

4.6.2 OTHER ISSUES AND MITIGATING FACTORS

The theory in the preceding sections makes several unrealistic assumptions. When the true physical situation is taken into account, several factors help to mitigate the potentially serious diffraction and coherence length-related issues.

4.6.2.1 Coherence Scrambling

As shown above, the magnitude of the Fresnel diffraction effects is modified by coherence length effects. Dainesi et al.[3] have shown that if the individual beamlets are delayed with respect to each other by more than the laser coherence time, then the diffraction effects from the individual beamlets will cancel each other. This is very important, since this coherence scrambling effect will not degrade the edge quality of the homogenized plane. A similar effect can be achieved by simply rotating the fly eye at 45° to the astigmatic beam as previously discussed. In this case, a mixing of the coherence of the two beam dimensions occurs.

The difference in thickness of an optical material with index of refraction, n, to spatially delay two emerging rays of light by a distance L_c with respect to each other is

$$\Delta_c = \frac{L_c}{n-1} = 0.1 \text{ mm} \qquad (n = 1.5, L_c = 0.05 \text{ mm}) \qquad (4.21)$$

Dainesi et al.[3] achieved this by placing a distributed delay device in the beam line. However, they were using short pulse-amplified dye laser beams. In this case, the coherence length is much greater than for standard multi-ns excimer laser pulses. It can be seen from Equation 4.18 that for standard excimer laser pulses, the variation in spherical lens thickness could alone result in significant coherence scrambling if the focal length is short enough.

Referring once again to our example, the planoconvex lens has a curved surface radius of:

$$r = (n-1) \times F = 7.5 \text{ cm} \quad (n = 1.5, F = 150 \text{ cm}) \tag{4.22}$$

For a 3.5 cm aperture, this results in a lens thickness variation of ~0.001 mm. From Equation 4.18, this is insufficient to cause significant coherence scrambling.

The impact of coherence scrambling on any particular system will depend entirely on the system-specific parameters. If diffraction fringes are observed at the homogenized plane when using a nonimaging fly's eye homogenizer, then they may be mitigated or significantly reduced. This is accomplished by implementing an imaging homogenizer, which is less prone to generating diffraction fringes, by applying the concept of Dainesi et al.[3] and by placing a distributed delay device in the beam path and thus scrambling the coherence effects, or, if necessary, by redesigning the homogenizer and beam delivery system.

4.6.2.2 Divergence

Standard excimer laser beams usually have several mrad beam divergence. Much of the divergent energy is lost in passage through the anamorphic beam expanders to the homogenizer, but the residual divergence can still have a large impact at the homogenized plane. Consider the simple geometry of two rays that are 1 mrad divergent. Over the distance of the focal length of the homogenizer spherical lens, even this small divergence can result in a significant separation between the two rays. In the case of our example, for instance, $F = 150$ cm, and the resultant ray separation is

$$\Delta \approx 1 \text{ mrad} \times 150 \text{ cm} = 0.15 \text{ cm} \tag{4.23}$$

For long working distance nonimaging fly's eye homogenizers, even a very small beam divergence can therefore result in significant *smearing* of the homogenized plane. This has the advantage of smoothing out any diffraction effects, but also has the disadvantage of softening the flanks of the homogenized beam, and thus wasting usable energy.

The actual divergence characteristics of any particular configuration will be highly dependent upon the specific laser used, the preconditioning optics, and the homogenizer configuration. It will be necessary to evaluate the impact of divergence for each specific implementation.

4.6.2.3 Spherical Aberrations

Another area with potential impact on the quality of the homogenized plane is spherical aberrations due to the spherical lens. For a single-element planoconvex lens operating in the optimum orientation, the longitudinal spherical aberration (LSA) for a lens of focal length F (see Reference 4) is given by

$$LSA \approx 0.272 \frac{F}{f\#^2} \approx F \times NA^2 \qquad (4.24)$$

For a large f-number (θ (radians) ~ $\sin(\theta)$ $\tan(\theta)$), the transverse spherical astigmatism (TSA) is related to the LSA by

$$TSA \approx NA \times LSA = F \times NA^3 \qquad (4.25)$$

Consider our example once again. The spherical lens has an aperture of 3.5 cm and a focal length of 150 cm. Thus, NA = 0.012 and (from Equation 4.20) the corresponding TSA is ~2.5 × 10⁻⁴ cm. Obviously, in this example, spherical aberrations are negligible. They will only be of concern for large homogenizer apertures and short spherical lens focal lengths (i.e., large NA).

4.6.2.4 Multielement Lenslet Effects

The analysis by Henning et al.[1] not only assumes fully collimated light, but it only considers the contribution from a single-array lenslet. If the homogenizer was perfectly assembled, then with every lenslet having the exact same focal length with perfect centration and zero wedge, and if the lenses were true thin lenses and there were no spherical aberrations, then the Fresnel diffraction patterns from each lenslet in the array would exactly overlap, and the analysis would hold true to the extent that the incoming beam is fully collimated.

In a *real world* configuration, these assumptions are not true, and the resultant effects on the homogenized plane must be evaluated.

Lenses always have some degree of variability in focal length. From Equation 4.1, a 5% variation in lenslet focal lengths, for example, will result in a 5% variation in the size of the individual beamlets at the focal plane. This will have the benefit of slightly smoothing out the diffraction patterns, but also has the disadvantage of softening the flanks of the homogenized plane. As long as the focal length variations are relatively small, the resultant homogenized plane smearing effect will be minimal.

Variations in lenslet centration and wedge must also be considered. Custom cylindrical lenses typically have centration tolerances ~0.5 mrad, and wedge tolerances ~1 mrad. Unless the arrays are carefully selected and matched, the variations in wedge and centration can have effects on the homogenized plane that are similar in magnitude to divergence effects.

In summary, any effects due to the multielements in the lenslet array will smooth out any diffraction patterns and soften the flanks of the homogenized plane. A little smoothing may help uniformity, but too much will be unacceptably detrimental. The lenslets should be matched for focal length, centration, and wedge as closely as possible.

Again, the specific impact of these effects on a particular system must be analyzed individually to derive lenslet tolerance specifications.

4.6.3 Imaging Fly's Eye Homogenizers

Although nonimaging fly's eye homogenizers are comparatively straightforward to design and implement,[5] situations will occasionally arise where the resultant quality of the homogenized plane is inadequate, in spite of all attempts to optimize the design.[6] This is particularly true if the homogenized plane is to be used as the application work plane, without going through a mask-imaging section. In these cases (planarization, surface annealing, deposition, etc.), the required homogenized plane specifications (edge steepness and uniformity) may necessitate the use of an imaging fly's eye homogenizer.

This type of homogenizer is illustrated in Figure 4.21. There are some observations that can immediately be made.

1. The homogenizer requires two lenslet arrays.
2. The lenslets in these arrays must be positive (convex).
3. The resulting homogenizer is longer than an equivalent nonimaging homogenizer with concave (negative) lenslets.

While these factors result in the homogenizer being more expensive and more difficult to set up and align, as well as having higher optical losses compared with nonimaging homogenizers, the resultant beam quality at the homogenized plane is always superior.

4.7 BASIC HOMOGENIZER DESIGN CONSIDERATIONS

The imaging fly's eye homogenizer uses the first lenslet array to divide the beam into multiple beamlets. The second lenslet array, in combination with the spherical lens, acts as an array of objective lenses that superimposes the images of each of the beamlets in the first array onto the homogenized plane. Because this system acts as an imager, as opposed to the illuminator of the nonimaging homogenizer, diffraction effects are negligible for all practical excimer laser coherence lengths, and for all practical lens focal lengths and dimensions.

However, the size of the homogenized plane is more complicated than the relationship for the nonimaging homogenizer given in Equation 4.1. Referring to Figure 4.20, and using standard thin lens equations, the dimension of the homogenized plane is

$$D = p \frac{F}{f_1 f_2} [(f_1 + f_2) - a] : (f_1 < a < f_1 + f_2) \tag{4.26}$$

Note that the size of the homogenized plane is now dependent not only upon the focal lengths of all three lenses, but also upon the separation of the two lenslet arrays. This is one advantage of imaging fly's eye homogenizers where varying the separation between the two arrays results in a variable homogenized plane size. Care must be taken, however, not to damage the second lenslet array by passing too much intensity into its elements or to overfill the second lenslet array.

The divergence for this homogenizer after the homogenized plane is (as given in Reference 2):

$$\tan \theta < \frac{1}{2} \frac{(W + D - 2p)}{F} \qquad (D > p) \qquad (4.27)$$

$$\tan \theta \approx \frac{1}{2} \frac{(W - D)}{F} \qquad (D < p) \qquad (4.28)$$

$$W \approx np \qquad (4.29)$$

Consider the same example as for the nonimaging homogenizer. First, we select the focal length of the spherical lens. As before, we want a divergence from the homogenizer such that $\tan (\theta) \leq 0.01$. With the same values as before for W, D, and p, we immediately get the result from Equation 4.27 that $F \geq 110$ cm. We will use a value of $F = 150$ cm for this example.

In order to keep the energy flux through the second array reasonable, and yet allow for some variability in the array separation, we will choose the condition that:

$$a \approx f_1 + 0.75f_2 \qquad (4.30)$$

Putting this relationship into Equation 4.26, along with the derived value for F(150 cm) gives the result:

$$f_1 \approx 25 \text{ cm} \qquad (4.31)$$

For simplicity, we have chosen the same values for f_1 and f_2.

Lenslet arrays	5 × 5 lenslets, 0.5 cm facet width, ~25 cm lenslet focal length, 2.5 × 2.5 cm total, ~4.4 cm separation
Spherical lens	~150 cm focal length, >3.5 cm clear aperture
Homogenized (mask) plane	150 cm from spherical lens, 0.71 × 0.71 cm, ~0.5 J/cm² energy density
Imaging system demagnification	3×
Objective lens	30 cm focal length, >4.2 cm clear aperture, 0.07 NA, 120 cm from mask plane
Image plane	40 cm from objective lens, >2 J/cm² energy density

This gives the following set of values for our imaging fly's eye homogenizer with the same system specifications as the previous nonimaging homogenizer (again excluding a field lens):

As with the nonimaging fly's eye homogenizer, this example shows how important it is to match the specific homogenizer parameters to the beam delivery requirements. There is no *one size fits all* solution, and every system is unique.

4.7.1 QUALITY OF THE HOMOGENIZED BEAM

Because this homogenizer uses an imaging process to fill the homogenized plane, there are no problems related to Fresnel diffraction and coherence length. Beam divergence is also much less of a concern. Some of the issues raised in the nonimaging homogenizer discussion are, however, still applicable.

Figure 4.23 illustrates the improved homogenized plane beam quality that can be achieved with an imaging homogenizer, as compared to a nonimaging homogenizer, under similar conditions.

The preceding discussions on multielement lenslet effects (focal length, centration, and wedge tolerances) are equally applicable. In fact, because the imaging homogenizer is capable of generating a much *crisper* homogenized plane than the nonimaging homogenizer, it is even more imperative that these parameter tolerances be kept as tight as possible. The use of a dual anamorphic beam expander is still normally recommended, and rotating the homogenizer by 45° can largely mitigate any residual astigmatism effects.

4.7.1.1 Imaging versus Nonimaging Fly's Eye Homogenizer Trade-Offs

It is evident from the preceding discussion that there are many trade-offs between imaging and nonimaging fly's eye homogenizers, and the best choice is not always immediately apparent. Here, we will attempt to summarize the various trade-offs as concisely as possible.

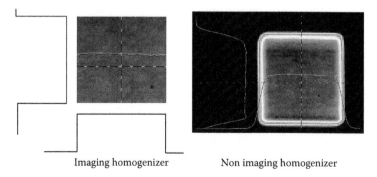

Imaging homogenizer   Non imaging homogenizer

FIGURE 4.23 Sample beam profiles for imaging and nonimaging fly's eye homogenizers.

Note that both types of homogenizer are equally dependent upon the quality of the optical components, in particular the focal length, centration, and wedge tolerances of the lenslet arrays.

Nonimaging Fly's Eye Homogenizer		Imaging Fly's Eye Homogenizer	
Advantages	Disadvantages	Advantages	Disadvantages
Fewer optical elements (less expensive, lower optical losses)	Prone to generating Fresnel diffraction patterns at homogenized plane	Comparatively immune to Fresnel diffraction effects	More optical elements (more expensive, higher optical loses)
Well matched to large area mask imaging systems with low NAs	Flanks of homogenized plane are generally quite *soft*	Well matched to applications where the homogenized plane is the work plane	More complicated to set up and align
Easier to set up and align	Quality of homogenized plane highly dependent on beam divergence	Gives a *crisp* beam shape at the homogenized plane with sharp edges	
Astigmatism effects partially mitigated by rotating homogenizer by 45°	Size of homogenized plane cannot be varied	Quality of homogenized plane mostly independent of beam divergence	
		Variable size of homogenized plane (adjusting the two lenslet array separations)	

Although it is apparent that imaging fly's eye homogenizers will always give superior technical results, our practical experience in having designed and manufactured many beam delivery systems over the years has shown us that the nonimaging fly's eye homogenizer is the preferred choice for many applications, particularly for large area mask imaging systems where the resultant homogenizer designs have large Fresnel numbers.

4.7.2 AXICON ARRAY

Occasionally, there is an application that requires an annular (ring) pattern. The typical method is to use mechanical stages or galvanometers to *steer* a small laser spot around the circumference of the desired annulus. This approach has the advantage of being highly flexible, but tends to be somewhat slow and beam astigmatism can result in varying line widths around the annulus.

The same annular pattern can also be achieved, however, by use of an axicon array, as illustrated in Figure 4.24. The axicon array is very similar to the nonimaging fly's eye homogenizer, except that it uses an array of very small angle cones. If the corresponding F numbers are sufficiently large, or if a multielement field lens is

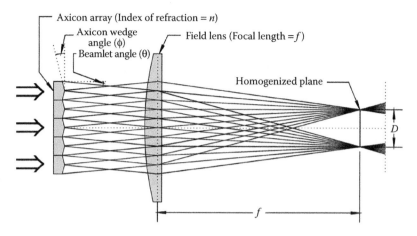

FIGURE 4.24 Schematic illustration of axicon array homogenizer.

used, then the resultant annulus can be close to diffraction-limited. This is contingent upon the same concerns for collimation, divergence, and centration/alignment of the axicon array as the nonimaging fly's eye homogenizer. The axicon array is typically used to illuminate a mask, which is then imaged onto the image surface.

Referring to Figure 4.24, the angle of the beamlets exiting the axicon array with respect to the optical axis is

$$\theta = a\sin\left(n\sin(\Phi)\right) - \Phi \qquad (4.32)$$

where:
Φ is the axicon cone angle
n is the index of refraction of the axicon material

For small angles, this relationship can be approximated by

$$\theta \approx (n-1)\Phi \qquad (4.33)$$

The diameter of the homogenized annulus is then given by

$$D = 2f\,\tan(\theta) \qquad (4.34)$$

An immediate observation is that the size of the annulus is fixed for any given cone angle and field lens focal length. Therefore, unless a variable focal length field lens is used, the only means to vary the diameter of the annulus on the work surface is to vary the demagnification of the mask to image stage.

4.7.3 COLLIMATED BEAM HOMOGENIZERS

One of the major disadvantages of most types of homogenizer is that the emerging beam is decollimated. This typically necessitates long beam delivery systems or high NA optics, both of which add cost and complexity. Furthermore, these homogenizers are extremely difficult to integrate into spot separators or spot generators.

However, it is possible to implement a collimated beam homogenizer wherein the emerging beam remains collimated. The Samshima-type homogenizer5 can in fact improve the useful fraction of the laser beam, but these types of homogenizers only operate on one axis of the laser beam. In addition, there is a limit to how a dual-fold type homogenizer can correct the beam. These devices can be used as illuminators for various beam utilization efficiency enhancers that geometrically split the beam up into a number of parts to increase production but provide no beam homogenization.

4.8   EFFICIENCY ENHANCEMENT (A FORM OF ILLUMINATION)

4.8.1   BEAM SPLITTING FOR HIGHER BUF

Optimum beam utilization is essential to quality part manufacturing at affordable manufacturing costs. To improve the BUF, various techniques are employed to precisely match the open area of the mask to the available laser light. The concept is easy to understand when observing Figure 4.25. A raw beam is subdivided into parts that slightly overfill the mask vias. The mask is then reimaged and de-magnified onto the part.

4.8.2   GEOMETRIC SPOT SEPARATOR

Note that, in reality, the individual beamlets cross paths between the imaging lens and the part, and there is usually a degree of magnification/demagnification.

Geometric beam dividing is a powerful technique whereby part throughput is increased by using one laser beam to illuminate multiple imaging systems. Systems that use one laser beam to feed more than 10 separate imaging systems are currently running in high-volume parallel production. For geometric splitting systems with high split factors, proper alignment of each beamlet to its respective imaging system is critical. These alignment issues have spurred the development of innovative miniature optic mounts. A large number of beam splitting techniques have been

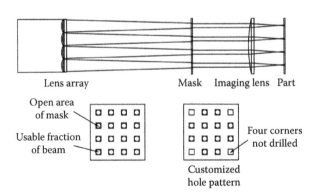

FIGURE 4.25   Schematic illustration of lens array mask illuminator.

engineered and proven successful at JPSA. These include incorporating prisms, mirrors, lenses, lenslet arrays, and diffractive phase plates. Prism beam separators have been used with a high degree of success in 24-hour production environments. Spot separation is adjustable by altering the separation of the prisms. The concept can be extended to multifaceted prisms such as quad prisms and prism arrays.

4.8.3 SPOT ARRAY GENERATORS

Laser spot array generators can be constructed of spherical or cylinder lenses (Figure 4.26). Crossed cylinder lenses are a convenient method for creating an equivalent array of spherical lenses. In addition to crossed cylinder lens arrays, many types of premask-conditioning lenses are used to increase the BUF of a laser application.

Arrays of spherical lenses and groups of cylinder lenses can be used. It is not necessary to overfill each via coordinate on the mask because a group of vias that are closely packed can be flooded with intensified light. For example, a line of vias may be imaged by a mask that is illuminated by a cylinder lens that produces an intensified line of laser light onto the row of vias located on the mask.

Spot array generators are frequently very similar to the various homogenizers already described, but usually without the field lenses that are used to generate the homogenized plane. Compare, for example, Figure 4.25, the lensarray mask illuminator with, Figure 4.20, the nonimaging fly's eye homogenizer, reproduced again later.

Spot array generators such as the fly eye type do not homogenize the laser beam and therefore the arrays of spots will vary in intensity depending on the illumination beam profile. Using a collimated beam homogenizer prior to the lens array spot array generator illuminator can serve to provide a more uniform energy distribution over the spot array.

Where a field lens is implemented, it is for the purpose of matching the generated spot size to the mask spot dimensions, or matching the exiting beam to the NA of the imaging system, not for any homogenization purposes.

Figure 4.27 is a clear example of an industrial implementation that has been shown to increase the productivity of a system by an order of magnitude. We took the beam and split it up geometrically into 10 or more different parts. Each section

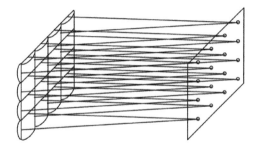

FIGURE 4.26 Nonimaging fly's eye homogenizer.

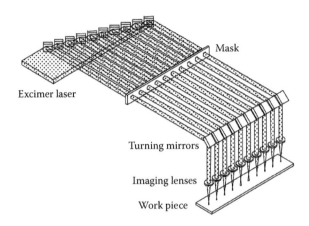

FIGURE 4.27 Line focus ganged into 10 separate parts.

of the beam drills a single hole. This involves a mirror assembly that geometrically splits the beam up into 10 different parts, hitting 10 different masks, and 10 different imaging lenses to parallel process a number of parts. That works very well in production where using 10 separate intensity measurement devices, as well as 10 different attenuators to keep all of the light balanced, satisfactory results were achieved.

In Figure 4.28, we see a classic example of a mask illuminator efficiency enhancer. In this instance, the laser beam is shaped to overlay the open area of a slit mask. This illustrates how if one is using a simple cylinder lens to compress the beam in one direction and is imaging a slit, then the overfill of the mask is reduced. Thereby, you get more energy on-target, and process efficiency is increased considerably. Taken a step further, we have built such units using the basic idea of Figure 4.25 combined with Figure 4.28. In this case, a cylinder lens array created a series of line focuses to expose a multislit mask. Four and five slits were used to gang dice small die on a wafer. Due to geometric limits, the application actually used interdigitation to further reduce the die size effectively by scanning grids at one half of the beamlet separation on-target.

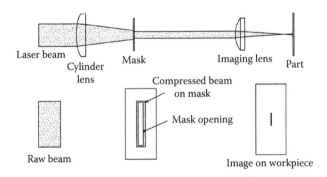

FIGURE 4.28 Mask illuminator efficiency enhancer.

4.8.4 ANGULARLY MULTIPLEXED

Figure 4.29 illustrates a technique whereby one can eliminate a number of optics and use special segmented mirrors to geometrically and angularly split the beam and irradiate a mask. This is a powerful technique because, in this particular case, it shows that you are imaging three spots that are farther apart than a normal beam would image (i.e., you are covering a much larger area with a single lens). This brings the added advantage that because you are angularly multiplexing these beams, you are eliminating the need for a large-diameter (and costly) field lens, and are also thus further eliminating optics.

Computer generated holograms (CGH) or holographic optical elements (HOE) are useful means of dividing the beam geometrically to expose a complex target[11] (Figure 4.30). These devices are fabricated from fused silica and are made up of etched submicron features. These precision-etched features can be binary (two levels) or of higher order (eight or more) designs. The etched features cause phase shifts in the incident laser beam that effectively redirects the laser energy through constructive or destructive interference.

Using a relatively incoherent source such as an excimer laser will degrade the theoretical efficiency of the HOE in the form of signal to noise. Noise results in light focused to areas not desired. If a part is located at the image plane of the hologram, then the excess light will drill into the material to be processed in unwanted areas and also limit the feature resolution.

Higher numbers of phase level designs are more efficient (and more costly), but are still susceptible to noise if an imperfect beam is used to irradiate the HOE. Cleaning up the beam through the use of unstable resonators and or oscillator amplifier lasers improves CGH performance. CGH efficiency is greater, but the overall laser output to process efficiency result is still low due to a less-efficient laser source and a more complicated laser design.

If the CGH is used as a mask illuminator optic, then the mask will spatially filter the intensity distribution. Then, using a lower beam quality laser, or a lower phase

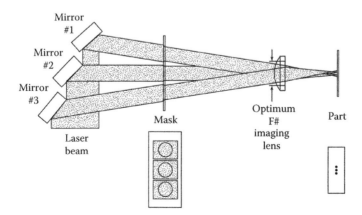

FIGURE 4.29 Segmented mirror illuminator.

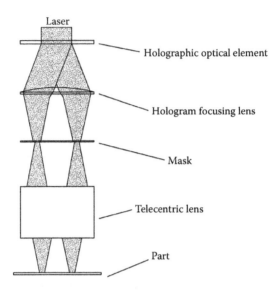

FIGURE 4.30 Holographic elements as mask illuminators and homogenizers.

level CGH, results in overall improved efficiency and high image quality at the part
being processed.

These multilevel phase plates show great promise for beam homogenizers. New
designs that eliminate the central order from the homogenized plane are evolving. The advantages of this kind of homogenizer are the reduced number of optical
surfaces in the beam line and therefore lower losses, and the ability to design the
units simultaneously as a spot array, a pattern generator, and a beam homogenizer.
Improved designs and higher phase levels allow better signal to noise even with relatively poor laser beam quality.

4.9 CONCLUSION

The illumination section of an excimer laser system is an important optical device
for the performance of a system. This section can shape the beam, control the beam's
divergence, and control the beam profile through the use of different homogenizers.
It can also be used to geometrically increase the efficiency of a process by only filling the open area of the mask. All of the techniques shown here have been proven
in devices that are currently used in production around the world, 24 hours per day
and seven days a week.

REFERENCES

1. Henning, T., Unnebrink, L., and Scholl, M., UV laser beam shaping by multifaceted
   beam integrators: fundamental principles and advanced design concepts, *SPIE.*, 2703,
   62–73, 1996.
2. LIMO Product Application Notes, www.limo.de/Application_Notes-preliminary.pdf

3. Dainesi, P., Ihlemann, J., and Simon, P., Optimization of a beam delivery system for a short-pulse Krf laser used for materials processing, *Appl. Opt.*, 36(27), 7080–7085, 1997.
4. Melles Griot Optics Guide Tutorial, www.mellesgriot.com/products/optics/toc.htm
5. Sameshima, T., and Usui, S., Laser beam shaping system for semiconductor processing, *Opt. Commun.*, 88, 59–62, 1992.
6. Dickey, F.M., and O'Neil, B.D., Multifaceted laser beam integrators: general formulation and design concepts, *Optical Eng.*, 27(11), 999–1007, 1988.
7. Latte, M.R., and Jain, K., Beam intensity uniformization by mirror folding, IBM Research Report, RJ 3935 (44535), 6/14/83, 1983.
8. Kalamura, Y., Itagaki, Y., Toyoda, K., and Namaa, S., A simple optical device for generating a square top intensity irradiation from a Gaussian beam, *Opt. Commun.*, 48(1), 44–46, 1983.
9. Sercel, J.P., Optimized beam delivery for industrial excimer lasers, *Photon. Spectra*, 84 1991.
10. Sowada, U., Sercel, J., Kahlert, H.J., Basting, D., and Austin, L., Improved beam uniformity in excimer lasers, *SPIE*, 889, 13, 1991.
11. Veldkamp, W.B., Technique for generating focal plane flat top laser beam profiles, *Rev. Sci. Instruments*, 53(3), 294, 1982.

5 Micro-Optics for Illumination Light Shaping in Photolithography

Reinhard Voelkel

CONTENTS

5.1   INTRODUCTION

Photolithography is the engine that empowered microelectronics and semiconductor industry for more than 50 years. Photolithography allows very complex
micro- and nanostructures to be built by copying a pattern from a photomask to a
wafer. Photolithography is the key enabling technology (KET) behind the powerful concept of *shrinkage*, also referred to as *die shrink*, the ability to reduce the
minimum feature size of transistors, electronic wires and other components of a
microchip from some 50 microns (μm) in the early 1960s to some tens of nanometers today. Die shrink allows more chips on a wafer to be manufactured, thereby
reducing manufacturing costs, minimizing power consumption, and improving the
performance in terms of speed, storage capacity, and customer convenience. In a
photolithographic process, a thin layer of photosensitive resist is coated on a wafer,
the photomask is illuminated by light, and the light propagates from the mask to
the wafer surface, transferring the pattern from the mask onto the resist. As shown
schematically in Figure 5.1, there are two different ways to accomplish the pattern
transfer by light propagation. Light could propagate in free space, which is referred
to as shadow printing lithography; or through an imaging system, which is referred
to as projection lithography.

In a mask aligner, the mask pattern is transferred to the wafer by shadow printing
as shown in Figure 5.1a. If the mask is in direct contact with the wafer, this is referred
to as contact lithography. If the mask is not in contact, this is referred to as proximity
lithography. Contact lithography allows large mask fields to be printed with submicron resolution. Practical problems, such as mask contamination and resist sticking,

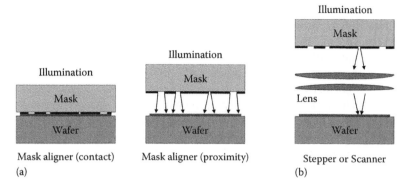

FIGURE 5.1   (a) Photolithography: In a mask aligner, the light propagates in free space from the mask to the wafer. (b) In projection lithography, the mask pattern is projected onto the wafer by an imaging system.

require a frequent cleaning and inspection of the mask that reduces the lifetime of the mask. Contact lithography is used for small wafer series in research labs, but not applicable for high-volume production. Proximity lithography, where the mask is located at a distance of typically 20 to 70 μm above the wafer, provides a resolution of some 2 to 5 μm, limited by diffraction effects. Proximity lithography is used for uncritical lithography processes, for example, in advanced packaging and micro-electromechanical systems (MEMS) production.

In a state-of-the-art projection system, the mask pattern on a reticle is projected onto the wafer by a very sophisticated catadioptric projection lens with a numerical aperture (NA) of up to NA = 1.35 for immersion systems. An optical 4× reduction system provides a resolution below 40 nm (half-pitch) for image fields of typically 20 mm × 30 mm and an overlay accuracy better than 2 nm. Today high-end projection systems are able to pattern up to 250 wafers of 300 mm wafer size per hour, equivalent to 16 m² per hour.

For both methods, shadow printing lithography in a mask aligner and projection lithography, the shaping of the illumination light is of high importance for the resolution and fidelity of the resulting resist pattern. During exposure, the illumination light impinging onto the photomask is diffracted at micro- and nanostructures of the mask pattern. Changing the illumination light influences the interaction of the incident light with the mask pattern, and thus changes the light propagation in free space or within the projection optics. In modern photolithography systems, the light is carefully shaped to minimize diffraction effects. In the case of projection lithography, it is also possible to reduce residual lens aberrations by modifying the illumination light. Shaping the illumination light is often referred to as customized illumination (CI). For CI, the mask pattern is illuminated with a well-defined angular spectrum, often referred to as *illumination settings*. Another possibility is to modify the mask pattern itself to minimize or precompensate for diffraction effects. This is referred to as *optical proximity correction* (OPC). If both illumination light and the mask pattern are optimized, this is referred to as *source mask optimization* (SMO). Today, lithography simulation software allows the complete lithography process to be simulated and optimized on a computer. As lithography simulation is quite time-consuming, this is usually done for the critical features of a mask pattern only. Simulation then provides the ideal mask pattern and the corresponding illumination settings. E-beam writing technology allows manufacturing almost perfect photomasks with very high accuracy of the structures and the positioning. Implementing desired illumination settings in a lithography system—the topic of this book chapter—is much more challenging. For both shadow printing and projection lithography systems, the shaping of the illumination light is usually based on micro-optical components. Micro-optical components like refractive microlens arrays and diffractive optical elements (DOEs) allows the illumination light to be freely shaped with little loss. A major part of this book chapter is dedicated to the design, manufacturing, testing, and system integration of the micro-optical key components used for light shaping in photolithography. The following section is about the early days of photolithography and its decisive role for semiconductor industry.

5.1.1 Pioneers in Photolithography

Jay W. Lathrop and James Nall from the Diamond Ordnance Fuze Laboratory (DOFL) are reported to be the first to use the term photolithography.[1] They used a standard microscope to project a pattern from a photographic film onto a slice of germanium. The germanium was painted with a thin layer of Kodak photoresist and exposed in the microscope. After wet-chemical resist development, an inverted copy of the mask pattern appeared on the substrate. The resist pattern serves as masking layer for additional processing like etching or ion implantation. Both left DOFL soon after their invention. Jay W. Lathrop joined Texas Instruments (TI); James Nall went to Fairchild Semiconductor and both implemented their photolithography approach of microstructuring semiconductor material at their new companies. These two companies were the technology leaders at that time and thus photolithography became the new standard technology for microstructuring on planar substrates.

The next decisive step towards modern photolithography was Fairchild's revolutionary *planar process*, invented by Jean Hoerni in 1957 and transferred to production two years later.[2] In Hoerni's planar process, a thin silicon oxide (SiO_2) film on a silicon wafer was coated with photoresist and microstructured by photolithography using a photographic film containing the layout of the circuit in a mask aligner. Subsequent SiO_2 etching, heat diffusion, and metal-layer deposition were applied to manufacture the transistors and to connect them electrically. Introducing planar silicon substrates, referred to as *wafers*, the planar process with several subsequent exposure steps also initiated the name *mask aligner*, a photolithography tool in which the transferred pattern is aligned with a previously microstructured pattern on the wafer using alignment marks.

5.1.2 Illumination Systems in Early Mask Aligners

In early mask aligners, simple light bulbs were used for photomask illumination. A very popular light source for mask aligners in the 1960s was the Sylvania Sun Gun, a white light lamp designed for 8 mm movie cameras and photography, shown

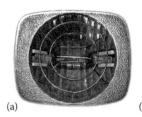 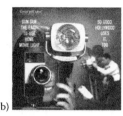 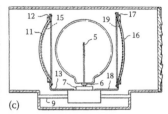

(a)   (b)   (c)

FIGURE 5.2   (a) Sylvania Sun Gun, (b) a light source for 8 mm home movies, photography, was widely used in first contact mask aligners, and (c) Scheme of a similar photographic spotlight with a light-concentrating reflector (11), a second heat-directing reflector (16) with dichroic film (16) for heat confinement and a Fresnel lens (3) for collimation (U.S. patent 2,798,943.)

in Figure 5.2. Such lamps were cheap, easy to handle, and provided bright white light with a sun light spectrum. For contact lithography, the angular spectrum of the illumination light is rather uncritical and uncollimated light could be used.

When industry moved from contact lithography to proximity lithography in the 1970s, light collimation became more critical. For proximity lithography, the exposure pattern blurs with increasing mask-to-wafer gap due to diffraction. Thus, collimated illumination light with an angular divergence ranging from ±1° to ±5° is required. In addition to the technology transition from contact to proximity lithography, the size of the wafers and the exposure wavelengths also changed in the 1970s. The semiconductor industry moved from 1″ to 2″ wafer size in 1969 to 3″ in 1972 and to 4″ in 1976. The filament lamps used in early mask aligners did not provide enough light. They were replaced by more powerful high-pressure gas-discharge lamps like mercury short-arc plasma lamps.

5.1.3 Mercury Short Arc Plasma Lamps and Ellipsoid Reflectors

High-pressure gas-discharge lamps contain a gas, typically a noble gas, such as argon, neon, krypton, and xenon, liquid mercury, and two tungsten electrodes at a distance of typically 2 to 7.5 mm. The liquid mercury is changed into a gaseous state by the high temperature in the electric arc. Mercury short-arc plasma lamps are designed for maximum output in the violet and ultraviolet wavelength region, particularly at 435.8 nm (g-line), 404.7 nm (h-line), 365.4 nm (i-line), 253.7 nm, and 184.5 nm. Mercury short-arc plasma lamps reach a luminous flux of up to 10,0000 lumen and luminance of more than 100 candela per m^2. Compared to most high-power light sources, mercury short-arc plasma lamps provide a relatively high light efficiency of some 40 to 60 lumen per watt. Nevertheless, they are high power consumers emitting typically more than 95% of the electric power as heat. Mask aligners are typically operated in a cleanroom environment where very tight tolerances for temperature and humidity are mandatory. Especially for production mask aligners, using 1,000 and 5,000-watt mercury short-arc plasma lamps, the heat removal is always critical. The lamp house of a mask aligner should therefore be placed outside the cleanroom, for example, in the technical zone.

For efficient collection of the light from a mercury short-arc plasma lamp, a parabolic reflector providing well-collimated light from extended sources would be a perfect solution.[11] Unfortunately, high-power mercury short-arc plasma lamps, which have a temperature of some 800°C at the surface of the Quartz glass bulb, are too hot to be placed in parabolic reflectors. For parabolic reflectors, the focal point is close to the reflector vertex and the mirror coating will be damaged by the heat of the lamp. Figure 5.3 shows the focal point position for an ellipsoidal reflector. The emitting characteristics of a mercury short-arc plasma lamp allow most of the light from the lamp to be collected and to have a large hole for cooling by ventilation or gas flow.

Ellipsoidal reflectors perfectly refocus the light emitted from an ideal point source located at the primary focal point at the secondary focal point, as shown schematically in Figure 5.3. However, the imaging quality dramatically decreases for any other object points. This is a severe problem for extended light sources like mercury short-arc plasma lamps with some 2 to 5 mm arc length. As shown schematically in Figure 5.3, the cross section of the light in the secondary focal point of the ellipsoid reflector is very large. For lamps with some 2 to 5 mm arc length, a cross section of the light at the secondary focal point of typically 30 to 40 mm is found. The propagating light has an angular spectrum of typically ±15°. The geometrical optical flux is proportional to the product of maximum light angle and size of the illuminated field. For such a high geometrical optical flux, the light emitted from the ellipsoid is quite difficult to collimate and homogenize in an optical integrator.[12] This is a fundamental drawback of mercury plasma lamps. In addition, these lamps require a very precise alignment of the lamp position and any misalignment will immediately lead to light loss. Despite these issues, mercury short-arc plasma lamps with ellipsoidal mirrors remain the only available high-power light sources for full-field shadow lithography tools like mask aligners.

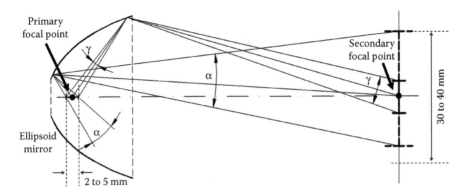

FIGURE 5.3 Scheme of an ellipsoidal reflector as used for collecting light from a mercury short-arc lamp. For plasma sources of some 2 to 5 mm arc length, the cross section of the light at the secondary focal point of the ellipsoid reflector is typically 30 to 40 mm.

5.1.4 GLASS DIFFUSER PLATES SERVING AS OPTICAL INTEGRATORS

In early mask aligners, a ground glass diffuser was placed in or near the secondary focal point of the ellipsoid mirror for light mixing. These ground glass diffusers just smeared out the light, but did not provide uniform illumination. Thus the ground glass diffusers were soon replaced by array integrator plates, as shown in Figure 5.4, manufactured, for example, by glass molding, that is, heating and press-forming glass using metal tooling.

Figure 5.4 shows an optical integrator plate with a matrix of pyramids as used for mask aligners to smear out the light after the ellipsoid mirror. Whereas glass molding in standard low-T_g glass has long been understood, molding in fused silica is much more difficult, as the glass-softening point is 1,600°C. The achievable profile quality of molded array plates in fused silica is typically quite limited. Thus the achievable uniformity using molded integrator plates as shown in Figure 5.4 was typically some ±5% or worse.

5.1.5 LENS ARRAY PLATES SERVING AS OPTICAL INTEGRATORS

A significant advance on the glass-molded integrator was lens array integrators, comprising some 20 to 40 individual lenses mounted in a frame, as shown in Figure 5.5.

In a lens array integrator, as shown in Figure 5.5, each lens generates an image of the source under a different angle. In the mask plane, these images overlap and form a quasi-uniform illumination. Lens array integrators with a higher fill-factor are shown in Figure 5.6, (a) two sets of cylindrical lenses crossed, and (b) square rod lenses mounted in a metal frame.

FIGURE 5.4 Glass-molded optical integrator plate (array of pyramids) made of fused silica, as used in early mask aligners for light homogenizing. (Courtesy of SUSS MicroTec, Germany.)

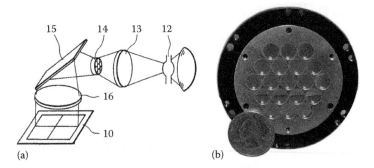

FIGURE 5.5 (a) Scheme of an illumination optical system from Canon described in U.S. Patent 4,530,587 comprising a mercury lamp (12), a condenser lens (13), a lens array serving as an optical integrator (14), a reflecting mirror (15), and a collimator lens (16), and (b) a photograph of an optical integrator plate from a Canon 501F mask aligner.

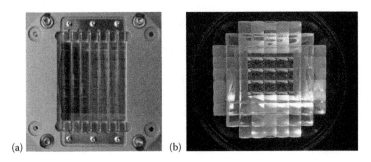

FIGURE 5.6 (a) Lens array generator comprising two sets of seven individual plano-convex cylindrical lenses (90° twisted) forming a 7 × 7 array of square aperture lenses and (b) square rod lenses mounted in a metal frame as used in early Nikon g-line steppers. (Courtesy of (a) SUSS MicroTec Photonics Systems Inc., and (b) Mark Loreto, both Corona, CA, USA.)

5.1.6 MASK ALIGNER ILLUMINATION SYSTEM

The optical system of a modern mask aligner is shown schematically in Figure 5.7. Light from a mercury short-arc plasma lamp is collected by an ellipsoidal mirror and collimated by a condenser lens. A light integrator, typically a fly's eye condenser or optical integrator, is used for homogenizing and light shaping. A field lens, referred to as the *front lens*, collimates the light. The optical systems provide collimated light with a defined angular spectrum of typically ±1° to ±5° divergence.

As discussed in Section 5.1.3, a large cross section of 30 to 40 mm and a large angular spectrum of ±15° make it difficult to collimate the light from the ellipsoid by a condenser lens. Especially for proximity lithography, where well-collimated illumination light is needed, a very significant amount of light is lost. This light needs to be blocked in the optical system. Another consequence of the high-geometrical optical flux is the long path length of the optical systems. The focal length of the front lens is typically some 1 to 1.5 m to achieve well-collimated mask illumination. Usually, only 10% to 20% of the light emitted by the lamp is used for mask

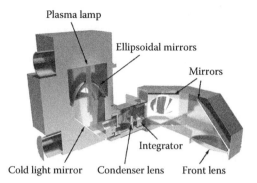

FIGURE 5.7 Scheme of the optical system of a mask aligner for proximity lithography. Light is collected by an ellipsoidal mirror and collimated by a condenser lens. An integrator unit provides uniform illumination of the photomask with a well-defined angular spectrum.

illumination. At the mask, the incident light is diffracted at micro- or nanostructures of the photomask and propagates in free-space towards the photosensitive resist on the wafer surface, as shown in Figure 5.40. After the exposure, the photoresist is developed in wet chemistry. Depending on the type of photoresist, either the exposed (positive resist) or the un-exposed (negative resist) parts of the resist layers are dissolved by the wet chemistry. After a baking process, a three-dimensional resist relief, a copy of the mask pattern, is obtained on the wafer.

5.1.7 PROXIMITY LITHOGRAPHY IN A MASK ALIGNER

The light propagation from the mask to the wafer in proximity lithography, as shown schematically in Figure 5.1, is described by near-field (Fresnel–Kirchhoff) diffraction.[3] In the general case, near-field diffraction can be investigated using so-called rigorous numerical methods solving Maxwell's equations. For mask feature sizes significantly larger than the wavelength of the illuminating light, and for sufficiently large proximity gaps, approximate methods such as scalar diffraction theory yield satisfactory results.[49] The achievable resolution for lines and spaces, half-pitch, for proximity lithography is deduced from the Fresnel integral formula and given by the following expression:

$$\text{Linewidth (half-pitch)} = \frac{3}{2}\sqrt{\lambda\left(g + \frac{d}{2}\right)} \approx \sqrt{\lambda g} \tag{5.1}$$

where:
    λ is the wavelength
    g is the proximity gap
    d is the resist thickness[4]

The resolution degrades with the square root of the proximity gap. The exposure pattern blurs with increasing mask-to-wafer gap due to diffraction. A major problem for larger proximity gaps is the interference of the diffraction pattern from adjacent mask features. Higher diffraction orders propagating from adjacent mask features may interfere and affect the desired features. These effects, also referred to as optical proximity effects, may lead to linewidth or shape variations of a desired feature.

5.1.8 OFF-AXIS ILLUMINATION

With the introduction of proximity lithography in mask aligners in the mid-1970s, the angular characteristics of the illumination light started to play a more decisive role than that for contact lithography. For proximity lithography, well-collimated illumination is mandatory to enable larger proximity gaps. According to the van Cittert–Zernike theorem,[5] the degree of partial coherence is related to the degree of collimation of the illumination light. Thus, for better-collimated mask illumination light, the diffraction effects even get worse. Unwanted secondary diffraction orders, also referred to as side lobes, become stronger and the features printed in photoresist on the wafer are blurred for increasing proximity gaps.

Illumination techniques developed for microscopy, like off-axis illumination[6] and apodization, were applied to enhance the resolution in proximity photolithography. Off-axis illumination turned out to be a very useful strategy to reduce the diffraction effects for proximity lithography significantly. Figure 5.8 shows the optical system described in U.S. Patent 3,941,475 and filed for the usage in Tamarack Scientific proximity mask aligners[7] in 1974. A fly's eye integrator configured for annular illumination was introduced to reduce the diffraction effects, especially side lobes.

Figure 5.8a describes how the resulting aerial image in the near-field of a mask is dominated by secondary diffraction orders, referred to as side lobes. Figure 5.8b describes how the diffraction effects could be reduced by off-axis illumination. Figure 5.8c describes a two-lens-array optical integrator, also referred to as fly's eye condenser; and Figure 5.8d describes different kinds of optical settings (ring, multiple rings, multipole) for diffraction-compensated illumination. In the 1970s and 1980s, it was not possible to simulate the light propagation for shadow printing lithography, where extended light sources illuminate complex mask features. At that time, off-axis integrator configurations for proximity lithography were derived by trial and error. For a better understanding, the basic types of illumination systems for microscopy will be explained.

5.2 OPTICAL INTEGRATORS

5.2.1 CRITICAL OR NELSON ILLUMINATION (ABBE)

In early microscopes, the so-called critical or Nelson illumination shown in Figure 5.9 was used. The critical illumination, introduced by Edward Nelson, relies on Ernst Abbe's fundamental work[6] on image formation and resolution published in 1873. Nelson called this illumination *critical* when the illumination aperture fills at least ¾ of the objective pupil ($\sigma > 0.75$). Nelson also discovered that the aberration

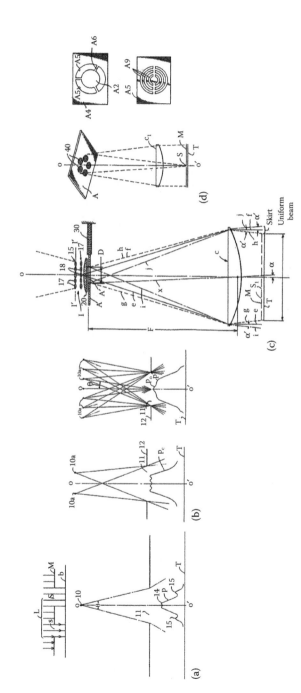

FIGURE 5.8 Illumination system for proximity mask aligner lithography,[6] as described in U.S. Patent 3,941,475, filed by Tamarack Scientific in 1974. Diffraction effects at a narrow slit for illumination with (a) plane wave, (b) dipole and multipole illumination. (c) A two-lens-array integrator (fly's eye), and (d) off-axis ring illumination to reduce diffraction effects as shown in (b).

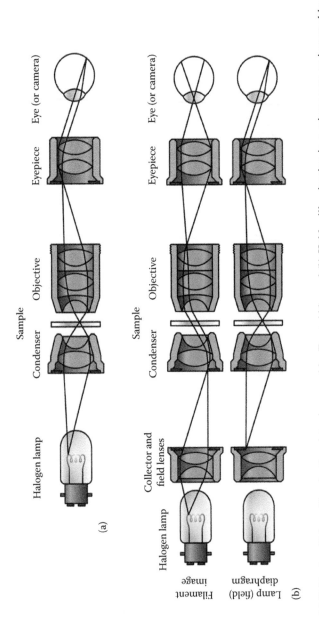

FIGURE 5.9 (a) Critical or Nelson illumination, later improved by Ernst Abbe and (b) Köhler illumination in a microscope as invented by August Köhler in 1893. (Courtesy of Richard Wheeler, http://en.wikipedia.org/wiki/User:Zephyris/Gallery.)

level of imaging instruments depends on the illumination.[5] For critical illumination, the light source is imaged in the sample plane. Preferably, an extended and homogenous light source is used.

5.2.2 KÖHLER ILLUMINATION

Köhler illumination, proposed by August Köhler[8] in 1893, provides uniform illumination of the object plane independent of shape, extension, and angular field of the light source. Each source point can be treated as generating a coherent plane wave of spatial frequency determined by the position of the source point relative to the optical axis. In other words, using Köhler illumination, each point at the target area is illuminated by the entire source, so that irradiance variations across the source do not affect the target illumination. However, if a single-lens element is used to collect the flux of the source, intensity variations of the source limit the achievable uniformity for Köhler illumination.[9] This problem could be overcome using multiple Köhler illumination channels, as shown schematically in Figure 5.10. The superposition of the light from the parallel Köhler illumination channels is done by a large lens, referred to as Fourier lens. Multiple-channel Köhler illumination systems are often referred to as Köhler integrators,[9] fly's eye uniformizer (see Chapter 2), or imaging beam integrator,[10] and will be discussed in detail in the next section.

5.2.3 KÖHLER INTEGRATORS

Figure 5.10 shows a scheme of a Köhler integrator, consisting of multiple Köhler illumination channels. The Köhler illumination is realized using two identical lens arrays, located at a focal length's distance of each other's. A large Fourier lens is used for light integration. For each channel of a Köhler integrator, the entrance pupil of the first lens is imaged by the second lens and the Fourier lens to the Fourier

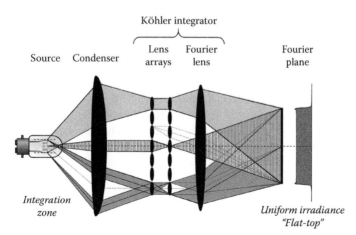

FIGURE 5.10 Scheme of a fly's eye or Köhler integrator comprising two lens arrays. A uniform light distribution referred to as *flat-top* is generated in the focal plane of the Fourier lens.

plane. The outer boundary of the uniform illumination area is a superposition of these individual images of the lens array subapertures and provides a sharp cut-off, often referred to as *flat-top* profile. If the integration zone is larger than the source, the source can be moved within the integration zone without affecting system performance, which helps to stabilize the flux of the illumination light on the sample.[9]

To achieve optimum irradiance uniformity, the subapertures of the lens array should be sufficiently small to ensure that the incoming light from the source is constant over each subaperture. On the other hand, good imaging quality of the individual lens channels is required to ensure aberration-free subimages of the lens array's individual entrance pupils in the Fourier plane.[10] This imaging quality requirement is a severe limit of both the maximum acceptance angle and the achievable irradiance uniformity of Köhler integrators. If the array lenses are too small, then diffraction is the limitation.[11] For lenses with a higher NA, the lens aberrations are the limitation. In both cases, the image formation deteriorates, so that the resulting integrated pattern in the Fourier plane becomes fuzzier, less uniform, and the light integration is less efficient.[12] As discussed in the next sections, the proper choice of the lens array parameters, namely the lens diameter, focal length, and number of lenses in the array, is very important for the performance of Köhler integrator.

5.3 LENS ARRAYS FOR KÖHLER INTEGRATORS

Köhler integrators were introduced for the mask illumination of lithography systems in the 1970s, replacing simpler one-element diffusers and integrators.[7] At that time, it was not possible to manufacture high-quality microlens arrays. Thus, most lens arrays used for Köhler integrators were based on some large lenses mounted in a metal frame, as shown in Figures 5.5 and 5.6. Although these Köhler integrators significantly improved the light uniformity, they had some drawbacks. Manufacturing a series of identical lenses is quite demanding and expensive. Mounting or gluing the individual lenses as arrays with a high fill factor is difficult. For high-power light sources, the lenses in the array need to be fixed by screws and springs as optical glues or cements will not survive the thermal heating. Mechanical clamping is not very stable against vibrations and the numerous cycles of heating and cooling-down of the array integrator during operation. As costs for manufacturing and mounting rise with the number of lenses within the array, single-lens-mounted optical integrators typically comprise only some 10 to 20 individual lenses. The limited lens-to-lens uniformity in the array, grid displacement errors, and the low number of lenses typically limit the performance of such Köhler integrators to some ±5% light uniformity or worse. These limitations were overcome using monolithic microlens arrays manufactured with wafer-based technology in the 1990s, which will be discussed in the next section.

5.3.1 MICROLENS-BASED OPTICAL INTEGRATORS

Microlens arrays were already proposed more than a century ago, for example, for integral photography or light diffusing in cinema projectors. Many researchers published and patented inventions, where microlens arrays were the decisive key elements.[18] At that time, microlens arrays were engraved or polished, for example, on a lathe. Later, glass molding, casting, and pressing were used, for example, to manufacture fly's eye condensers for slide and film projectors. Often, the quality of these microlens arrays was quite poor. Surface roughness, defects, lens profile accuracy, and non-uniformities in the array constrained their field of applications to simple illumination tasks. For more demanding applications, requiring, for example, two or three micro-optical layers, the lateral mismatch (grid imperfections), and the array-to-array alignment were often problematic. This situation changed with the rapid progress of microstructuring technology in the second part of the last century.

5.3.2 PLANAR OPTICS AND WAFER-BASED MANUFACTURING TECHNOLOGY

The planar manufacturing technology, initiated by Jean Hoerni's planar process (see Section 5.1.1), also had an impact on micro-optics. Dennis Gabor's invention of holography in 1947 allowed the recording of complex optical functions in a planar photographic plate. In 1963, at the dawn of Silicon Valley, Adolf W. Lohmann worked at IBM Research Laboratory in San Jose, when he was approached by Byron Brown, a summer student, who asked for a project that would combine holography and computers. Adolf W. Lohmann and Bryon Brown then developed a method to calculate holograms in a computer and to print them with the plotter.[13] The black-and-white pattern printed by the plotter was then photographed in a special repro-camera, reducing the plot pattern by a factor of 20:1 to 40:1. The camera image was recorded on high-resolution photographic plates. This was the invention of the computer generated hologram (CGH), the very beginning of laser beam shaping using DOEs. Lohmann's CGHs were binary amplitude holograms (two levels of amplitude transmittance), still suffering from low diffraction efficiency, parasitic diffraction orders, and a high noise level. The next step to improve DOEs was to use wafer-based technology, like photolithography and etching,[14] allowing binary phase holograms to be manufactured. For this process, the high-resolution plates were used as photomasks and contact-copied onto glass wafers using mask aligner lithography and subsequent sputter or plasma etching of the microstructured resist layer into the bulk material. In 1977, Mike Gale et al. manufactured multi-level DOEs serving as color filters on wafer-level.[15] In 1985, Popovic[16] proposed a microlens fabrication technology, which is based on microstructuring of photoresist by photolithography and a subsequent resist melting and reflow process. Reactive ion etching (RIE) is used to transfer the resist microlens into wafer bulk material like fused silica, silicon, or borofloat glass.[18]

5.3.3 MELTING RESIST AND REFLOW TECHNOLOGY

Figure 5.11 shows the different steps of manufacturing microlens arrays by melting resist and reflow technology.[16–18] A uniform layer of photoresist is spin-coated onto a glass wafer. The resist is microstructured by photolithography in a mask aligner and a subsequent wet-chemical development step. The resulting resist structures are melted in an oven or on a hotplate at temperatures of typically 150°C to 180°C. The reflow procedure itself is quite simple. Above the softening temperature, the

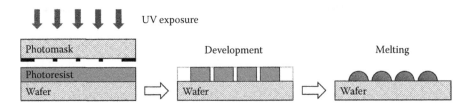

FIGURE 5.11 Manufacturing of microlens arrays by melting and reflow photoresist technology. A thick layer of photoresist is spin-coated onto a wafer and exposed by a photomask in a mask aligner. The exposed resist is developed and the resist structures are melted in an oven or on a hotplate at temperatures of typically 150°C to 180°C.

edges of the resist structure start melting. Above the glass transition temperature, the amorphous resist polymer changes into a glass-state system. The surface tension tries to minimize the surface area by rearranging the liquid masses inside the drop.

Ideally, the resist melts completely, the masses are freely transported, and surface tension forms a microlens with a spherical lens profile. After melting, the sag of the molten lens is typically significantly higher than the height of the resist layer before exposure. In practice, the lens-melting process needs careful process optimization and precise control of all process parameters to obtain good lens-to-lens uniformity within one wafer and from wafer-to-wafer. Repeatability and uniformity of molten resist lenses are key factors for the following etch process.

5.3.4 REACTIVE ION ETCHING TRANSFER

In the next step, the microlens structures are transferred into the bulk wafer material, for example, fused silica, silicon, or borofloat, using RIE technology. RIE is a dry etching technology, where the chemical reaction of etch gases like SF_6 and CHF_3 is enhanced by an ion bombardment. The ions are generated under low pressure in a plasma and accelerated by directional electric fields to achieve an anisotropic etching of the molten resist and the bulk material, as shown in Figure 5.12.

The etching process removes atoms from the resist and wafer surface at different etch rates. Surface areas covered by resist are protected until the covering resist layer is removed. Typical etch rates range from less than 0.01 μm per min to more than 1 μm per min, depending on the ion energy and the reactive etch gases in the plasma chamber. The molten resist lenses are usually very close to a spherical lens profile with a conic constant around $k \approx 0$ after melting. The transfer of the molten

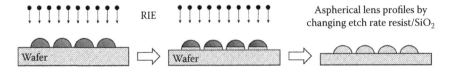

FIGURE 5.12 Scheme of the reactive ion etching (RIE) transfer process of resist microlenses in a wafer bulk material. A correction of the lens slope is obtained by changing the etch rate between the resist and the bulk material during the etching process.

resist lens by RIE allows the lens profile to be changed from spherical to aspherical profiles. This is done by varying the mixture of the etch gases and oxygen during the etch process. If the etch rate for resist is higher than that for the wafer bulk material, the resulting lens profile will have a flatter slope than the resist lens profile. A continuous change of all etch parameters allows one to obtain aspherical lens profiles.

Figure 5.13 shows the comparison of the measured lens profile (solid line) to the desired lens profile (dotted line) of a microlens with 1.08 mm lens diameter and 93 µm lens sag height etched in Silica. The profile was measured in a KLA-Tencor P15 mechanical profilometer.[19] Figure 5.14 shows the deviation of the measured lens profile, expressed by a 12th degree polynomial fit, from the ideal lens profile. For a microlens of 1.08 mm lens diameter, 93 µm lens sag, 1.8 mm radius of curvature, and a conic constant of $k = -1$, a deviation of only 155 nm (rms) is obtained.[18]

Melting resist technology and subsequent RIE allow us to manufacturing aspherical microlenses with excellent profile accuracy and lens-to-lens uniformity

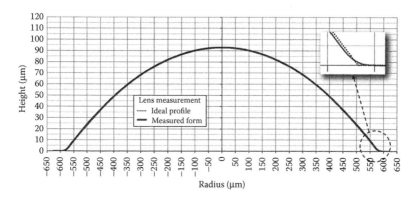

FIGURE 5.13 Comparison of measured lens profile (solid line) to ideal lens profile (dotted line) for a microlens of 1.08 mm lens diameter and 93 µm sag height etched in fused Silica.[18]

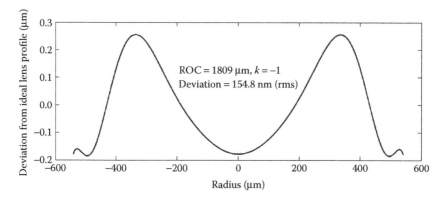

FIGURE 5.14 Deviation of measured lens profile, expressed by a 12th degree polynomial fit, versus the ideal lens profile shown in Figure 5.13. For a refractive microlens of 1.08 mm lens diameter, 93 µm lens sag, 1.8 mm radius of curvature, and a conic constant of $k = -1$, an excellent profile quality with only 155 nm (rms) deviation is obtained.[18]

on full wafer level. A lens-to-lens profile uniformity within one array of typically ±1% to ±3% is obtainable.[18] Using photomasks and wafers made of fused silica, an overlay and grid accuracy below ±0.5 µm is achievable for 8″ wafer manufacturing. Additional alignment marks allow different arrays, for example, on the front and backside of the same wafer to be aligned with an accuracy of better than ±1 µm.

5.3.5 CHALLENGES AND LIMITATIONS OF WAFER-BASED MANUFACTURING TECHNOLOGY FOR MICROLENS ARRAYS

One of the challenges of melting photoresist manufacturing technology is to manufacture lenses larger than 1 mm in diameter (or width) with a high fill factor. During the melting and reflow process, the resist forms a droplet with an ideal spherical lens profile. A fundamental limit for the melting process is the contact angle α of the resist lens in the liquid state as shown in Figure 5.15.

The contact angle is a function of the balance between the properties of the wafer surface, the liquid resist, and the atmosphere above the droplet. To ensure a proper lens profile, the contact angle α must be higher than the minimum contact angle α_{min}, which is defined by the surface energy of the wafer-resist atmosphere system. If this condition is not maintained, the lens will have a dip in the center and the resulting lens profile will show strong aberrations.[20] In practice, the reflow process works best for 1:15 to 1:20 aspect ratio (lens sag to lens diameter). Thus, for the manufacturing of microlenses with a lens diameter of more than 1 mm, a photoresist layer thicker than 50 µm is required. For the melting resist technology, the uniformity of the resist layer is directly related to the uniformity of the microlenses after melting. Thus, the resist-coating process needs to be perfectly optimized to achieve uniform resist layers. Not only is the resist coating a critical process for melting resist lens technology, but also the lithography for very thick resist layers is quite critical. For contact lithography in mask aligners, the aspect ratio of resist height versus the minimum obtainable feature size is typically around 5:1. For a 50 µm thick resist layer, a minimum gap of 10 µm between adjacent resist cylinders is required to ensure full separation of the exposed structures after the resist development process.[21] Although the RIE process is highly anisotropic, the remaining minor isotropic component of the etching process might further increase the gap in-between adjacent microlenses in a densely packed array. These

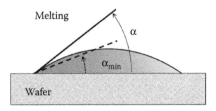

FIGURE 5.15 Scheme of a photoresist lens during the melting and reflow process. To ensure a continuous and quasi-spherical lens profile, the resulting contact angle α must be larger than the minimum contact angle α_{min} determined by the surface energy of the wafer-resist atmosphere system.

process limitations make it difficult to manufacture microlens arrays with a 100% fill factor by melting resist and RIE technology.

5.3.6 FILL FACTOR OF MICROLENS ARRAYS

Figure 5.16 shows basic configurations of closely packed periodic arrays consisting of identical circular (type A, C), rectangular (type B), or hexagonal microlenses (type D).

The fill factor $\xi_{A,C}$ for circular shaped lenses (type A, C) is as follows:

$$\xi_{A,C} = \frac{r^2\pi}{p_x p_y} \tag{5.2}$$

where:

p_x, p_y are the pitches of the lenses

r is the radius of the clear lens aperture

The gap between two adjacent lenses is $gap_{x,y} = p_{x,y} - \varnothing$. The fill factor ξ_B for regular arrays of rectangular lenses (type B) and hexagonal lenses (type D, $d_y = \sqrt{3}/2\,d_x$) is given as follows:

$$\xi_{B,D} = \frac{d_x d_y}{p_x p_y} \tag{5.3}$$

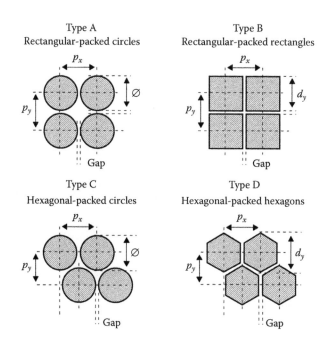

FIGURE 5.16 Different types of microlens array configurations.

where:

d_x, d_y are the lens widths
p_x, p_y are the pitches

The gap between two adjacent lenses is $gap_{x,y} = p_{x,y} - d_{x,y}$. Table 5.1 shows numerical values of the fill factor ξ as a function of the gap in relation to p_x.

It is difficult to manufacture type B and D microlens arrays using melting resist and reflow technology. The sharp corners in the lens base will lead to different lens profiles along the short and long axes after melting. This problem could be overcome using a two-step melting and etching procedure. The more elegant solution to obtain high fill factor microlens arrays for Köhler integrators is to use two stacked microlens arrays of densely packed cylindrical lenses for x- and y-directions as shown schematically in Figure 5.17.

TABLE 5.1

Fill factor ξ for lens arrays. Ø is the lens diameter and p_x, p_y are the pitches in x- and y-direction.

Gap in [%] of Pitch	Fill Factor ξ in [%]		
	Type A	Type C	Type B, Type D
$(p_x - \varnothing)/p_x$	$p_x = p_y$	$p_y = \sqrt{3}/2\, p_x$	
Contact	78.5	90.7	100
1	77.0	88.9	98.0
2	75.4	87.1	96.0
3	73.9	85.3	95.1
5	70.9	81.9	90.0
10	63.6	73.5	81.0

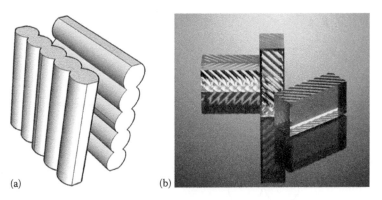

(a)   (b)

FIGURE 5.17 Double-sided cylindrical microlens arrays as used for optical integrators: (a) Scheme of two microlens arrays at a focal length distance and (b) photography of a similar microlens array, where front and backside arrays are twisted by 90°. (Courtesy of SUSS MicroOptics, Switzerland.)

5.4   MICROLENS-BASED KÖHLER INTEGRATORS

5.4.1   INTRODUCTION

The basic concept of a Köhler integrator has been introduced in Section 5.2.2. The advantages and disadvantages of microlens-based Köhler integrators will be discussed in more detail within this section. Wafer-based microlens manufacturing technology provides high quality and densely packed microlens arrays with lens diameters ranging from some 10 µm to several millimeters.[18,22] This gives the optical designer an almost free choice regarding the number of lenses within an array, the lens diameters, the lens profiles (spheres or aspheres), and the NA of the lens arrays for microlens-based optical integrators. However, it is not always easy to choose the optimum parameters for a microlens-based optical integrator system.

On the one hand, the subapertures of the microlenses should be sufficiently small to ensure that the incoming light is constant over each subaperture. The superposition of the intensity distribution within the subapertures will then provide uniform light distribution, the so-called *flat-top* light profile, in the focal plane of the Köhler integrator, as shown schematically in Figure 5.10. For most light sources, some 20 to 40 microlenses will provide acceptable uniformity of the light intensities in the flat-top. On the other hand, the microlenses of a Köhler integrator should not be too small. Otherwise, the optical performance of the integrator is deteriorated by diffraction effects and the limited space-bandwidth-product (SW) of the secondary microlens array, as discussed in the following sections in more detail.

5.4.2   FRESNEL NUMBER

As discussed in 5.2.3, the optical performance of a microlens-based Köhler integrator is influenced by two lens parameters, the size, and the NA of the microlenses.[10,53] If the lens diameter is too small, the optical performance of the Köhler integrator is deteriorated by diffraction effects. The Fresnel number (FN) is used to describe the ratio of the geometrical and diffraction effects on beam propagation for light focused by a microlens.[23] The FN is as follows:

$$\text{FN} = \frac{a^2}{\lambda f_E} = \frac{a^2(n-1)}{\lambda R_c} \tag{5.4}$$

where:
   $2a$ is the lens diameter
   λ is the wavelength
   R_c is the radius of curvature
   n is the refractive index of the lens material
   f_E is the paraxial focal length derived from geometrical optics as shown schematically in Figure 5.18

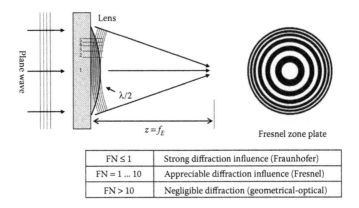

FN ≤ 1	Strong diffraction influence (Fraunhofer)
FN = 1 ... 10	Appreciable diffraction influence (Fresnel)
FN > 10	Negligible diffraction (geometrical-optical)

FIGURE 5.18 Definition of the Fresnel number (FN): A convergent wave front is moving from a microlens toward an observation point at $z = f_E$. The lens aperture can be broken into Fresnel zones, each indicating an optical path difference of $\lambda/2$.

The FN indicates the number of Fresnel zones representing the microlens, as shown schematically in Figure 5.18. A microlens with a small lens aperture $2a$ and a long focal length f_E has a low FN near one or below, and is more dominated by diffraction effects at the lens aperture than by refraction of the incident light at the lens profile.[24,25] For microlenses with low FNs, the focal length cannot be calculated by geometrical optics anymore. The dominance of diffraction at the lens aperture leads to a significant focal length shortening, referred to as *focal shift*. Figure 5.19 shows the propagation of the light behind a microlens of 150 μm lens aperture. The light distribution was calculated by Gaussian beam decomposition for coherent fields.[26]

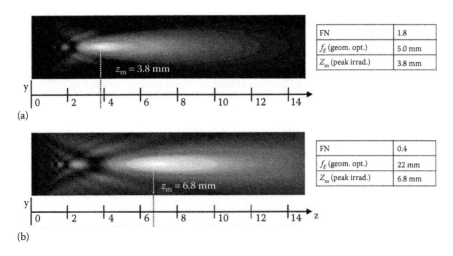

FN	1.8
f_E (geom. opt.)	5.0 mm
Z_m (peak irrad.)	3.8 mm

FN	0.4
f_E (geom. opt.)	22 mm
Z_m (peak irrad.)	6.8 mm

FIGURE 5.19 Intensity (arbitrary units) distribution behind a microlens of Ø = 150 μm, and (a) radius of curvature R_c = 2.03 mm and (b) R_c = 10 mm, illuminated by a plane wave at 633 nm. The lens stands in the x, y plane, while z corresponds to the propagating axis.

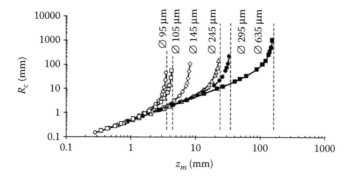

FIGURE 5.20   Radius of curvature R_c versus position z_m of the peak irradiance, namely the best focus for different microlens diameters, illuminated by a plane wave at 633 nm. (Graph: Ruffieux et al.[24])

For both examples shown in Figure 5.19(a) FN = 1.8 and (b) FN = 0.4, the best focus spot, namely the position z_m of the peak irradiance, is significantly closer to the lens than the paraxial focal length f_E would indicate. For microlenses with very large radius of curvature, the light distribution converges with the light distribution of a pinhole of the same diameter. In other words, for every lens diameter and a specific wavelength, the maximum obtainable focal length is derived from a pinhole corresponding to a lens with no optical power, that is, $R_c = \infty$, as shown in Figure 5.20.

As shown in Figure 5.20, for a low radius of curvature R_c, corresponding to a FN >> 10, the position z_m of the peak irradiance corresponds well to the paraxial focal length f_E derived from geometrical optics. For increasing values of R_c, that is, for low FNs near one and below, the peak irradiance position z_m converges to a maximum value of z_m illustrated by vertical dash lines in Figure 5.20. These values correspond to the peak irradiance position obtained for pinholes with no optical power.

This example demonstrates that for the design of microlens-based devices or systems, it is mandatory to evaluate the influence of diffraction on the optical performance, for example, by calculating the FN. For nonimaging optical integrators with a single-microlens array, as shown schematically in Figure 5.25(a, left), the FN of the microlenses corresponds roughly to the number of ripples in the flat-top, shown in Figure 5.25(a, right). This correlation is explained if the light inside a subaperture of the first lens array is regarded as coherent. Thus, the far-field distribution shows a typical Fresnel structure with diffraction ripples.[5,10]

5.4.3   SPACE-BANDWIDTH-PRODUCT LIMITS THE PERFORMANCE OF MICROLENS-BASED KÖHLER INTEGRATORS

Besides the FN, another important key parameter for microlens-based Köhler integrators is the so-called SW parameter, also referred to as space-bandwidth-product. The SW *beta* parameter is treated in detail for beam integrators in Reference 10. Reference 10, Chapters 10 and 12 of this book also treat channel integrators and gives a unique Fresnel formula for them. For microlens-based Köhler integrators, the SW is related to the ability to obtain a flat-top light distribution with well-defined sharp edges. The SW is the number of object points that an imaging system can transport to the image plane.[27] In other words, the SW is the useable field-of-view (FOV) of an imaging system divided by the pixel size. It describes how many individual object points the imaging system can transfer to resolvable image pixels in the image plane. The size of an image pixel is derived from the point spread function (PSF) of the imaging system. The PSF describes the response of an imaging system to an ideal point source. Due to the low-pass characteristics of optical system, an infinitesimal small point source in the object plane will be blurred out to a larger spot in the image plane. For aberration-free imaging systems, the minimum image pixel size corresponds to the diffraction-limited airy disk pattern, defined by NA = sin(u) of the lens and the wavelength λ as follows:

$$\varnothing_{Airy} = 1.22 \cdot \lambda / NA \qquad (5.5)$$

In the simplest case of a Köhler integrator, both lens arrays are identical and located at a focal length distance of each other as shown in Figure 5.21. Thus each lens of the second array images the corresponding subaperture of the first array to infinity. The Fourier lens collects the light and superimposes the subimages in its Fourier plane.

Usually the bottleneck of the imaging part of a Köhler integrator is the microlens array. The off-axis imaging quality of the microlenses and their SW must provide enough resolution to form a sharp superposition image of the subaperture of the first lens array in the Fourier plane.

The ideal configuration of the microlens arrays is a trade-off between NA, which is large enough to resolve the subapertures, and the off-axis imaging properties, namely, the useable FOV of the microlens. In the ideal case, the microlens

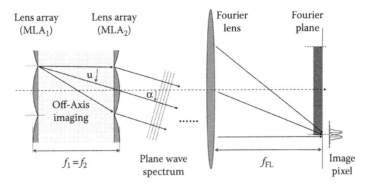

FIGURE 5.21 Scheme of a Köhler integrator with two identical microlens arrays located at a focal length distance of each other. The second microlens array (MLA$_2$) images the subaperture of the first microlens array (MLA$_1$) to infinity. The Fourier lens collects the plane wave spectrum generated by the second array (MLA$_2$). The SW of a Köhler integrator is the number of object points from the subapertures of the first lens array transported to the Fourier plane.

array would consist of lens doublets or triplets, well corrected for off-axis imaging. Unfortunately, no cost-efficient technology exists to manufacture arrays of micro-lens doublets or triplets on planar substrates. For reasons of simplicity and cost, most Köhler integrators are built using plano-convex microlenses with a spherical or aspherical lens profile. Typically, the microlens arrays are manufactured on the front and backside of a wafer made of fused silica, as shown in Figure 5.17. Fused silica is the preferred glass material due to its high transparency from 190 nm to the mid-infrared wavelength region and its robustness against high laser power. A drawback of fused silica is the relative low refractive index, ranging from 1.56 (193 nm) to 1.44 (2 μm). Due to the low refractive index, a plano-convex lens made of fused silica needs to have a relatively high lens sag to achieve a large NA. A high lens sag leads to severe spherical aberrations that can only be corrected using aspherical lens profiles. Microlenses with an aspherical lens profile show excellent performance for on-axis operation, but the aspherical profile might lead to strong aberrations for off-axis operation, as required for a larger FOV.

These restrictions make the design of a microlens-based Köhler integrator quite tricky. The optical design must take into account the divergence and coherence of the incident light, must optimize the asphericity of the lens profiles to minimize both on-axis and off-axis aberrations, and must consider diffraction effects at the micro-lens arrays. Simulation and optimization of microlens-based Köhler integrators have been investigated in many ways using all kind of simulation approaches, like ray tracing, beam or wave propagation, or rigorous optical diffraction analysis.[5,10,28,29,49] For most of these approaches, the major problem remains the definition of the light source. Optical design and simulation work very well for completely incoherent and coherent light sources. Unfortunately, most light sources deliver partial coherent light. Thus, it is very difficult or even impossible to derive the exact parameters of the light source for simulation. In practice, a Köhler integrator is often adapted itera-tively by simulation and experiments. As a starting point, microlenses with a slightly aspherical lens profile, for example, a conic constant of $k = -0.5$, are usually a good choice. In another approach, the lens array comprises three to five different aspheri-cal lens profiles well balanced to generate a complete set of plane waves for α_{min} to α_{max} with equal intensity.

5.4.4 Microlens-Based Köhler Integrators for Array Generation

In the case of a well-collimated and coherent light source, like a laser light source, a microlens-based Köhler integrator could be used as array generator.[28,30,31] Array generators were investigated in the 1990s for the illumination of optical and opto-electronic devices in digital optics. Streibl et al. describe the microlens array as a phase grating having many diffraction orders with equal intensity.[30] The second lens array in a standard Köhler integrator, similar to the scheme shown in Figure 5.21, is described as an array of field lenses. For an incident plane wave, the first micro-lens array focuses the light onto the vertices of the lenses of the second array. For a collimated laser beam, the foci represent multiple copies of the laser beam waist. A matrix of sharp spots is observed in the Fourier plane.

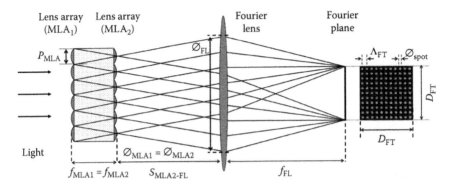

FIGURE 5.22 Scheme of a Köhler integrator with two identical microlens arrays located at a focal length distance of each other. For the illumination with well-collimated and coherent laser light, the Köhler integrator serves as an array generator.[30]

As shown in Figure 5.22, the width D_{FT} of the flat-top in the focal plane of the Fourier lens is given by the pitch P_{MLA}, and the focal length f_{MLA} of the microlens arrays, and the focal length f_{FL} of the Fourier lens as follows:

$$D_{FT} = P_{MLA} \frac{f_{FL}}{f_{MLA}} \tag{5.6}$$

For periodic microlens arrays with a pitch of P_{MLA}, the flat-top profile will be modulated by diffraction orders with a period as follows:

$$\Lambda_{FT} = \frac{f_{FL} \cdot \lambda}{P_{MLA}} \tag{5.7}$$

The number of diffraction orders $N_{x,y}$ within the flat-top correlates with the FN of the microlens and is given by the following:

$$N_{x,y} = 4 \cdot FN = \frac{P_{MLA}^2}{\lambda \cdot f_{MLA}} \tag{5.8}$$

For circular microlenses, the illuminated aperture of the Fourier lens is circular with a diameter $Ø_{FL}$, derived from the NA of the first microlens array NA_{MLA1} and the distance $s_{MLA2\text{-}FL}$ in-between the second microlens array and the Fourier lens, as shown schematically in Figure 5.22. Thus, the diffraction-limited diameter $Ø_{spot}$ of the individual spots (Airy disk) is given by the following:

$$Ø_{spot} \approx 2.44 \frac{\lambda \cdot f_{FL}}{Ø_{FL}} \tag{5.9}$$

Figure 5.23 shows a matrix of spots generated with a Köhler integrator consisting of two identical microlens arrays with pitch $P_{MLA} = 250 \ \mu m$ and $f_{MLA} = 1.6$ mm focal

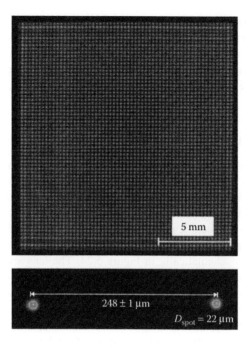

FIGURE 5.23 Array of 61 × 61 spots of 22 μm diameter at 248 μm pitch generated with a microlens-based Köhler integrator illuminated by a collimated laser beam of 633 nm (HeNe). (Courtesy of Maik Zimmermann, Bayerisches Laser Zentrum, Erlangen, Germany.)

length, located at a focal length distance of each other and a Fourier lens of $f_{FL} =$ 100-mm focal length.[31]

The Köhler integrator was illuminated with a collimated laser beam of 633 nm wavelength (HeNe). A spot matrix with 61 × 61 spots of $\varnothing_{spot} \approx 22$ μm and a period of $\Lambda_{FT} \approx 248$ μm is observed. The envelope of the spot array is a flat-top with a homogeneity of ±2%. Typical applications for array generators are laser ablation and medical applications, like laser epilation, rejuvenation, and tattoo removal.

As discussed above, for illuminating a Köhler integrator with a coherent and collimated laser beam, the resulting flat-top will be modulated by equidistant sharp intensity peaks. The size of the peaks correlates with the size of the beam waist in the laser, in other words, with the beam parameter product M^2. The peaks are related to multiple beam interference. The first microlens array generates multiple images of the light source located at the vertex of the second microlens array for perpendicular incidence of a collimated laser beam. The light distribution in the plane of the second microlens array could be interpreted as a Dirac comb function. The propagating plane wave spectrum is then transformed by the Fourier lens. The Fourier transformation of a Dirac comb function is also a Dirac comb, the observed matrix of spots. For a Köhler integrator with periodic arrays, the transmission function can be derived from the convolution of one lens channel and the comb function representing the positions of the individual lenses. For lasers with a $M^2 \gg 10$, the subimages of the source, that is, the beam waist, are getting larger and fill the subapertures of the second lens array. Thus, the individual peaks in the Fourier plane also get larger and overlap with the adjacent peaks. For lasers with a small $M^2 \approx 1$, additional measures need to be applied to obtain a smooth flat-top intensity profile. This is typically done by introducing nonperiodic or quasi-statistical optical components, like random diffuser plates[10,32–35] or using nonperiodical microlens arrays, as described in the following sections.

5.4.5 KÖHLER INTEGRATORS WITH CHIRPED MICROLENS ARRAYS

An interesting approach to overcome interference effects of microlens Köhler integrators for coherent laser illumination is to use chirped microlens arrays, as proposed by Wippermann et al.[36] and shown schematically in Figure 5.25. The width D_{FT} of the flat-top in the Fourier plane of a Köhler integrator is given by the NA of the microlenses and the focal length of the Fourier lens f_{FL} as explained in Section 5.4.4. The basic idea behind nonperiodic or chirped microlens arrays for a Köhler integrator[36] is shown schematically in Figure 5.24. The lens arrays consist of two series of microlenses with different diameters and sag heights, but identical NA. The second array is placed in the focal plane of the first array, as shown in Figure 5.24a. Köhler integrators using chirped microlens arrays can be manufactured as 1D arrays using arrays of chirped cylindrical microlenses or 2D arrays using elliptical microlens arrays. For reason of simplicity of manufacturing and mounting, the 1D solution is the preferred method. 2D Köhler integrators are then obtained using two chirped 1D integrators twisted by 90°.

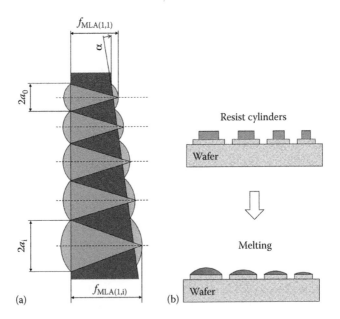

FIGURE 5.24 (a) Scheme of two monolithic chirped microlens arrays as used for Köhler integrators (Courtesy of Frank Wippermann, Fraunhofer IOF, Jena, Germany) and (b) schematic drawing of the reflow lens-manufacturing procedure.

Wippermann et al.[36] proposed manufacturing the chirped microlens arrays using a slightly modified melting resist and reflow process (see Section 5.3.3), shown schematically in Figure 5.24b. In a first step, pedestals are manufactured in resist and etched into the wafer. In a second step, resist cylinders or cuboids with a smaller diameter (width) are placed on top of the pedestal by lithography, as shown in Figure 5.24 (b, top). The volume of the resist cylinders is adapted to form microlenses with constant NA, but different diameters and sag heights, as shown in Figure 5.24 (b, bottom).[36]

Figure 5.25 shows a comparison of the resulting intensity distribution in the Fourier plane for: (a) nonimaging array integrator based on a single microlens array, (b) Köhler integrator based on two identical periodic arrays, and (c) Köhler integrator with two chirped arrays. The resulting intensity distribution is plotted for two cases: (top) a single-lens channel with no interference effects and (bottom) the multiple interference of many lens channels of an array, which is illuminated by coherent light.

As shown in Figure 5.25, the nonperiodic chirped microlens arrays in a Köhler integrator result in a nonregular intensity distribution. Due to the remaining coherence, light from different channels still interferes and form a speckle pattern. Figure 5.25c shows the speckle pattern for an ideal point source providing perfectly collimated and coherent illumination. For laser light sources with $M^2 \gg 1$, the sharp delta peaks shown Figure 5.25c will broaden and overlap.

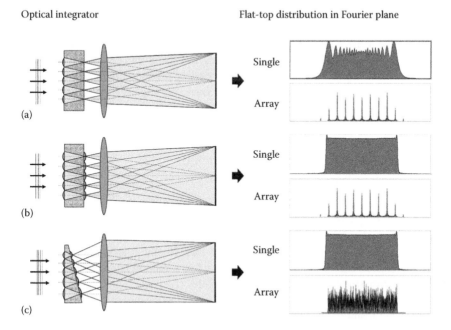

Optical integrator   Flat-top distribution in Fourier plane

FIGURE 5.25 Comparison of fly's eye condenser types and the resulting intensity distribution in the Fourier plane for: (a) nonimaging fly's eye integrator based on a single microlens array, (b) Köhler integrator based on two identical periodic arrays, and (c) Köhler integrator with two chirped arrays. (Courtesy of Frank Wippermann, Fraunhofer IOF, Jena, Germany.)

Compared to Figure 5.25b with periodic microlens arrays, the peaks in Figure 5.25c are located much more densely, which improves homogenization, especially for lens arrays with small lens pitch. Since all channels have the same NA, Figure 5.25a Köhler integrator with two chirped microlens arrays will provide a flat-top with sharp edges. Like all Köhler integrators, the systems using chirped microlens arrays will benefit from low levels of stray light, broad spectral range, relaxed alignment tolerances, and insensitivity to temporal changes of the incident laser beam.

The major problem of the chirped Köhler integrator approach is the manufacturability of the chirped microlens arrays. The proposed method[36] of manufacturing these arrays using the melting resist and reflow technology is difficult to be implemented in production. Photoresist chemistry is very sensitive to process parameters like humidity and residual solvents in the air and temperature. The critical issue for the proposed technology is to choose the right volume of the resist cylinders or cuboids before melting. In addition, as described in 5.3.3, the melting resist and reflow technology only allow spherical lens profiles ($k \approx 0$) to be manufactured. Aspherical profiles are obtained by changing the etch gases during the transfer of the resist lenses to the wafer material using plasma etching, as described in Section 5.3.4. This is only possible for identical microlenses. For etching chirped lens arrays with many different lens diameters and lens sags side-by-side, a profile correction is not possible. Another issue is the alignment of the two-chirped arrays. Wafer-based manufacturing technology only works for planar substrates, thus both arrays must be manufactured separately, wedge-polished, and mounted together to obtain a monolithic double-sided array as shown in Figure 5.24(a). A similar approach to reducing the coherence and interference effects of microlens-based Köhler integrators is partially delaying the input wavefront using, for example, a set of two stair mirrors, as demonstrated by Erdmann.[34] The stair mirrors split the incoming wavefront into beamlets. The different optical path lengths in-between the corresponding mirror pairs reduce the temporal coherence of the light.[53] Other methods for wave splitting are based on a bundle of optical fibers with different lengths or the combination of two gratings.[5]

5.4.6 RANDOM DIFFUSER FOR THE REDUCTION OF MULTIPLE INTERFERENCE AND SPECKLES

As discussed in the previous sections, the temporal and spatial coherence of light sources is a significant problem for optical integration. Diffraction effects, multiple interference, and speckles might have a strong influence on the flat-top profile, especially for microlens-based optical integrators. Thus to obtain a smooth flat-top profile, the preferred choice is light sources with a low temporal and spatial coherence, like lasers with $M^2 \gg 20$ or light-emitting diodes (LEDs). If such light sources are not applicable, then moving random diffusers and time averaging are the good alternative to obtain smooth flat-top profiles.[5]

Figure 5.26 (a) shows schematically the basic optical set up of a microlens-based Köhler integrator equipped with a rotating random diffuser (a), located in, or near, the common focal plane of a telescope. The z-position of the rotating diffuser in the telescope is usually optimized to fill the subapertures of the second microlens array

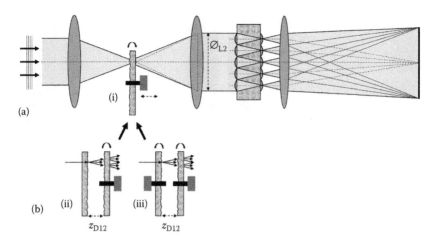

FIGURE 5.26 (a) Optical set up of a microlens-based Köhler integrator equipped with a rotating random diffuser (i) located in or near the common focal plane of a telescope; (b) shows alternative set ups using (ii) a combination of a static and a rotating diffuser and (iii) two diffusers rotating in opposite directions separated by a distance of Z_{D12}.

in the Köhler integrator. Figure 5.26 (bottom) shows alternative set ups using (b) a combination of a static and a rotating diffuser, or (c) two diffusers rotating in opposite directions separated by a distance of Z_{D12}. Using two sequential scattering plates, of which one or both are moving, provides a significant improvement of the speckle and interference-effect suppression.[5,37] The drawback of using random diffusers is a potential loss of light due to diffraction into angles too large to be collected by the second telescope lens and the Köhler integrator. This problem could be avoided using random DOEs,[38] or shaped diffusers,[35] as explained in more detail in Section 5.4.7 (Figure 5.28).

Rotating random diffusers are well suited to smear out interference and speckle effects for microlens-based Köhler integrators if the integration time at the sensor or in the application is long enough. This is typically not the case for pulsed lasers, like Nd:YAG and excimer lasers. In the best case, the application allows multiple pulses to be recorded on the same substrate or detector. In this case, a rotating or moving diffuser would modulate the interference or speckle pattern differently for each pulse. The integration of multiple patterns will then lead to a smooth flat-top. This is, for example, the case for high-end projection lithography systems, where up to 50 pulses are recorded at each position on the wafer. For some applications of pulsed lasers, where throughput or cost requirements do not allow multiple pulses to be used, dynamic flat-top smoothing becomes very difficult. Rotating diffusers typically rotate with about 5000 rotations per minute, that is, for a disk of 50 mm diameter, the speed in the usable range is about 10 μm per nanosecond (ns). For a minimum feature size of typically 40 to 50 μm, the pulse length for the laser should be longer than 20 ns to allow sufficient randomization by the rotating diffuser. Preferably, a double-disk set up is used, where the two disks rotate in opposite direction, similar to Figure 5.26(b, ii).

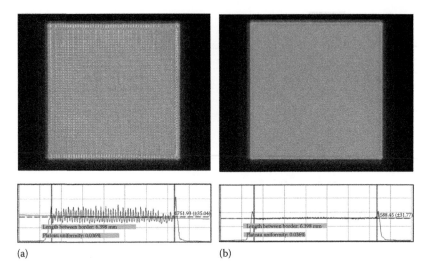

(a)   (b)

FIGURE 5.27 Flat-top profiles for a microlens-based Köhler integrator illuminated with coherent and collimated laser light from a laser diode (670 nm). A rotating diffuser, similar to the set up shown schematically in Figure 5.26 (a) is used to smooth the flat-top profile (b).

5.4.7 MANUFACTURING OF RANDOM DIFFUSERS

As discussed in Section 5.4.6, random diffusers are well suited to reduce interference effects or speckles for Köhler integrators used with coherent laser light. For these applications, the random diffusers are typically placed before the Köhler integrator, as shown in Figures 5.26 (a) and 5.27. Random diffusers could also be placed at different positions within a microlens-based Köhler integrator, after the Fourier lens, or in combination with diffractive beam shaping elements.

For these applications, the random diffusers allow residual noise, speckles, oscillations, and the zeroth or unwanted higher orders appearing in flat-top profile to be smoothed or flattened out.

The simplest case of a random diffuser is a ground-glass diffuser plate, as shown in Figure 5.28a. Ground-glass diffuser plates are usually manufactured by grinding, lapping, and sand blasting of glass plates. The finer the particles in the slurry, that is, the smaller the grits for the sand blasting are, the smoother is the surface and the smaller are the resulting diffusion angles. However, due to their rough surface structure, shown in the SEM image in Figure 5.28a, a significant amount of the incident light is diffracted, refracted, or reflected back or into very large angles and is thus lost in the optical system.

A surface smoothing is achieved by isotropic wet-etching of fused silica ground-glass diffusers in hydrogen flouride (HF), as shown in Figure 5.28b. A variety of manufacturing techniques, comprising holography, speckle recording, photolithography, e-beam writing, laser surface machining, volume diffusers based on micro or nanobeads, and so on have been developed for the manufacturing of all kinds of diffractive, refractive, or hybrid random diffusers. However, most of the commercially available light shaping or engineered diffusers are not suitable for high laser power

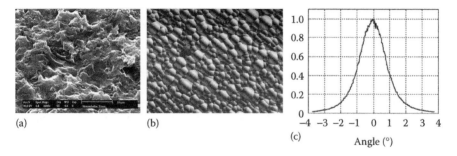

(a)   (b)   (c)

Angle (°)

FIGURE 5.28   Random diffuser plates: (a) SEM image of a ground-glass diffuser obtained by sand blasting; (b) microscope image of a diffuser plate obtained by successive etching of a ground-glass plate with hydrogen fluoride (HF); and (c) angular spectrum for plane wave incidence measured in a goniometer. (Courtesy of SUSS MicroOptics, Neuchâtel, Switzerland.)

in the ultraviolet wavelength range, as required for photolithography. An alternative manufacturing technique for 1D and 2D random diffusers allows shaped small-angle diffusers to be manufactured.

Figure 5.29 shows a scheme of the process flow for manufacturing of 1D or 2D random diffusers in fused silica (SiO_2) on wafer-level.[35] A poly-Silicon (p-Si) etch mask is deposited by low pressure chemical vapor deposition (LPCVD) on both sides of a fused silica wafer. The desired mask pattern is transferred into the poly-Silicon mask by photolithography and RIE. The wafer is dipped into nondiluted hydrofluoric acid (HF) at room temperature. During the wet-etch process, the poly-Silicon masks gets completely under-etched and falls off. Further etching leads to larger radius of curvature R_c of the random concave elements without changing their position or form. As shown Figure 5.30b, the resulting far-field pattern of the linear 1D random diffuser is a flat-top with no zeroth or higher orders (Figure 5.31).

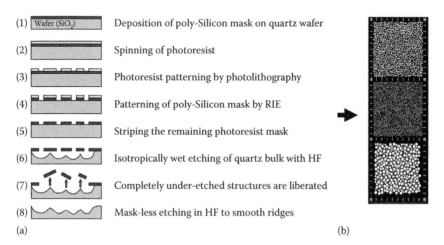

(1) Wafer (SiO₂)   Deposition of poly-Silicon mask on quartz wafer

(2)   Spinning of photoresist

(3)   Photoresist patterning by photolithography

(4)   Patterning of poly-Silicon mask by RIE

(5)   Striping the remaining photoresist mask

(6)   Isotropically wet etching of quartz bulk with HF

(7)   Completely under-etched structures are liberated

(8)   Mask-less etching in HF to smooth ridges

(a)   (b)

FIGURE 5.29   (a) Scheme of the process flow for manufacturing 1D or 2D random diffusers by photolithography, a poly-Silicon mask, and selective wet etching with hydrogen fluoride (HF) and (b) typical mask pattern for the process step 3 as shown in (a).[35]

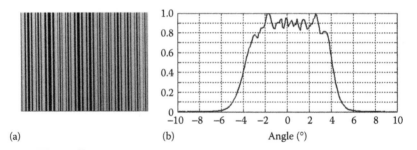

(a)   (b)   Angle (°)

FIGURE 5.30   1D random diffuser for smoothening a modulated flat-top profile: (a) Scheme of the 1D random slit pattern on the photomask for surface structuring and (b) far field pattern of the etched 1D random diffuser measured in a goniometer.[35]

FIGURE 5.31   (a) Spot pattern in the flat-top of a 1D Köhler integrator illuminated with coherent laser light and (b) smooth flat-top obtained by placing the linear 1D random diffuser (Figure 5.30) in the Köhler integrator in-between microlens arrays and the Fourier lens.

The wafer-level manufacturing technology described allows linear (1D) and two-dimensional (2D) random diffusers with diffusion angles with very small angles below 1° to be freely shaped. This allows flat-top profiles to be smoothed with only a small increase of the divergence of the laser beam. The drawback of the described technology[35] is the high sensitivity of the HF etching to process parameters like temperature, age of the chemical solution, and agitation during the process. Typically, the etching process is highly irreproducible and the far-field pattern of the resulting random diffusers manufactured with an identical process might vary significantly. In practice, the pattern from a master wafer is replicated by nano-imprint (or similar processes) to photoresist and then transferred by RIE into fused silica wafers. The replication process allows larger series of identical random diffusers to be manufactured.

5.4.8 DYNAMIC RANDOM DIFFUSERS

Rotating random diffusers are a very effective means for flat-top smoothing, but cannot be used for many applications. The rotating disk might introduce vibrations, and periodic maintenance of the motor is required. Other way to randomize interference and speckle effects from microlens-based Köhler integrators is to use MEMS actuators, like deformable silicon membranes,[39] electroactive polymer membranes,[40] and micromirror arrays.[41]

Figure 5.32 shows a scheme of a MEMS dynamic linear diffuser. The magnetic actuation works as follows: AC current flows in the membrane from one anchor to the other and a permanent magnet is located about 2 mm under the chip. A vertical Lorentz force is generated on the entire membrane. An upward or a downward force

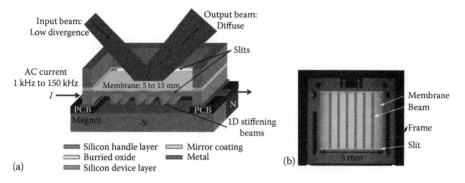

FIGURE 5.32   (a) Scheme of a MEMS dynamic linear diffuser based on a thin silicon membrane and magnetic actuators and (b) backside view (photo) of the membrane with stiffening beams onto the silicon membrane. The electrical connections are made on the PCB. A permanent magnet is located under the PCB. A vertical Lorentz force is created when the current flows in the membrane. The input beam is then diffused, when it reflects on the deformed mirror. (Courtesy of Wilfried Noell, formerly at EPFL Neuchâtel, Switzerland.)

is produced alternatively depending on the current flow direction. When the current frequency matches a mechanical mode, the membrane resonates and the largest deformation amplitude is reached.[41]

Figure 5.33 shows experimental results for flat-top smoothing using the MEMS dynamic linear diffuser shown Figure 5.32. Light from a He–Ne laser (633 nm) and a Köhler integrator with two identical cylindrical microlens arrays was used to generate a 1D flat-top. Illumination with a coherent and well-collimated laser beam generated a line of discrete spots, as discussed in Section 5.4.4. A static 1D linear random diffuser with ±7.5° diffusing angle was then placed behind the Fourier lens. As shown in Figure 5.33a, the random diffusers randomize the spot pattern from the 1D array generator to a statistical pattern of bright and dark lines. Figure 5.33b shows the flat-top pattern for the combination of the static and the dynamic MEMS diffuser. A combination of static and dynamic random diffusers allows the interference and diffraction effects to be smoothed out, but only slightly reduces speckles.

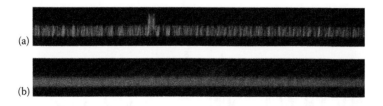

FIGURE 5.33   (a) Magnified image of a flat-top generated using a microlens-based 1D Köhler integrator and a static 1D linear random diffusers (±7.5°), as described in Section 5.4.7 and (b) flat-top after addition of a dynamic linear MEMS diffuser as shown in Figure 5.32. The dynamic diffuser further smoothens out the remaining diffraction pattern of the static device. Speckles are slightly reduced. (Courtesy of Wilfried Noell, formerly at EPFL Neuchâtel, Switzerland.)

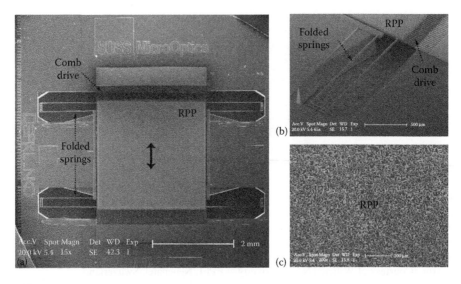

FIGURE 5.34 SEM images of MEMS translation stage with random phase plate for beam shaping and despeckling. (a) The translation stage consists of two large X-shaped torsion springs, a transparent membrane for 3 mm × 3 mm, and the comb drive. (b) Shows a close-up view on the springs and the comb drive. (c) Shows the random-phase plate.[42]

This is achieved by combining a MEMS actuator with a diffractive random phase plate, shown in Figure 5.34.

Figure 5.35 shows experimental results for despeckling of light from a diode laser (532 nm).

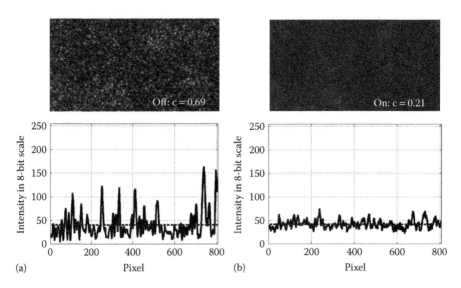

FIGURE 5.35 Speckle reduction (a) using the MEMS translation stage with random-phase plate (b) shown in Figure 5.34.

The pixel size of the random diffuser was about 5 µm; the step height of the phase pattern was 4 µm. For 314 Hz scanning, a lateral translation of 180 µm was achieved. A significant reduction of the speckle contrast was achieved. For diode lasers, speckle reduction could also be achieved electronically by decreasing the longitudinal coherence lengths of the emitted light. Commercially available dynamic speckle reduction systems are based, for example, on deformable silicon membranes, electroactive polymer membranes,[40] or micromirror arrays. They work very well for low-power laser beams, but it remains difficult to find solutions for high-power lasers, ultraviolet light, and 1D line generation.

5.5 ADVANCED MASK ALIGNER LITHOGRAPHY

5.5.1 Tandem Köhler Integrators for Illumination in Photolithography Systems

A major problem of mask aligner illumination systems based on high-pressure mercury lamps and ellipsoidal mirrors (see Section 5.1.3) is the tight tolerance for the lamp alignment within the ellipsoidal mirror. A misplacement of the plasma lamp directly influences the uniformity and the angular spectrum of the illumination light, and thus affects the shadow printing process. Thus, a daily control of the light uniformity is mandatory for mask aligners in a production environment. These limitations could be overcome using a tandem integrator configuration,[43,44,45] shown schematically in Figure 5.36.

In a tandem Köhler integrator, a second Köhler integrator is placed in the Fourier plane of the first Köhler integrator. The flat-top illumination at the entrance pupil of the second Köhler integrator ensures a uniform angular spectrum of the mask illumination light. For tandem Köhler integrators, the light uniformity and angular spectrum of the mask illumination light are completely decoupled from the light source.

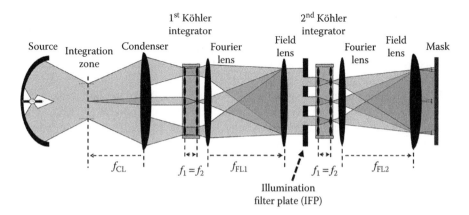

FIGURE 5.36 Scheme of a microlens-based tandem Köhler integrator configuration referred to as MO Exposure Optics® for mask aligners. In a tandem Köhler integrator, a second Köhler integrator is placed in the Fourier plane of the first Köhler integrator.

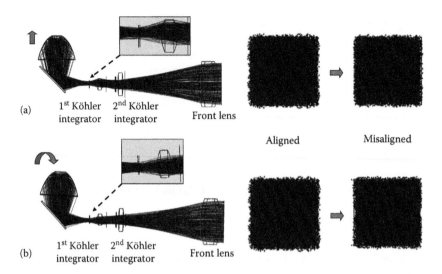

FIGURE 5.37 Raytracing simulation (Zemax®) of a tandem Köhler integrator system: (left-side) ray distribution for (a) a shift of the plasma lamp within the ellipsoid by +10 mm and (b) a tilt of the ellipsoid by +1°, and (right-side) the comparison of the flat-top light distribution of the aligned and the misaligned cases.

Figure 5.37 shows raytracing simulation (Zemax®) for (a) a misplacement of the plasma lamp within the ellipsoid by $\Delta z = +10$ mm along the symmetry axis of the ellipsoid, and (b) a tilt of the ellipsoid by +1°. As shown in Figure 5.37 (left), the misalignment of the lamp leads to a significant shift of the ray bundle in the Fourier lens of the first Köhler integrator. This misalignment is completely compensated by the second Köhler integrator and no difference between the flat-top light distribution of the aligned and the misaligned cases shown in Figure 5.37 (right) is observable. Microlens-based tandem Köhler integrators also improve the obtainable light uniformity in the flat-top. As shown in Figure 5.38, a uniformity of ±1% is obtainable for broadband illumination at 365 nm (i-line), 405 nm (h-line), and 435 nm (g-line). The two-step integration of a tandem Köhler integrator allows different light sources, such as arrays of diode lasers and LEDs, as well as solid-state and excimer lasers, to be used within one lithography tool.

Mask aligner illumination systems using a tandem Köhler integrator have been introduced in 2009[45–47] and are referred to as MO exposure optics® (MOEO). The primary advantages using a tandem Köhler integrator for mask aligner illumination are the constant light uniformity–independent from lamp alignment or aging issues, the identical angular spectrum over the full mask field, and telecentric mask illumination.

In addition, the tandem Köhler integrator allows the illumination light to be spatially filtered, by placing a filter plate in between the two integrators, as shown schematically in Figure 5.36. This illumination filter plate (IFP) partially blocks the light at the entrance of the second Köhler integrator, as shown in Figure 5.39a. Each (x,y) position on the filter plate corresponds to a specific angle of the mask illumination

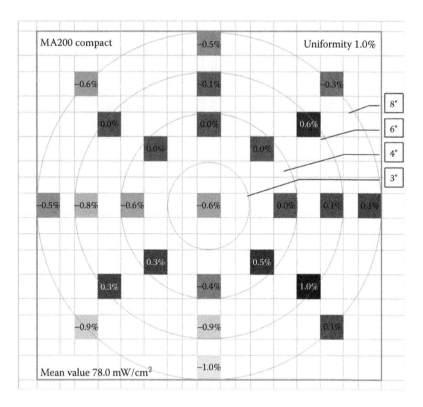

FIGURE 5.38 Measurement of the light uniformity in the mask plane of a mask aligner. A uniformity of ±1% is achieved for a mask field of Ø = 200 mm. (Courtesy of SUSS MicroTec, Garching, Germany.)

light. Figure 5.39b shows a library of different IFP configurations typically used for shadow printing lithography in a mask aligner.[47] Illumination settings providing a large angular spectrum are preferred for contact lithography or printing very uncritical mask patterns. Illumination settings providing a small angular spectrum, that is, well-collimated illumination light, are preferred for large gap proximity lithography. Other illumination settings like ring, quadruple, or Maltese cross are used to further reduce diffraction effects for specific mask patterns, as shown in Figure 5.39b.

Constant and telecentric illumination over the full mask field and the ability to shape the illumination light, referred to as customized illumination (CI) enable simulation and optimization of shadow printing lithography. Computational lithography software allows the shadow printing lithography in a mask aligner to be simulated from the light source to the resist pattern (SMO). Lithography enhancement methods from projection lithography are applied to shadow printing lithography in mask aligners.[48–50]

Using customized illumination (CI), source mask optimization (SMO), and other innovative light-shaping technologies for shadow printing lithography are usually referred to as advanced mask aligner lithography (AMALITH).[46] Targets for shadow

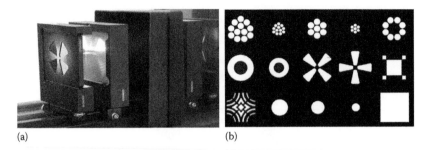

(a)   (b)

FIGURE 5.39 (a) Photograph of the second part of the tandem Köhler integrator in a mask aligner, referred to as MO Exposure Optics®. A 45° rotated Maltese cross illumination filter plate (IFP) is placed in front of the second microlens Köhler integrator and (b) a library of different IFPs used for mask aligners.[47]

printing lithography optimization are manifold: compensating for errors and irregularities like corner rounding, line-width narrowing and edge shortening; elimination of remaining diffraction effects, increasing the gap range of operation (minimum to maximum gap) and the proximity gap, as well as resolution enhancement.

5.5.2 SIMULATION OF SHADOW PRINTING LITHOGRAPHY FROM LIGHT SOURCE TO RESIST

Commercially available lithography simulation software like, for example, LAB from GenISys GmbH,[48] provides full 3D simulation of shadow printing lithography in mask aligners for multiple wavelengths and different illumination settings. The aerial image calculation is based on Kirchhoff scalar diffraction theory solving the Rayleigh–Sommerfeld integral. Propagation in the resist is simulated by a transfer matrix model (thin-film algorithm) including bleaching effects.[49] The light-induced modification and the development of the photoresist material are described by the Dill parameters (extinction in the unbleached/bleached state and photosensitivity of resist) and by the Mack 4 (development rate) parameters. The bulk-image intensities are transferred into inhibitor concentrations, which define the dissolution rate and the resulting resist profile after development. Lithography simulation software like Dr. Litho from Fraunhofer IISB[50] also includes the simulation of chemically amplified photoresists.

Figure 5.40a,c shows the intensity profiles for proximity lithography for two different illumination settings obtained from simulation in LAB lithography software.[48,51] Figure 5.40b,d shows the illumination settings as intensity plots of the angular distribution (α_x, α_y). As shown in Figure 5.40, a smaller angular spectrum ($\pm 1.4°$) allows printing at a larger proximity gap. However, the diffraction effects are stronger for better-collimated illumination, which leads to a more modulated light intensity distribution. To better visualize exposure latitude, the data were plotted as lithographic process windows for 5 µm CD with ± 1 µm tolerance, shown in Figure 5.41. A lithographic process window is defined as the set of values for proximity gap, exposure dose to control critical parameters, like CD, sidewall angle, or resist pits. The largest inscribed rectangle or ellipse in the plot represents the process window.

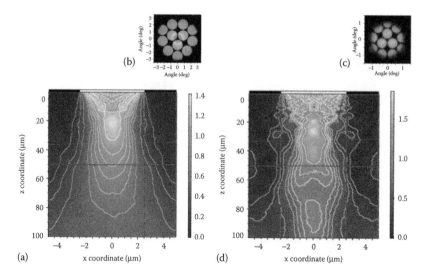

FIGURE 5.40   (a,c) Intensity profiles in air obtained by simulation in LAB lithography software for a photomask with lines and spaces, 5 μm half-pitch (i-line). Two different angular spectra are shown: (a,b) HR (±3°) illumination and (c,d) LGO (±1.4°). The contour plot isobars correspond to equal exposure for the same exposure time. The value 1.0 is equivalent to the dose-to-clear for a 1 μm thick layer of AZ® 1512HS positive photoresist on silicon.

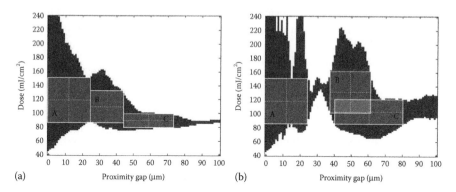

FIGURE 5.41   Process window for critical dimension, CD, of 5 μm with ±1 μm tolerance using (a) HR (±3°) illumination optics suitable for contact or small gap proximity lithography and (b) LGO (±1.4°) applied for large gap proximity, both referring to an illumination light intensity distribution as shown in Figure 5.40. The dark areas show combinations of proximity gaps and dose values, which produce the desired CD value. The exposure latitude is indicated by rectangles fitted into the process window.

The process windows "A," shown in Figure 5.41, are related to a hard- or soft-contact lithography. Process windows "B" and "C" represent proximity lithography. For wafers with an excellent flatness and mask aligners providing a gap-setting accuracy of better than ±5 μm, the process window "B" is the preferred choice. In production, where wafers are not perfectly flat or even slightly bended due to preprocessing, the process window "C" and LGO (±1.4°) are the appropriate choices. By comparing both process windows shown in Figure 5.41, it is interesting to see that for LGO (±1.4°) illumination (b), the process window is very narrow for proximity gaps from 25 to 35 μm. The process window narrowing corresponds to the stronger diffraction effects for better-collimated light, as shown in Figure 5.40c,d. This is contrary to equation derived in Section 5.1.7, predicting that the resolution degrades with the square root of the proximity gap. This is contrary to what the *good instinct* of a lithography engineer would expect. This example demonstrates the relevance of the proper illumination settings for yield enhancement and process stabilization in shadow printing lithography.[51]

5.6   ILLUMINATION SYSTEMS FOR PROJECTION LITHOGRAPHY

5.6.1   Laser Light Sources for Photolithography

For projection lithography systems, the mercury plasma lamps (i-line) were replaced by introducing KrF excimer lasers (248 nm) in the late 1980s and ArF excimer lasers (193 nm) in the late 1990. Excimer lasers are also widely used for many other applications, like medical applications (eye surgery), material processing, and for inspection systems. Whereas the bandwidth of an excimer laser is typically $\Delta\lambda = \pm 0.5$ nm, the ArF excimer lasers used in photolithography are optimized to $\Delta\lambda \leq \pm 0.25$ pm bandwidth to minimize the dispersion effects (chromatic errors) in the projection system. This extremely narrow bandwidth is usually achieved using a combination of prisms and diffraction gratings. Today, leading ArF lasers typically provide 100 to 120 watt and 10 to 20 mJ pulse energy at a maximum repetition rate of 6000 Hz.[52]

The introduction of a monochromatic and well-collimated light source also changed the requirements for the illumination systems fundamentally by allowing DOEs to be used for free-form light shaping. Optical integrators based on lens arrays and light-mixing rods are limited to the form of the lens or rod, and are typically used to generate round or rectangular flat-top profiles. DOEs can be used to create much more complex angular distributions for the illumination light. This will be explained in the following sections.

5.6.2   Illumination Systems Using an Angle-Defining System and a Köhler Integrator

Figure 5.42 shows a scheme of the illumination system as used for state-of-the-art projection lithography, comprising an angle-defining system, a Köhler integrator, and a projection system. The angle-defining system defines which individual lens channels of the Köhler integrator are illuminated. As shown in Figure 5.42, each lens channel corresponds to a discrete plane wave of the mask illumination light. For projection systems, the angular spectrum of the illumination light and the pupil function are conjugated.[5] Thus, shaping the illumination light of a projection system is also referred to as *pupil shaping*. The pattern shown for the pupil plane in Figure 5.42

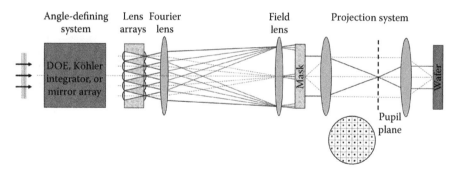

FIGURE 5.42 Scheme of the illumination of a state-of-the-art projection lithography stepper, comprising an angle-defining system, a Köhler integrator, and a projection system.

(right) corresponds to a plane wave illumination of the lens array generating a matrix of discrete spots in the pupil plane.

Illuminating an imaging system with a set of discrete plane waves or a grid is also referred to as *structured illumination* and used in microscopy.[5,34,53] Using a microlens-based Köhler integrator for illumination, the number of microlenses in the array defines the number of discrete pupil points. Increasing the number of pupil points is referred to as *pupil filling*. A higher degree of pupil filling will allow more precise pupil shaping and will significantly improve resolution and fidelity of the printed pattern for projection lithography.[34] The much higher degree of pupil filling triggered the change from light-mixing rods to microlens-based Köhler integrators for high-end projection lithography systems in the mid-2000s.

5.6.3 APODIZATION FOR PROJECTION LITHOGRAPHY

In the Section 5.1.8, we discussed the influence of off-axis illumination on shadow printing lithography. As described by Abbe,[6] off-axis illumination and apodization are suited to improve the resolution in imaging systems. Apodization, from Greek[23] *removal of the feet*, was first applied in astronomy to suppress the secondary maxima (rings) of the diffraction limited Airy pattern. In astronomy, it was established by introducing opaque filter plates in the pupil plane. Apodization is achieved by manipulating the illumination light. For projection lithography systems, as shown schematically in Figure 5.42, apodization is obtained by illuminating the Köhler integrator with an annular illumination, generated by the angle-defining system.

A very simple optical system for generating annular or ring illumination is axicon telescopes, as shown schematically in Figure 5.43a. Two different configurations of axicon telescopes could be used as angle-defining system. Figure 5.43b shows the resulting angular spectrum (simulation).[55] Axicon zoom telescopes in combination with a light mixing rod have been introduced for projection lithography systems by Wangler[56] in 1993. When industry changed from mercury plasma lamps to well-collimated and mono-chromatic KrF excimer lasers, the use of DOEs became industry standard.[57,58]

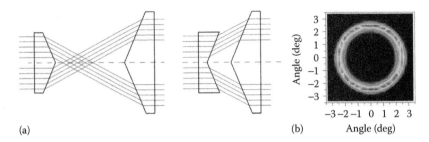

(a)   (b)   Angle (deg)

FIGURE 5.43   (a) Axicon telescopes[54] are widely used in illumination systems for projection lithography, as they are telecentric and allow the angles of annular illumination to be changed without loss. (b) Angular spectrum for annular illumination, as achieved using an axicon telescope.[55]

5.6.4   DIFFRACTIVE OPTICAL ELEMENTS FOR FREE-FORM BEAM SHAPING IN PHOTOLITHOGRAPHY SYSTEMS

A DOE is by definition an optical element which modifies the wavefront of the incident light by the means of diffraction.[59–62] DOEs are a class of optical elements that includes a large variety of different micro- and nano-optical components ranging from Fresnel lenses to volume holograms. This implies as well a large variety of different manufacturing methods from mechanical micromachining to e-beam writing in the subwavelength range. Figure 5.44 shows different refractive and diffractive micro-optical elements for laser beam shaping.

DOEs for applications in projection lithography are typically manufactured in fused silica or calcium fluoride, the only suitable glass materials providing very high transmission and robustness (lifetime) at 193 nm to 248 nm ultraviolet wavelengths. Blazed or kinoform DOEs, shown in Figure 5.44, are typically manufactured by

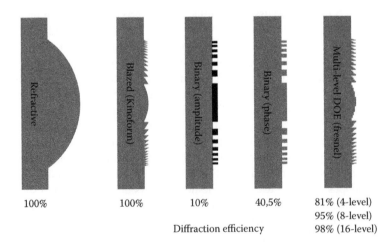

Refractive	Blazed (Kinoform)	Binary (amplitude)	Binary (phase)	Multi-level DOE (fresnel)
100%	100%	10%	40,5%	81% (4-level) 95% (8-level) 98% (16-level)

Diffraction efficiency

FIGURE 5.44   Schemes of different micro-optical components and the related diffraction efficiency for modifying the light of an incident laser beam.

mechanical micromachining, like diamond turning, fly cut, and micro or nanopolishing techniques. However, it is very difficult to meet the very high demands for profile accuracy and surface roughness required for beam and pupil shaping in projection lithography systems using micromachining. Thus, typically 8-level or 16-level DOEs are used for these tasks. These elements are typically manufactured using successive litho-etch steps. Projection lithography steppers or e-beam writing are used for the lithography. Sputtering or RIE is used for the transfer of the resist structures into the bulk material (Figure 5.45).

The manufacturing limit for DOEs is usually not the resolution of the lithography. The more critical parameters are the depth of focus of the lithography, the uniformity of the resist coating on pre-structured wafers, and proximity effects in lithography and etching. As resolution and depth of focus are related parameters, the best choice for the manufacturing of multi-level DOEs is a moderate resolution of 350 nm to 500 nm, and >0.5 μm usable depth of focus, if possible >1 μm. Obviously, an excellent planarity of the wafer surface over the full step field of typically 20 mm × 24 mm is mandatory. Especially for CGHs, as shown Figure 5.46a, where high- and low-level structures are side-by-side, the resist coating, lithography, and etching process are not trivial. The major problem for the etching process is to obtain the correct step height for all phase structures. Due to microloading effects, the etch rates for isolated microstructures and large plateau areas might differ significantly.

SEM images, as shown in Figure 5.47, allow typical manufacturing errors, such as a sidewall angle of 80° for a 470 nm high plateau (a), an overlay mismatch around 50 nm (b), and the degree of corner rounding (b) due to diffraction effects of the stepper to be observed.

Manufacturing errors influence the diffraction efficiency, the amount of light remaining in the zeroth order, the light diffracted into undesired higher orders or ghost images, and noise. To obtain excellent optical performance a correct etch depth (i.e., the step height) with less than 3% deviation from the ideal value is required for 8-level DOEs.

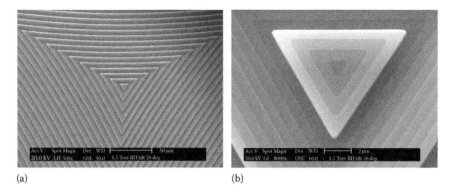

(a)   (b)

FIGURE 5.45 SEM images of 8-level diffractive optical elements (Fresnel lens) designed for laser beam shaping at 193 nm wavelength (a). The point of contact of three adjacent hexagon-shaped cells is shown at high magnification (b). (Courtesy of SUSS MicroOptics, Neuchâtel, Switzerland).

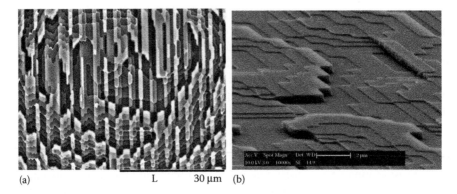

(a) L 30 μm (b)

FIGURE 5.46 SEM images of different 8-level diffractive optical elements (CGH) designed for laser beam shaping (a) at 193 nm and (b) 248 nm wavelength. (Courtesy of SUSS MicroOptics, Neuchâtel, Switzerland.)

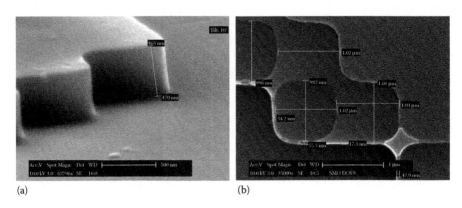

(a) (b)

FIGURE 5.47 SEM images showing details of the diffractive optical element shown in Figure 5.46b. A sidewall angle of 80° for 470 nm high steps (a) and an overlay error about 50 nm is observed (b). (Courtesy of SUSS MicroOptics, Neuchâtel, Switzerland.)

Figure 5.48 shows the phase profile of a densely packed array of 16-level DOEs for focus spot generation, (a) measured in a white light profilometer Wyko NT3300; (b) scan of x-profile from same measurement shows a total phase depth of 320 nm subdivided to 16 phase levels. The ideal phase step for 16 phases is 322.5 nm, thus the etch depth error is less than 1%. A diffraction efficiency of 98% and less than 0.1% in the zeroth diffraction order is achieved.[63] Even for perfect DOEs, light diffraction at sharp edges, reflection at steep sidewalls, and quantization errors might lead to stray light and noise, especially for oblique incidence of light with a larger angle. Typically, the most critical problems using DOEs are the zeroth order, artifacts or ghost images, and noise. Diffractive beam shapers usually have a higher noise level than refractive components, and the available range of diffraction angles decreases with the number of phase levels.

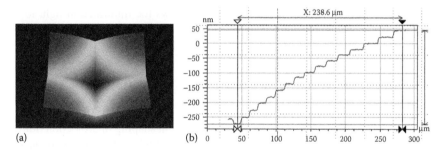

(a) (b)

FIGURE 5.48 Phase profile of a densely packed array of 16-level diffractive optical elements for focus spot generation (a) measured in a white light profilometer Wyko NT3300; (b) scan of x-profile from same measurement shows a total phase depth of 320 nm subdivided to 16 phase levels.

The big advantage of diffractive optics is the freedom to design almost any far-field light distribution. In projection lithography systems, diffractive beam-shaping elements are widely used to implement annular, dipole, quadrupole, and other more sophisticated illumination settings (pupil shaping). They are used, for example, as angle-defining elements to illuminate the subapertures of a microlens-based Köhler integrator.

5.6.5 FREE-FORM BEAM SHAPING IN PHOTOLITHOGRAPHY SYSTEMS

The introduction of DOE-based free-form beam shaping in the illumination systems of projection lithography systems helped to improve the resolution from some 500 nm below 40 nm (half-pitch). The progress in lithography simulation allowed special DOEs to improve the lithography process for critical mask features to be manufactured. Figure 5.49 shows the evolution of pupil pattern as used in projection lithography systems to improve the resolution and depth of focus. The pupil pattern is evolved from annular illumination, provided by axicon telescopes to very sophisticated multipole patterns, and generated by DOEs.

The combination of diffractive or reflective (micromirror) angle-defining elements in combination with a microlens-based Köhler integrator, as shown in Figure 5.42, allows a very fine grid pattern in the pupil plane to be obtained. In Figure 5.49 (2009), every *pixel* of the pupil plane pattern corresponds to a microlens channel of the Köhler integrator.

In 2009, ASML introduced FlexRay,[64] a pupil-shaping system based on an array of some thousands of individually addressable MEMS micromirrors, shown

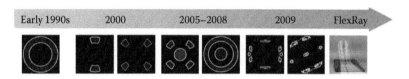

FIGURE 5.49 Evolution of pupil patterns from (left) axicon telescope annular illumination in the early 1990s to (right) FlexRay micromirror-based free programmable illumination today. (Courtesy of Johannes Wangler, Carl Zeiss SMT GmbH, Oberkochen, Germany.)

schematically in Figure 5.49 (at right). FlexRay allows the illumination light to be shaped on demand, avoiding a delay of several weeks for the manufacturing of beam-shaping DOEs, and it has become the industry standard today.

5.7 MICRO-OPTICS IS A KEY ENABLING TECHNOLOGY FOR PROJECTION LITHOGRAPHY

First attempts to use micro-optics for illumination light shaping in projection lithography started in the early 1990s, when wafer-based technology became available for manufacturing planar micro-optics in UV-transparent fused silica. DOEs are the first choice to combine a high degree of freedom in light shaping and a high diffraction efficiency. Refractive microlens arrays are the first choice to provide uniform flat-top illumination for light mixing of divergent light and pupil filling. Micro-optical elements for pupil shaping soon evolved from simple flat-top illumination to annular, dipole, quadrupole, multipole, and very sophisticated free-form illumination settings, as shown in Figure 5.49. Micro-optical illumination became standard in optical lithography. The next major step forward was to combine refractive and DOEs with a programmable micromirror array, referred to as FlexRay (ASML) shown in Figure 5.49 (right). The programmable micromirror array allows illumination settings to be changed on the fly. Similar micromirror concepts are now realized in EUV lithography systems. Interestingly, besides the illumination tasks, discussed within this book chapter, micro-optical components and technology are also used for other tasks in a modern projection stepper, as shown schematically in Figure 5.50.

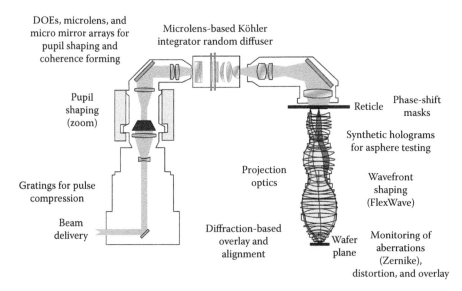

FIGURE 5.50 Fields of applications for micro-optics in state-of-the-art projection lithography system. (Courtesy of Robert Brunner, formerly at Carl Zeiss AG, Germany.)

Blazed gratings are used in the excimer laser light sources for linewidth narrowing without wavefront distortion. As the remaining material dispersion is a very critical parameter for the lens design, achieving ultra-narrow bandwidth and high wavelength stability[65] allows the optical designer to eliminate residual chromatic aberrations. Today, lens manufacturing for projection lithography systems has achieved its highest level with aspheric lens of atomic-scale profile accuracy. For testing of aspheres at this precision, special interferometers using synthetic holograms as proposed by Johannes Schwider in 1976 are used.[66] These synthetic holograms, masterpieces of micro-optics, are manufactured by e-beam writing at the fabrication limits. For ultrafast and ultraprecise alignment and overlay control in a state-of-the-art projection stepper, a multitude of micro-optical elements are used on wafers, reticles, wafer-stages, and reticle-stages. These elements allow precise alignment of reticle to wafer on the single digit nanometer level. MEMS mirror arrays will also play a decisive role in future extreme ultraviolet (EUV) lithography for illumination light shaping. Micro-optics has proven to be decisive KET for optical lithography over the last 20 years.

REFERENCES

1. Jay, W., Lathrop, "The diamond ordnance fuze laboratory's photolithographic approach to microcircuits," *IEEE Ann. Hist. Comput.,* 35(1), 48–55, 2013, doi:10.1109/MAHC.2011.83.
2. Hoerni, J.A., Method of Manufacturing Semiconductor Devices. U.S. Patent 3,025,589 (1959).
3. García-Sucerquia, J.I., Castaneda, R., Medina, F.F., and Matteucci, G., "Distinguishing between Fraunhofer and Fresnel diffraction by the young experiment", *Opt. Commun.,* 200, 15–22, 2001.
4. Rai-Choudhury, P., ed., *Handbook of Microlithography, Micromachining, and Microfabrication*, SPIE Press, Bellingham, WA, 1997.
5. Wolfgang Singer, Michael Totzeck, Herbert Gross, "Handbook of Optical Systems, in ed. Herbert Gross, *Physical Image Formation*", 2, Wiley-VCH, Verlag 2005.
6. Ernst Abbe, "Beiträge zur Theorie des Mikroskops und der mikroskopischen Wahrnehmung", *M. Schultze's Archiv für mikroskopische Anatomie*, IX, 413–468, 1873.
7. Roland, E., Sheets, "Optical microcircuit printing system", U.S. Patent 3,941,475 (1976).
8. August Köhler, Zeitschrift für wissenschaftliche Mikroskopie, Band X, Seite 433–440, 1893.
9. Dross, O., Mohedano, R., Hernández, M., Cvetkovic, A., Benítez, P., Carlos and Miñano, J., "Illumination optics: Köhler integration optics improve illumination homogeneity", *Laser Focus World*, 45, 2009.
10. Fred M. Dickey, "Laser Beam Shaping: Theory and Techniques, Second Edition", CRC Press, ISBN 9781466561007, 2014.
11. Voelkel, R., Singer, W., Herzig, H.P., and Dändliker, R., "Imaging properties of microlens array systems", MOC'95 Hiroshima, The Japan Society of Applied Physics, Tokyo, pp. 156–159, 1995.
12. Dross, O., Mohedano, R., Hernández, M., Cvetkovic, A., and Benítez, P., Carlos Miñano, J., "Illumination optics: Köhler integration optics improve illumination homogeneity", Laser Focus World, 45, 2009.

13. Brown, B.R., Lohmann, A.W., "Complex spatial filtering with binary masks", *Appl. Opt*, 5, 967ff, 1966.
14. U.S. Patent 3733258, Joseph John Hanak, John Patrick Russell, "Sputter-etching technique for recording holograms or other fine-detail relief patterns in hard durable materials", filed in February 1971.
15. U.S. Patent 4155627, Michael T. Gale, Hans W. Lehmann, Roland W. Widmer," Color diffractive subtractive filter master recording comprising a plurality of superposed two-level relief patterns on the surface of a substrate", filed December 21, 1977.
16. U.S. Patent 4689291, Z.D. Popovic, R.A. Sprague, G.A. Neville Connell 1985.
17. Philippe Nussbaum, Reinhard Voelkel, Hans Peter Herzig, Martin Eisner, Stefan Haselbeck, "Design, fabrication and testing of microlens arrays for sensors and microsystems", Pure Appl. Opt. 6, 617–636, 1997.
18. Voelkel, R., "Wafer-scale micro-optics fabrication", Adv. Opt. Technol. (AOT), 1, 135–150, 2012.
19. P10, P15 Stylus Profilometer, KLA-Tencor, www.kla-tencor.com.
20. Daniel Tomicic, "Adhesion measurement of positive photoresist on sputtered aluminium surface", PhD thesis, Linköping University, Norrköping, Sweden, August 2002.
21. Cullmann, E., Loechel, B., Maciossek, A., and Rothe, M., "Advanced resist processing for thick photoresist applications", Microelectr. Eng. 30, 551–554, 1996.
22. Voelkel, R., "Micro-Optics: Enabling Technology for Illumination Shaping in Optical Lithography", SPIE Advanced Lithography 9052–67 Opt. Microlithography XXVII, 2014.
23. Herbert Gross, "Handbook of Optical Systems, Volume 1: Fundamentals of Technical Optics", Edited by Herbert Gross, Wiley-VCH Verlag 2005.
24. Ruffieux, P., Scharf, T., Herzig, H.P., Voelkel, R., and Weible, K.J., "On the chromatic aberration of microlenses", Opt. Express, 14(11), 4687–4691, 2006.
25. Li, Y., Platzer, H., "An experimental investigation of diffraction patterns in low Fresnel-number focusing systems", *Opt. Acta*, 30, 1621, 1983.
26. FRED Software by Photon Engineering, Tucson (AZ), USA, www.photonengr.com
27. Lohmann, A.W., "Scaling laws for lens systems", *Appl. Opt.*, 28, 4996–4998, 1989.
28. Streibl, N., "Beam shaping with optical array generators," *J. Mod. Opt.* 36, 1559–1573, 1989.
29. Lindlein, N., Bich, A., Eisner, M., Harder, I., Lano, M., Voelkel, R., Weible, K.J., and Zimmermann, M., "Flexible beam shaping system using fly's eye condenser", *Appl Opt*, 49(12), 2382–90, 2010, doi: 10.1364/AO.49.002382.
30. Streibl, N., Nölscher, U., Jahns, J., and Walker, S., "Array generation with lenslet arrays", *Appl. Opt.*, 30(19), 2739–2742, 1991.
31. Zimmermann, M., Schmidt, M., Bich, A., and Voelkel, R., "Refractive micro-optics for multi-spot and multi-line gneration", Proc. LPM2008, 9th Int. Symposium on Laser Precision Microfabrication, 2008.
32. Wangler, J., Sickmann, H., Weible, K., Scharnweber, R., Maul, M., Deguenther, M., Layh, M., Scholz, A., Spengler, U., and Voelkel, R., "Illumination system for a microlithographic projection exposure apparatus", U.S. 20090021716 A1, Jan 22, 2009.
33. Kopp, Ch., Ravel, L., Meyrueis, P., "Efficient beamshaper homogenizer design combining diffractive optical elements, microlens array and random phase plate", *J. Opt. A: Pure Appl. Opt.* 1, March 11, 1999398–403, doi:10.1088/1464-4258/1/3/310, 1999.
34. Erdmann, L., Burkhardt, M., and Brunner, R., "Coherence management for microlens laser beam homogenizers," *SPIE*, 4775, 145–154, 2002.

35. Bitterli, R., Scharf, T., Herzig, H.P., Noell, W., de Rooij, N., Bich, A., Roth, S., Weible, K.J., Voelkel, R., Zimmermann, M., and Schmidt, M., "Fabrication and characterization of linear diffusers based on concave micro lens arrays," *Opt. Exp*, 18(13), 2010.
36. Wippermann, F.C., Dannberg, P., Braeuer, A., and Sinzinger, S., "Improved homogenization of fly's eye condenser setups under coherent illumination using chirped microlens arrays", *SPIE*, 6466, doi: 10.1117/12.700127, 2007.
37. Schertler, D.J., George, N., "Uniform scattering patterns from grating-diffuser cascades for display applications", *Appl. Opt.*, 38(2), 1999.
38. Wang, L., Tschudi, T., Halldorsson, T., and Petursson, P.R., "Speckle reduction in laser projection systems by diffractive optical elements", *App. Opt.* 37(10), 1998.
39. Masson, J., Bich, A., Herzig, H.P., Bitterli, R., Noell, W., Scharf, T., Voelkel, R., Weible, K.J., and de Rooij, N.F., "Deformable silicon membrane for dynamic linear laser beam diffuser," Proc. SPIE 7594, MOEMS and Miniaturized Systems IX, 75940F, (February 16, 2010), doi:10.1117/12.842015.
40. Blum, M., Büeler, M., Grätzel, C., and Aschwanden, M., "Compact optical design solutions using focus tunable lenses," Proc. SPIE 8167, Optical Design and Engineering IV, 81670W (September 22, 2011); doi:10.1117/12.897608.
41. Noell, W., Weber, S., Masson, J., Extermann, J., Bonacina, L., Bich, A., Bitterli, R., Herzig, H.P., Kiselev, D., Scharf, T., Voelkel, R., Weible, K.J. Wolf, J.-P., and de Rooij, N. F., Shaping light with MOEMS. Proc. SPIE 7930, MOEMS and Miniaturized Systems X, 79300P (February 14, 2011); doi:10.1117/12.873051.
42. Masson, J., Bich, A., Noell, W., Voelkel, R., Weible, K.J., and de Rooij, N.F., "Tunable MEMS-Based Optical Linear (1D) Diffusers for Dynamic Laser Beam Shaping and Homogenizing," IEEE International Conference on Optical MEMS & Nanophotonics, Clearwater Beach, 1, 129–130, 2009.
43. Konno, K., Okada, M., "Light illumination device," U.S. Patent 4'497'015, Jan. 29, 1985.
44. Vogler, U., "Optimierung des Beleuchtungssystems für Proximitylithographie in Mask Alignern", Diploma Thesis, Technische Universität Ilmenau, 2009.
45. Voelkel, R., Vogler, U., Bich, A., Weible, K.J., Eisner, M., Hornung, M., Kaiser, P., Zoberbier, R., and Cullmann, E., "Illumination system for a microlithographic contact and proximity exposure apparatus," EP 09169158.4, 2009.
46. Voelkel, R., Vogler, U., Bich, A., Pernet, P., Weible, K.J., Hornung, M., Zoberbier, R., Cullmann, E., Stuerzebecher, L., Harzendorf, T., and Zeitner, U.D., "Advanced mask aligner lithography: New illumination system", *Opt. Express,* 18, 20968–20978, 2010.
47. SUSS MicroTec AG, Garching, Germany.
48. LAB, Lithography Simulation Software for Proximity Lithography, GenISys GmbH, Germany, www.genisys-gmbh.com.
49. Péter Bálint Meliorisz, "Simulation of Proximity Printing", PhD Thesis, Friedrich-Alexander University, 2010.
50. Litho, DR., *Lithography Simulation Software*, Fraunhofer IISB, Germany, www.drlitho.com.
51. Voelkel, R., Vogler, U., Bramati, A., Erdmann, A., Ünal, N., Hofmann, U., Hennemeyer, M., Zoberbier, R., Nguyen, and D., Brugger, J., "Lithographic process window optimization for mask aligner proximity lithography", Proc. SPIE 9052, Optical Microlithography XXVII, 90520G (31 March 2014); doi: 10.1117/12.2046332, 2014.
52. Gigaphoton, Oyama-shi, Tochigi-ken, Japan 2015.
53. Fred, M. Dickey, and O'Neil, D., "Multifaceted laser beam integrators: general formulation and design concepts", *Opt. Eng.* 27, 999–1007, 1988.
54. John H. McLeod, "The Axicon, a new type of optical element", *JOSA*, 44(8), 1954.

55. Vogler, U., "Optimierung des Beleuchtungssystems für Proximitylithographie in Mask Alignern", Diploma Thesis, Technische Universität Ilmenau, 2009.

56. Wangler, J., "Illumination device for an optical system with a reticle masking system", EP 0658810 B1, December 13, 1993.

57. Stanton, S., Oskotsky, M., Gallatin, G., and Zernike, F., "Hybrid illumination system for use in photolithography", EP 0744664 B1, 1995.

58. Singer, W., Herzig, H-P., Kuittinen, M., Piper, E., and Wangler, J., "Diffractive beam-shaping elements at the fabrication limit," *Opt. Eng.* 35, 2779–2787, 1996.

59. Bernard, C., Kress, P., Meyrueis, *"Applied Digital Optics: From Micro-optics to Nanophotonics"*, John Wiley & Sons Ltd., UK, ISBN 978–0–470–02263–4, 2009.

60. Bernard C. Kress, "Field Guide to Digital Micro-Optics", SPIE Field Guides Vol. FG33; SPIE Bellingham, Washington, USA; ISBN 978-1-62841-183-6, 2014.

61. O'Shea, DC., Suleski, T.J., Kathman, AD., Prather, DW., *"Diffractive Optics: Design, Fabrication, and Test"*, SPIE Tutorial Texts; SPIE Bellingham, Washington, USA; ISBN 0-8194-5171-1, 2003.

62. Soskind, Y.G., *"Field Guide to Diffractive Optics"*, SPIE Field Guides, SPIE Bellingham, Washington, USA, Vol. FG21, ISBN 978-0-8194-8690-5, 2011.

63. Voelkel, R., Weible, K.J., and Eisner, M., "Wafer-level micro-optics: Trends in manufacturing, testing and packaging," *SPIE* 8169–12, Conf. on Optical System Designs, 5–8 Sept 2011, Marseille, France 2011.

64. Bert Koek, SVP Litho Applications, "ASML Holistic Lithography", Semicon West 2009; July 14, 2009, URL: www.asml.com/doclib/productandservices/semicon_west_2009_FINAL.pdf.

65. Takashi Saito et al., Gigaphoton Inc, "Ultra-narrow bandwidth 4-kHz ArF excimer laser for 193-nm lithography", SPIE Vol. 4346, Optical Microlithography XIV 2001.

66. Schwider, J., Burow, R., "Testing of aspherics by means of Rotational-Symmetric synthetic hologramss", *Optica Applicata (Wroclaw)*, VI, 83, 1976.

6 Beam Shaping for Optical Data Storage

Edwin P. Walker and Tom D. Milster

CONTENTS

6.1   INTRODUCTION

One application of beam shaping is to improve signal quality in optical data storage systems.[1] Since the light source in an optical data storage device is a coherent laser beam, simple beam-shaping techniques can greatly influence the performance of the system. Beam-shaping elements take the form of spatially dependent optical filters that are strategically placed in the light path. This chapter discusses theory, simulations, and practical implementation for beam-shaping techniques used in optical data storage. The storage and retrieval of data on optical disks are described in two simple steps. First, data marks are recorded on a surface. The second step is retrieval of the information from the disk, where a focused light beam scans the surface. Modulation in the reflected light is used to detect the data-mark pattern under the scanning spot.

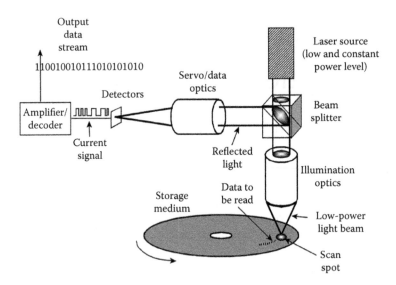

FIGURE 6.1 Readout of data marks utilizes a low-power focused laser spot to scan the data tracks on the spinning disk. Reflected light is collected by the optical system and directed to the servo and data detectors with a beam splitter. The light modulation due to the data pattern is converted into current modulation by the detectors, which is then converted to the user data signal by electronic circuits.

Readout of data marks on a disk is illustrated in Figure 6.1, where the laser is used at a constant output power level that does not heat the data surface beyond its thermal-writing threshold. The laser beam is directed through a beam splitter into the illumination optics, where the beam is focused onto the surface. As the data marks to be read pass under the scan spot, the reflected light is modulated. Modulated light is collected by illumination optics and returned by the beam splitter to servo and data optics, which converge the light onto detectors. The detectors change the light modulation into current modulation that is amplified and decoded to produce the output data stream.

Signal quality in an optical storage device depends greatly on the optical system transfer function that is used to read data patterns. A typical optical data storage system transfer function exhibits reduced contrast at high frequencies relative to low frequencies, where contrast is defined by the ratio of the signal amplitude at frequency v to the maximum signal amplitude in the frequency range of interest, that is, $H(v) = I_{sig}(v)/I_{max}$. Equalization of the system transfer function to obtain the desired frequency response is the basis of improved resolution, whether it is accomplished optically, electronically, or with a combination of techniques. Beam shaping is used in optical data storage to improve the quality of the electrical readout signal, rather than the more common uses of beam shaping to improve image quality or to provide uniform illumination.

One way to equalize the system transfer function is to apply optical beam shaping in the illumination path and/or the return path, as shown in the simplified optical path of Figure 6.2. Illumination-path filters shape the profile of the spot focused onto the disk, and return-path filters limit the extent of diffracted orders that reach the detector.

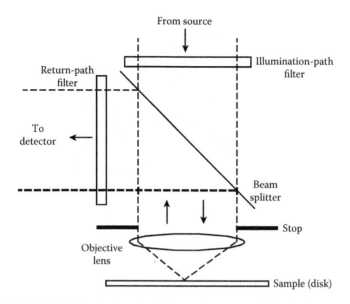

FIGURE 6.2 Simplified optical path of the readout system. Illumination-path filters or return-path filters can be placed in the optical system to change the system transfer function.

Section 6.2 of this chapter describes several simulation and experimental parameters used in the chapter, such as the type of test patterns and storage media. Section 6.3 describes the optical system transfer function and how it relates to optical data storage devices. Section 6.4 describes beam-shaping filters, including illumination-path filters, return-path filters, electronic filters, and filter combinations. Section 6.5 presents a short summary.

6.2 SIMULATION AND EXPERIMENTAL PARAMETERS

6.2.1 SIMULATION PARAMETERS

Simulation parameters for the optical data storage system are illustrated in this section. Simulations are performed in the OPTISCAN quasi-vector diffraction work environment.[2,3]

Eleven single-tone patterns are generated in order to build up the system transfer function from the amplitudes of the detector signals. The patterns range in pitch from 200 to 1300 lines/mm and are chosen to include typical commercial mark sizes. The smallest mark size used is 0.4 μm. A low-frequency single-tone pattern is shown as the top row in Figure 6.3, and a high-frequency pattern is shown as the center row. The optical system uses a 785 nm laser beam focused onto the disk with a 0.5-numerical aperture (NA) objective lens. Illumination on the objective lens is typically a Gaussian profile with the $1/e^2$ irradiance level equal to the diameter of the lens aperture.

The action of spinning the disk transforms a spatial frequency into a temporal one. The linear velocity v of the medium under the focused spot is $v = 2\pi r\omega$, where ω is the rotation rate in revolutions per second and r is the radius of the disk at which the

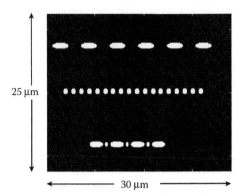

25 µm

30 µm

FIGURE 6.3 Data patterns used in the simulations. Top row: low-frequency data pattern; middle row: high-frequency data pattern; and bottom row: worst-case scenario data pattern. The long mark is 2.7 µm long and the short mark is 0.6 µm long.

spot is focused. For a given mark pitch Λ along the track, the corresponding temporal frequency of the data is given by $f = v/\Lambda$. Spatial frequencies $v = 1/\Lambda$ are used in the following discussions, because the simple relation $f = vv$ exists for conversion to a temporal frequency. The conversion from electric field amplitude to detector output current is accomplished internally in the OPTISCAN program.

Another data pattern of interest in data storage is the so-called *worst-case* scenario pattern, as shown in the bottom row of Figure 6.3. This pattern consists of the longest mark in a given data pattern followed by the shortest space, shortest mark, shortest space, and longest mark again. This combination is a worst case pattern, because it is one that is most likely to cause a detection error. It produces a signal response that is very much like a two-point response for the system.

6.2.2 EXPERIMENTAL PARAMETERS

The system used to collect experimental data was designed to work with magneto-optic (MO) media. MO media store information in small magnetic data marks, which are about the same size as pits on a compact disc (CD). The recording layer is initially erased so that all magnetic domains are aligned in one direction perpendicular to the recording surface, as shown in Figure 6.4a. In this configuration, the magnetic domains are extremely stable. A large magnetic field of several thousand oersteds is required to overcome the magnetic moment of the domains. The magnetic field required to reorient domains is called the coercivity.[4] To record data marks, a high-power focused spot is used to locally heat the recording surface. Heat reduces coercivity, and magnetic domains in the region of the focus spot are reoriented with an external magnetic field. When the laser beam is switched to low power between data marks, the recording layer is not heated and the external magnetic field has no effect on domain orientation. The laser beam is modulated between high power and low power as the disk spins in order to write a pattern of data marks along each track. Each mark contains magnetic domains oriented in the opposite direction compared to the magnetic domains of the background.[5]

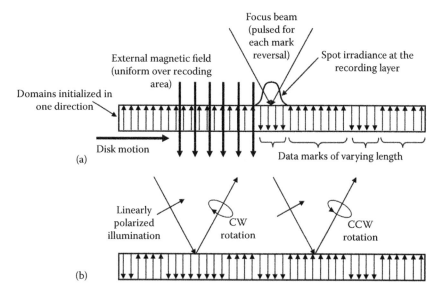

FIGURE 6.4 Recording and readout of magneto-optic (MO) media. (a) The recording process heats the medium to reduce coercivity in small areas that can change their orientation under the influence of an applied magnetic field, and (b) the readout process uses the polar Kerr effect to sense the orientation of the magnetic-domain mark pattern.

The marks have the property that, as a low-power focused light spot passes over it, polarization of the reflected light is rotated as shown in Figure 6.4b. Polarization rotation on reflection is due to the polar Kerr effect.[6] When the laser beam illuminates a data mark with domains oriented away from the laser beam, linear incident polarization is rotated slightly in the counterclockwise direction. When the laser beam illuminates the region between data marks, linear incident polarization is rotated slightly in the clockwise direction. In order to detect the data signal, a detector is used to sense change in the polarization of the reflected light.[*] For example, an indication that the reflected light is rotated in the counterclockwise direction implies that the laser spot illuminates a data mark. In order to erase data, the external magnetic field is reversed, Figure 6.4a, and the laser beam heats an entire section of the track. A major difference between CD and MO products is that the MO marks are produced in a track with an almost undetectable change in the topology of the track. In other words, there is almost no mechanical deformation of the track as the marks are recorded or erased. This property enables MO products to exhibit over one million erase cycles with little, if any, degradation in performance.[7,8] For the purpose of the discussion in this chapter, the results obtained with MO media can be considered similar to those obtained with other commercial media, such as CDs and digital versatile discs (DVDs).

---

[*] Differential detection is the most common detector geometry for MO media. The reflected light is directed through a polarizing beam splitter, where the surface of the beam splitter is oriented to split light equally for unrotated light. Clockwise or counterclockwise rotation of the polarization imbalances the split. The difference between light levels after the split can easily indicate the change in polarization.

Some of the experiments measured the signal distribution at the pupil plane of the objective lens. In these cases, signal distributions are obtained by writing a single-tone data pattern on the disk at the frequency of interest. After writing, the laser power is reduced to a constant read power level. A pinhole is placed in the pupil in order to pass only signal energy from a limited region to the detectors. The pinhole is scanned in a rectilinear pattern to sample the signal energy distribution in the pupil. The pinhole diameter is 400 µm, and the pupil diameter is 4.3 mm. By sampling at 100 µm intervals, signal distributions are measured as a function of position in the pupil. The disk velocity is 6.6 m/s. The scanning pinhole is used to obtain the signal power in a 30 kHz bandwidth as a function of the position in the stop. The data are then erased, and then a different frequency is written.

6.3 TRANSFER FUNCTION CHARACTERISTICS OF OPTICAL DATA STORAGE SYSTEMS

The data marks along a track can be thought of as a collection of spatial frequencies from which the diffracted orders reflected from the track are observed. There are often many diffracted orders observed, depending upon the spatial frequency content of the reflectivity variations, but the discussion in this section is limited to single-tone sinusoidal variations. Two diffracted orders are produced in the return path from each spatial frequency component as shown in Figure 6.5 for a low spatial frequency and a high spatial frequency. The phase difference between the zero-order reflected light and the ±1 diffracted orders changes as the object is scanned, which produces modulation in the data signal received at the detectors. Interference only occurs in the region of overlap between the diffracted orders and the zero order. The amount of modulation depends on the overlap area. For low-frequency components, there is a large overlap area. For high-frequency components, the overlap area is

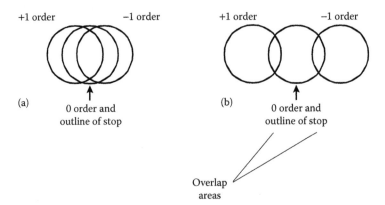

FIGURE 6.5   Diffracted orders produced in the return path for (a) low spatial frequency and (b) high spatial frequency. As the object is scanned, a phase shift is introduced between the zero order and the ±1 diffracted orders that produces a modulation in data signal received at the detectors. Modulation occurs in the overlap areas, which are clearly separated for the high spatial frequency.

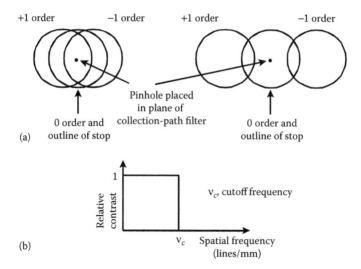

FIGURE 6.6 Theory of coherent-like transfer function (contrast vs. frequency) for a pinhole in the return-path filter plane. (a) ±1 and zero orders for low and high spatial frequencies and (b) the associated transfer function.

decreased and spreads toward the edges of the pupil. The reduction of overlap area explains the falloff in the transfer function, as the spatial frequency of the data pattern is increased.

Consider the return-path filter arrangement of Figure 6.6 with uniform illumination in the pupil and no illumination filter. Only light passing through a small circular pinhole in the center of the return-path filter is passed to the detectors. Low spatial frequencies exhibit the first diffraction order overlapping the area of the pinhole, as shown in Figure 6.6a. As the spatial frequency increases, the first diffraction orders separate, but the overlap area on the pinhole is not affected. As the spatial frequency increases further, there is a frequency such that none of the light from the first diffraction order overlaps the pinhole area. The result is a transfer function $H(\nu)$ that has a uniform response out to a cutoff frequency ν_c, as shown in Figure 6.6b. This behavior is a characteristic of coherent imaging systems.

Now consider the arrangement of Figure 6.7, where the pinhole is removed and all light deflected by the beam splitter passes to the detector. Low spatial frequencies completely overlap the aperture. As spatial frequency increases, the first diffraction orders no longer completely overlap the aperture, as illustrated in Figure 6.7a. As spatial frequency increases further, the amount of overlap area decreases until there is no overlap. The cutoff frequency is now $2\nu_c$, as shown in Figure 6.7b. This behavior is somewhat a characteristic of an incoherent imaging system and assumes that the collection and spot-forming lens diameters are equal, which is the usual case in optical data storage.

In Figure 6.8, the experimentally obtained signal current distributions for several different spatial frequencies are displayed. At low frequencies, the signal current is concentrated in the center of the pupil, as predicted by the theory described above. At high spatial frequencies, the signal current is concentrated near the edges of the

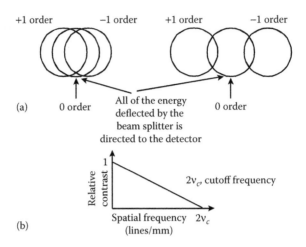

FIGURE 6.7 Theory of incoherent-like transfer function (contrast vs. frequency) for an open aperture in the return-path filter plane. (a) ±1 and zero orders for low and high spatial frequencies and (b) the associated transfer function.

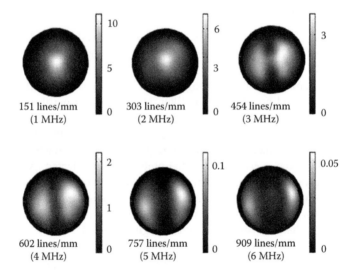

FIGURE 6.8 Signal current distributions in the pupil for conventional illumination after reflection from the disk. Increasing frequency left to right and top to bottom. The gray scale indicates the magnitude of the RMS signal current in mA. Bars on the right-hand side of each distribution indicate the relative scale.

pupil. Notice the result for 602 lines/mm, which is near the coherent cutoff frequency $v_c = 640$ lines/mm. Significant signal energy passes to the detectors at this frequency. As the spatial frequency increases toward $2v_c = 1280$ lines/mm, which is the cutoff frequency for this type of optical system, the overlap area is significantly reduced. The measured signal distributions are in excellent agreement with simulated signal distributions.

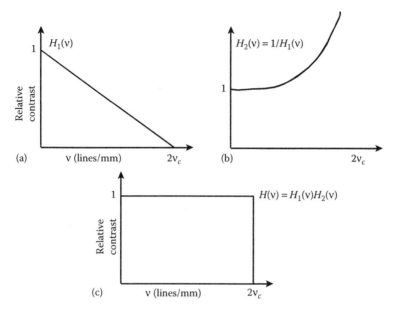

FIGURE 6.9 Equalization of the system transfer function: (a) Original system transfer function, (b) equalizing system transfer function, and (c) combined transfer function.

Altering the relative transmission of the low data frequencies relative to the transmission of the high data frequencies can influence the detector signal dramatically. In data storage systems that are used at, or near, the limit of resolution, it is desirable to increase the high-frequency content so that small features can be more easily resolved. The process of making the system transfer function uniform is known as *equalization*. Equalization can be accomplished by cascading a second linear system $H_2(v)$ with the optical transfer function $H_1(v)$ in order to produce a uniform response out to $v = 2v_c$, as shown in Figure 6.9. For example, the cascaded system $H_2(v)$ can be a simple electronic filter that boosts the high frequencies relative to the low frequencies. This type of electronic filtering has the desired benefit of increasing the relative amplitude of the high spatial frequencies. However, the noise is increased along with the signal, so the carrier-to-noise ratio (CNR) remains unchanged.[*] Optical filters can also be used to modify the system transfer function for the purpose of equalization.

6.4 BEAM SHAPING EXAMPLES

6.4.1 ILLUMINATION-PATH FILTERING

A significant contribution to the field of beam shaping was made by Frieden,[9] who showed how, in theory, there was the possibility of aperture synthesis where a given small aperture may be made to form images as if it were from a larger uniform

---

[*] The CNR is similar to signal-to-noise ratio, except that the signal strength is compared to the noise in a narrow frequency band close to the signal frequency, rather than integrating the noise power.

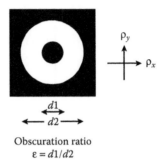

Obscuration ratio
$\varepsilon = d1/d2$

FIGURE 6.10 Illumination-path amplitude filtering using an opaque disk.

aperture. Frieden derived a pupil function to achieve arbitrarily perfect imagery over a given field of view. His example showed the central core width of the point-spread function reducing to 20% of the unfiltered case, with sidelobe levels comparable to the unfiltered focused spot. The derived pupil function had an amplitude variation that behaved as a cosine function with increasing frequency and amplitude at the edges of the pupil. The negative portions of the cosine were realized as π-phase changes.

Significant disadvantages of Frieden's approach are that peak irradiance is dramatically reduced and also that fabrication of the filter is difficult. The following discussion includes practical realizations of optical filtering that produce narrow focused spots. However, as is described in Frieden's theoretical work, the filters result in decreased peak irradiance and modified sidelobe irradiance.

6.4.1.1 Amplitude Filtering

One method of amplitude filtering is to place an opaque disk in the middle of a pupil plane, which creates an annular region of illumination, as shown in Figure 6.10. Figure 6.10 also shows the definition of the obscuration ratio $\varepsilon = d1/d2$, which is the ratio of the inner opaque circle diameter to the outer aperture diameter. Coordinates ρ_x and ρ_y represent the transverse dimensions in the plane of the filter. When illumination-path amplitude filters are placed before the objective lens, as shown in Figure 6.2, they generally have the effect of narrowing the central core of the light spot focused on the disk. Amplitude filters also lower the peak irradiance and redistribute the energy into the first few rings of the Airy pattern. This redistribution has the desired action of narrowing the central core of the focused spot, which increases resolution in the Rayleigh sense,[*] and the undesired consequence of lowering the peak irradiance and increasing the sidelobe energy. Even though amplitude filters narrow the central core of the focused spot, the cutoff frequency of the system transfer function remains unchanged. There are a number of papers published on obscured optical systems that describe the effect of the obscuration ratio on the focused spot properties.[10]

---

[*] Rayleigh two-point resolution is based on the minimum resolvable separation between two point-like objects, where the minimum separation is directly proportional to the width of the point-spread function of the optical system.

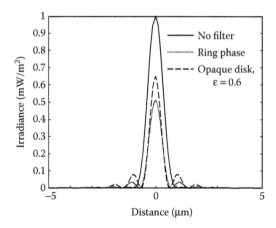

FIGURE 6.11 Irradiance profiles of focused spots with different illumination-path filters.

An example of the focused spot distribution simulated with OPTISCAN is shown in Figure 6.11, where the total power incident on the filter is the same as without the filter. The opaque disk (dot–dash line) with $\varepsilon = 0.6$ exhibits a smaller central core width than the unfiltered system, but sidelobe energy is increased. The peak of the focused spot is approximately 35% less than that without the filter.

Figure 6.12 shows the focused spot distributions as a function of ε. Qualitatively, as ε increases, the peak irradiance decreases and the sidelobes increase. The theoretical on-axis peak irradiance follows the expression $I_{peak} = (1 - \varepsilon^2)$. The trend for these data is that, for increasing s, the first sidelobe level becomes larger relative to the on-axis peak irradiance. In the limit as ε approaches unity, the filter simplifies to an infinitely thin ring, and the focused spot irradiance follows a squared J_0 Bessel function, with a theoretical sidelobe-to-peak ratio of 0.1621. The full-width-at-half maximum

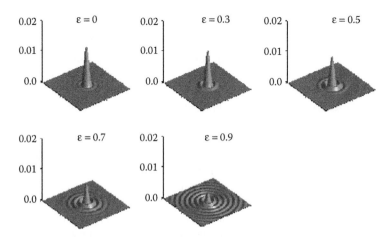

FIGURE 6.12 Focused spot irradiance distributions as a function of obscuration ratio ε for the opaque disk illumination filter.

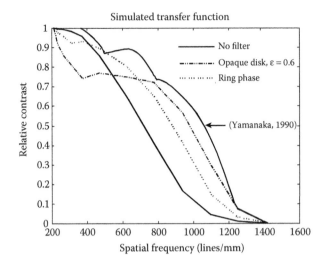

FIGURE 6.13 Simulated system transfer functions for several filter combinations, including the opaque disk and ring-phase illumination filters. Yamanaka's combination filter result is also shown.

(FWHM) of the central core irradiance decreases as ε increases, which is a desirable feature because contrast at high spatial frequencies will be increased with a reduced FWHM. However, the trade-off for reduced FWHM is a reduced peak irradiance and higher sidelobe levels.

Figure 6.13 shows the simulated transfer function curve for the opaque disk filter. The contrast is improved at high frequency relative to the unfiltered source. The shape of the ε = 0.6 transfer function is significantly different from the shape obtained without the filter. This shape is understood by observing that the amplitude filter blocks the central portion of the pupil, resulting in low spatial frequency attenuation. The nearly flat response in the midfrequency region indicates that the system is well equalized in this region. Figure 6.14 shows the two-point response of the ε = 0.6 source. There is a pronounced 53% central dip observed for this filter. It is tempting to use an even higher obscuration ratio. However, the effects of sidelobes become significant for ε > 0.6, such as the intersymbol interference and crosstalk from neighboring tracks. Intersymbol interference is characterized by the additional features present in the opaque-disk two-point response at low-signal amplitude ($i < 0.1$). Any corruption of the data signal reduces the system's ability to interpret the data signal reliably.

Researchers of scanning optical microscopy and optical data storage have investigated this type of amplitude filtering in some detail. Z. Hegedus[11] investigated amplitude filters having multiple annular zones of transmittance applied to a confocal scanning microscope. He found that using multiple zones, the sidelobe levels could be decreased from the case of a single opaque annular disk, but it was at the expense of increasing the central core width when compared to the system using the single opaque annular disk.

Y. Yamanaka[12] investigated amplitude filters having an opaque stripe (rectangle) down the center of the objective lens applied to an optical data storage system.

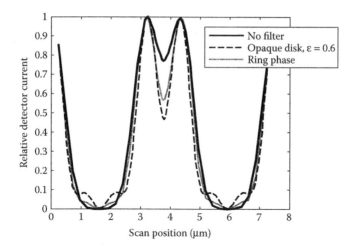

FIGURE 6.14   Data signal current for the worst-case scenario (two-point response). Each signal is normalized to the signal's maximum.

A rectangular amplitude filter breaks the circular symmetry in the focused spot. The consequence of this type of filter is that the central core reduction occurs in only one dimension (along the track), while the other dimension remains unchanged. The result of this work was a decrease in the central core width to 80% of the unobscured case, and this resulted in an increase in the linear recording density by a factor of 1.2. The disadvantage of this system is that to provide an acceptable signal quality, a more complex optical system must be used, one where the sidelobes are spatially filtered out of the reflected light by a slit before the signal reaches the detectors.

T. Tanabe[13] investigated the same amplitude filtering technique as Yamanaka while using an alternative technique to suppress the effects of the sidelobes on the readout signal quality. Rather than using a slit to spatially filter the sidelobes out of the reflected light, an electrical equalizer was used to pass the signal read from the main central core while suppressing the sidelobe signals. This technique of sidelobe elimination has the advantage of no additional optics to readout the signal, at the cost of more complicated electronic circuits.

6.4.1.2   Phase Filtering

Phase filtering, as with amplitude filtering, improves the optical data storage system in its ability to write and read smaller marks. Phase filters can also be used to partially equalize the system transfer function. Phase filtering is accomplished by placing optical path difference (OPD) masks in the optical system. Illumination-path phase filters are placed before the objective lens, as shown in Figure 6.2. They have the effect of narrowing the central core of the focused spot while lowering the peak irradiance and redistributing the energy into the first few rings of the Airy pattern. This redistribution has the desired action of increasing resolution in the Rayleigh sense and the undesired consequence of lowering the peak irradiance and increasing sidelobe energy. The consequences are very similar to those of amplitude filtering with an opaque disk. However, Sales and Morris[14]

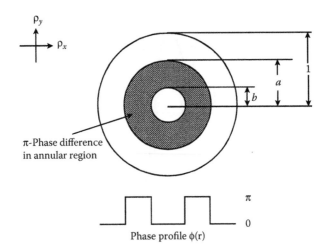

FIGURE 6.15   Three-portion ring-phase filter.

have shown that phase filters perform better in the sense that the magnitude of the undesirable consequences is less than that for amplitude filters. They also showed that the least reduction in peak irradiance occurs for phase filters.

H. Ando[15] discussed the optimization of phase filters to achieve focused spot performance criteria that have the desired effects of narrowing the central core and also reasonable sidelobe levels. Ando's three-portion phase filter is shown in Figure 6.15, where the phase is shifted by π in the ring area. The radii $a = 0.45$ and $b = 0.29$ correspond to the normalized outer and inner radii of the π-phase ring, respectively. Ando showed that, for a given reduction in the central core width, the reduction in the on-axis irradiance was not as much as for the amplitude filters, and the sidelobe level was not as high. Also, the sidelobe level experienced a further reduction by increasing the number of annular regions of phase shift.

The focused spot profile for the ring-phase filtered source is shown in Figure 6.11. Illumination for the phase filter is slightly different than with the opaque disk, in which the $1/e^2$ irradiance of the Gaussian is located at 70% of the pupil diameter, instead of at 100%. The phase filter produces a peak irradiance reduction of approximately 50% and a sidelobe level at 6% of the peak value. The peak reduction is understood from the central ordinate theorem of Fourier optics: the area of a function is equal to the value of the Fourier transform at the origin in the Fourier domain. The peak reduction is a function of the radii of the phase annulus and it is also a function of the illumination on the filter. The simulation results shown in Figure 6.11 used constant ring parameters, as well as constant total power in the pupil. A favorable aspect of the filter is that the diameter of the central core is reduced by about 12%, which results in better contrast for the high spatial frequencies. Notice that the sidelobe energy for the phase filter is much lower than that for the opaque disk.

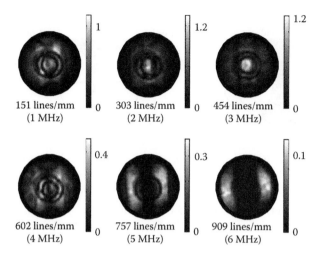

FIGURE 6.16 Signal current distributions in the pupil with the ring-phase illumination-path filter. Increasing frequency left to right and top to bottom. The gray scale indicates the magnitude of the signal current in mA. Bars on the right-hand side of each distribution indicate the relative scale.

Figure 6.16 displays the measured signal current distributions for the ring-phase illumination-path filtered system. The additional structure, compared to Figure 6.8, is due to the presence of the phase filter. Dark bands occur at the edges of the phase filter, and they closely follow the outline of the annulus. The shape of the dark region, when the distribution becomes bimodal, is no longer "x" shaped, as with the unfiltered case shown in Figure 6.8. Instead, the shape of the dark region is more rectangular in form.

Figure 6.13 shows the transfer function curve for the ring-phase filtered system. Contrast is indeed improved at high frequencies. For instance, at 900 lines/mm, the contrast is 0.2 for the unfiltered case and 0.5 with the ring-phase filtered source. In Figure 6.14, the two-point response curve of the ring-phase filtered system is shown. There is a pronounced 43% central dip observed for the ring-phase source without the undesirable low-amplitude features of the opaque disk filter.

6.4.2 RETURN-PATH FILTERING

6.4.2.1 Amplitude Filtering
Amplitude filters can also be placed in the return path, as shown in Figure 6.2, where they have no effect on the shape of the focused light spot. One return-path filter is a shading band that blocks the central portion of the aperture, as shown in Figure 6.17.

The effects of the shading band can be qualitatively understood by recognizing that the transfer function of Figure 6.7 can be derived from the collection of diffracted orders reflected from the data pattern. The amount of contrast, or modulation,

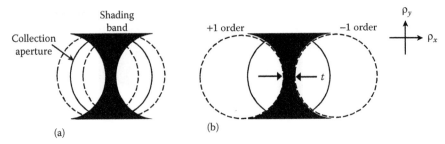

FIGURE 6.17　Action of the return-path x-band optical filter on the diffracted orders. (a) Low frequency and (b) high frequency.

depends on the overlap area of the ±1 diffracted order with the zero order, as described in Section 6.3. At low spatial frequencies of the data pattern, some of the overlap area is blocked, which reduces the signal amplitude at the detectors, as shown in Figure 6.17a. At high spatial frequencies, the overlap area is unaffected, so the signal amplitude is unchanged as shown in Figure 6.17b. When placed in the return path, amplitude filters have the effect of partially equalizing the system transfer function by attenuating the low frequencies relative to the high frequencies. The central width t of the filter relative to the aperture diameter affects the filter performance. Since the shape of the filter resembles an "x," it is called an x-band filter.

A shading band in the return-path also alters the CNR of the system. This effect can be understood as follows. For low spatial frequencies shown in Figure 6.17a, light is attenuated that contributes to the signal by the nature of the large overlap region of the diffracted orders. The attenuated light would also contribute to noise if it were allowed to pass through the detectors, since most optical noise sources are distributed in the pupil according to the irradiance distribution. At low frequencies, the signal power P_{sig} is attenuated more than the noise power P_n. As a consequence, there is reduced contrast and reduced CNR at low frequencies. However, high spatial frequencies experience no attenuation, because the diffracted orders for this frequency do not overlap with the shading band. The shading band blocks the light that contributes to P_n, so there is the desired action of increasing the CNR at the high spatial frequencies.

The simulated system transfer function for the unfiltered system and for a $t = 0.15$ x-band return-path filter is shown in Figure 6.18. Observe the reduction in amplitude of the low-frequency portion of the signal, relative to the unfiltered case. Also, the return-path x-band filter introduces no phase distortion into the signal, unlike electronic filters that usually have residual phase distortion. Figure 6.19 displays the measured two-point response for a worst case scenario data pattern with a 1.65 μm long mark and a 0.67 μm short mark. Figure 6.19a displays the signal trace with no filter, and Figure 6.19b displays the signal trace with $t = 0.15$ x-band filter. The modulation contrast is improved from 25% to 43% with the x-band filter. Figure 6.20 displays the measured CNR versus frequency for the x-band filter. Notice that the CNR is slightly lower in the low-frequency region, but it is up to 6 dB higher in the high-frequency region.

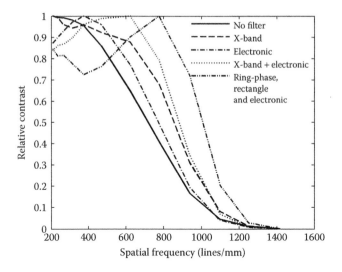

FIGURE 6.18 Simulated system transfer functions for several return-path filters and combinations.

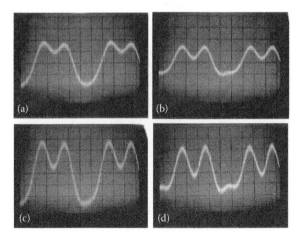

FIGURE 6.19 System two-point response for a worst-case scenario data pattern with long mark of 1.65 μm and short mark of 0.67 μm. (a) Conventional system (no filtering), 22% modulation, (b) return-path optical amplitude filter only, 43% modulation, (c) electronic boost filter only, 44% modulation, and (d) combination of the two filters, 60% modulation.

6.4.2.2 Return-Path Electronic Boost Filter

Another return-path filter of interest is the electronic boost filter. It is used to boost the high-frequency response of the system relative to the low-frequency response, as shown in Figure 6.9. This equalization is the opposite action of the return-path x-band amplitude filter discussed in Section 6.4.2.1. The electronic filter has a variable gain and a variable peak gain frequency. Figure 6.21 shows the transfer function of the electronic filter, as well as the measured transfer function of the electronic

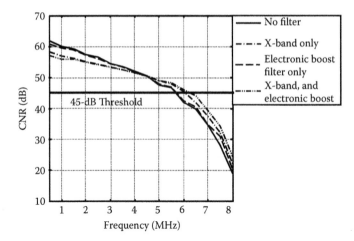

FIGURE 6.20 Measured carrier-to-noise ratio (CNR) for the x-band and electronic boost return-path filters and the combination. The 45-dB threshold is considered a rule-of thumb minimum threshold for reliable recording.

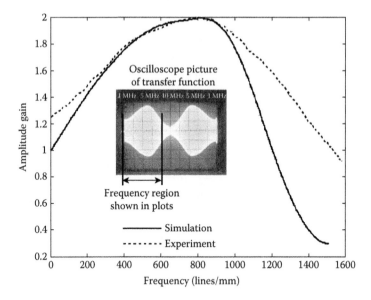

FIGURE 6.21 Simulated and measured transfer function of the electronic filter.

circuit when modeled in a MATLAB® program. The two curves agree reasonably well. The inset of Figure 6.21 shows the oscilloscope display of the measured transfer function as the filter is swept in frequency by a function generator. The modeled electronic filter can be used to predict the response of the system. The electronic boost filter only has 0.5° of phase distortion. Since the phase distortion is very small, the timing of the edges on the data signals is essentially unchanged. Figure 6.18 shows the simulated transfer function for the electronic boost filter, which exhibits slightly improved response at high frequencies. Figure 6.19c shows the effect of the electrical boost filter on the two-point response, where the modulation contrast is improved from 22% without a filter to 44% with the electronic boost filter. Figure 6.20 shows that there is a negligible effect of the electronic boost filter on CNR.

6.4.3 COMBINATION FILTERING

6.4.3.1 X-band Return-Path Filter and an Electronic Boost Filter

Figure 6.18 shows the simulated system transfer functions for five cases: the unfiltered system, electronic boost-filtered system, $t = 0.15$ return-path x-band filtered system, the combination of the $t = 0.15$ return-path x-band with the electronic boost filter, and the combination of a $t = 0.15$ rectangle band ($t = 0.15$ return-path x-band filter without the curved top and the bottom section) with the electronic boost filter and the ring-phase illumination-path filter. For the unfiltered system, a classic roll-off of the system transfer function is observed. For the electronic boost-filtered system, the transfer function exhibits improved contrast at high frequencies, and the frequency at which the maximum contrast occurs is shifted from 0 lines/mm to 375 lines/mm, or a mark pitch of 2.75 μm. The $t = 0.15$ return-path x-band system transfer function also exhibits improved contrast at high frequencies, which is better than that of the electronic filter. For the combination of the optical and electronic filter, the contrast improvement is greater than that of either filter alone, and the frequency of maximum contrast is shifted to 625 lines/mm which corresponds to a 1.6 μm mark pitch. It is interesting to note that simulation of the $t = 0.15$ return-path x-band filter gives better performance than the electronic filter. The combination of the ring-phase, rectangle band, and electronic boost filter provides the most dramatic increase in contrast at high frequencies. The frequency of maximum contrast shifts to 770 lines/mm in this case.

It is clear that the two-point response for the combination of filters behaves in the following manner. The low-frequency portion is reduced when the return-path optical filter is used. The high-frequency portion is increased when either the ring-phase illumination-path filter or the electronic boost filter is used. The percentage modulation of the two-point response for the filter combinations can be predicted from the transfer functions for the filter combinations by looking at the spatial frequency of 940 lines/mm, which corresponds to the small mark of the two-point mark pattern in Figure 6.19. The modulation contrast is improved from 22% without filtering to 60% with the combination filter.

Figure 6.20 shows measured CNR as a function of frequency for the return-path filtering schemes. At frequencies less than 4 MHz, the CNR for any of the filtering combinations is 1 to 4 dB less than the CNR of the conventional unfiltered system. However, the reduction is not significant enough to cause unacceptable signal quality. The CNR is comfortably above the 45 dB line, which is a rule of thumb for indicating a threshold for reliable recording.[16] For frequencies above 4 MHz, the CNR increases by 1 to 3 dB when optical filtering is used alone or in combination with the electronic filter. The electronic filter alone does not increase the CNR, because the electronic amplification increases the noise as well as the signal. The shape of the return path filter is designed to increase CNR by considering the distributions of signal and noise in the pupil of the optical system as a function of frequency.[17] For frequencies less than 4 MHz, the filter blocks some of the signal light and some of the noise. For frequencies greater than 4 MHz, the filter does not block any signal light, but it blocks light in the central portion of the pupil that contributes to noise.

6.4.3.2 Tapped-Delay-Line Electronic Filter with the Opaque Disk Illumination Filter

One way to deal with the corruption of the data signal due to the illumination-filter-induced sidelobes is to use an electronic filter called a tapped-delay-line filter. This aids in reducing the effect of the sidelobes on the data signal. The design philosophy in this section is reviewed from the work of Tanabe.[13] The filter is simulated in MATLAB. The filter's effectiveness is demonstrated as combined with the $\varepsilon = 0.6$ opaque disk illumination filter.

The block diagram for a tapped-delay-line filter with three taps is shown in Figure 6.22a. Each tap has a gain coefficient associated with it and is separated by a delay. The delay is usually expressed in units of time T, but the delay can be converted to units of length l, through the relation $l = vT$, where v is the velocity of the disk. The choice for the coefficients of the taps is based on properties of the focused spot. Specifically, the center tap gain $C_0 = 1$ is the normalized peak irradiance of the focused spot. The outer taps are $C_1 = C_{-1} = -K_0$, where K_0 is the ratio of the sidelobe peak level to the central core peak level. For this system, $K_0 = 0.122$. The choice for the tap spacing is based on the distance between the two sidelobe peaks. For this system, $l = 1.02$ μm.

Figure 6.23a shows a signal representing the profile of the focused spot before and after the application of the three-tap filter and the corresponding compensation waveform of the filter. The sidelobe levels are reduced but not completely. This is understood by observing the compensation waveform used to cancel the sidelobes. The compensation waveform is thought of as summed shifted replicas of the central core. Here, it is evident that the central core width is not the same as the first sidelobe width; in fact, the central core is a little larger. Tanabe did not use a centrally obscured source but rather the same rectangular shading band that Yamanaka used.

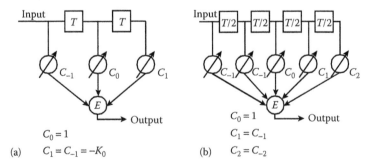

FIGURE 6.22 Tapped-delay-line electronic filter. (a) Three-tap filter and (b) five-tap filter; the delay of the five-tap filter is half the delay of the three-tap filter.

Tanabe's rectangular shading band width is chosen to provide a central core width very nearly equal to the sidelobe width, thus having better sidelobe reduction performance than indicated by Figure 6.23. Figure 6.24 shows the simulated transfer function for the three-tap delay-line filtered signals. The benefit of the three-tap filter is to equalize the transfer function in the midfrequency region.

The next step in the tap-delay-line filter design is to investigate whether or not going to a five-tap delay-line filter improves the system further. The block diagram for a five-tap delay-line filter is shown in Figure 6.22b. Each tap has a gain coefficient associated with it and is separated by a delay. The delay for the five-tap filter

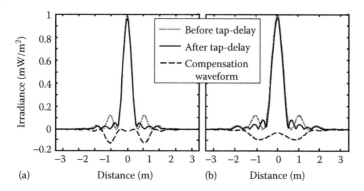

FIGURE 6.23   (a) Signal representing the focused spot before and after the application of the three-tap filter and (b) signal representing the focused spot before and after the application of the five-tap filter.

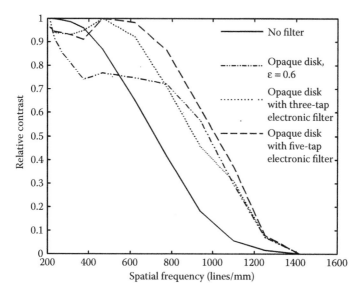

FIGURE 6.24   Simulated transfer functions for no filtering, the opaque disk illumination-path filter, and combinations of the opaque disk and tapped delay lines.

is half that of the three-tap filter, or $l = 0.501$ μm. The choice for the coefficients of the taps is based on properties of the focused spot. Specifically, the center tap gain $C_0 = 1$, the normalized peak irradiance of the focused spot. The outer taps $C_1 = C_{-1}$ and $C_2 = C_{-2}$ are designed by iterating until ringing in the data signal is reduced, and good equalization of the system transfer function is achieved. Figure 6.23b shows a signal representing the profile of the focused spot before and after the application of the five-tap filter and the corresponding compensation waveform of the filter. The sidelobe levels are reduced but not completely. Figure 6.24 shows the transfer function for the five-tap delay-line filtered signals. The benefit of the five-tap filter is to equalize the transfer function and reduce ringing in the data signal when combined with illumination-path filters. More taps may be used in the filter which leads into filter structures that are FIR (finite impulse response) digital filters. This increase in the number of taps leads to more complex electronic filter designs that are usually implemented in a digital signal processor (DSP). DSP's are an attractive option, but they lead to an increased cost of the data channel.

6.4.3.3 Yamanaka Combination Filter

Yamanaka[12] used a rectangular shading band as an illumination filter, which produced considerable sidelobe energy in the focused spot. To reduce the effect of the sidelobes in the detector signal, Yamanaka arranged a simple slit just before the plane of the detector, as shown in Figure 6.25. The slit was conjugate to the focused spot on the disk, and the slit width was designed to pass only the central lobe energy to the detector. In essence, this arrangement is similar to a 1D confocal microscope. The simulated transfer function for the Yamanaka system is shown in Figure 6.13 which indicates excellent equalization characteristics.

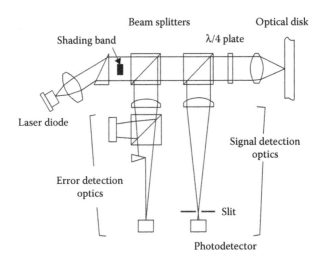

FIGURE 6.25 Yamanaka's combination system with an illumination-shading band and a slit filter in the return path. This system uses separate optics in the return path for the data signal and the servo signal. Also, the optics are designed for write-once media, like compact disc-recordable (CDR).

6.5 SUMMARY

Beam-shaping techniques for data storage applications fall into two general catego-ries: those that implement filters in the illuminating path and those that implement filters in the return path. The techniques may be combined with each other and with electronic filtering. For illumination-path filtering, a modest 10%–20% reduction of spot size can be achieved with either amplitude or phase filters. Smaller spot diame-ters can be achieved but at the expense of increased sidelobe levels that can affect the ability of the drive to write data. For return-path filtering, the system transfer func-tion can be significantly equalized without the need for changing the focused spot. A powerful combination of techniques is a shading band in the return path followed by a simple high-frequency boost electronic filter. Although the potential for beam shaping in optical data storage devices has been demonstrated in various laboratories, consumer devices have not implemented this technology. To date, the performance improvements do not outweigh the cost and complexity of the techniques.

REFERENCES

1. Walker, E.P., Superresolution applied to optical data storage systems, Ph.D. Dissertation, Optical Sciences Center, University of Arizona, Tucson, U.S., 1999.
2. Milster, T.D., A user-friendly diffraction modeling program, *IEEE Lasers and Electro-Optics Society* 1997, Optical Data Storage Topical Meeting Conference Digest, Apr. 7–9, Tucson, Arizona, 60–61, (IEEE Catalog Number 97TH8273), 1997.
3. Milster, T.D., Physical optics simulation in MATLAB for high-performance systems, *Opt. Rev.*, 10(4), 246–250, 2003.
4. Mee, C.D., and Daniel, E.D., *Magnetic Recording Volume I: Technology*, McGraw-Hill, New York, 1987.
5. Hurst, J.E. Jr., and McDaniel, T.W., Writing and erasing in magneto-optical recording, In *Handbook of Magneto-Optical Data Recording*, McDaniel, T.W. and Victora, R.H., eds., Noyes Publications, New Jersey, 1997, chap. 7.
6. Wright, C.D., The Magneto-Optical Readout Process, In *Handbook of Magneto-Optical Data Recording*, McDaniel, T.W. and Victora, R.H., eds., Noyes Publications, New Jersey, 1997, chap. 8.
7. Kryder, M.H., Magneto-optic recording technology, *Jpn. J. of Appl. Phys.*, 57(8), 3913–3918, 1985, pt. 2B.
8. Brucker, C., Magneto-optical thin film recording materials in pratice, In *Handbook of Magneto-Optical Data Recording*, McDaniel, T.W. and Victora, R.H., eds., Noyes Publications, New Jersey, 1997, chap. 5.
9. Frieden, B.R., On arbitrarily perfect imagery with a finite aperture, *Opt. Acta*, 16(6), 795–807, 1969.
10. Boyer, G.R., Pupil filters for moderate superresolution, *Appl. Opt.*, 15(12), 3089–3093, 1976.
11. Hegedus, Z.S., Annular pupil arrays application to confocal scanning, *Opt. Acta*, 32(7), 815–826, 1985.
12. Yamanaka, Y., Hirose, Y., Fujii, H., and Kubota, K., High density recording by super-resolution in an optical disk memory system, *Appl. Opt.*, 29(20), 3046–3051, 1990.
13. Tanabe, T., Superresolution readout system with electrical equalization for optical disks, *Appl. Opt.*, 34(29), 6769–6774, 1995.
14. Sales, T.R.M. and Morris, G.M., Fundamental limits of optical superresolution, *Opt. Lett.*, 22(9), 582–584, 1997.

15. Ando, H., Yokota, T., and Tanoue, K., Optical head with phase-shifting apodizer, *Jpn. J. Appl Phys.*, 32(11B), 5269–5276, 1993.
16. Marchant, A., *Optical Recording*, Addison-Wesley, Reading, MA, 1990.
17. Walker, E.P. and Milster, T.D., High-frequency enhancement of magneto-optic data storage signals by optical and electronic filtering, *Opt. Lett.*, 20(17), 1815–1817, 1995.

7 Laser Isotope Separation with Shaped Light

Andrew Forbes and Lourens Botha†

CONTENTS

7.1 INTRODUCTION

Laser-induced chemistry is an exciting and expanding field, which has led to commercial spin-off opportunities, such as the separation of isotopes of a given atom by means of selective laser-induced dissociation of a molecular structure containing those isotopes. This process, sometimes referred to as isotope enrichment, or just plain *enrichment*, is often the result of the molecule absorbing multiple photons, usually from an intense laser source. When a molecule is highly excited, it absorbs laser radiation by resonance, leading to dissociation of the weakest bonds. When the absorbed energy exceeds the dissociation energy of the weakest bond, the molecule undergoes decomposition. The trick is to make the absorption *selective*, so that only those molecules containing a particular isotope undergo this decomposition. One usually chooses the initial molecule so that after decomposition, the final molecule differs from the original in some chemical or physical way. For example, one might choose the initial molecule to be in gas form, which forms solid decomposition products. The isotopes of a given atom are then separated by conventional techniques, which, in our example, would be a phase separation (separating the solids from the gases).

Since the suggestion first appeared that the absorption of laser radiation and the subsequent decomposition of molecules could be isotopically selective, a good deal of work has gone into demonstrating this. The focus has been on several isotopes of commercial interest, with somewhat divergent absorption properties, and consequently multiple-laser systems have been investigated as the source of the radiation, covering the spectral range from the vacuum ultraviolet to the far infrared. Infrared lasers have, for example, been used in investigating the enrichment of uranium for nuclear fuel, and the separation of ^{12}C isotopes for better thermal management in electronic circuitry. Although laser-induced chemistry and laser isotope separation have received much attention in the past, very little of the initial expectations have been realized commercially. The obstacles to successful commercialization to date have been as follows:

  i. The cost of producing laser photons (or laser energy)
  ii. The utilization of these photons on an industrial scale

The first obstacle has more or less been removed in the case of CO_2 laser photons: running and capital costs have decreased over the past decade to a level that the industrial applications of infrared selective isotope separation have become feasible and economically viable (although photons are still very expensive as compared to electrical energy). This reduction in cost is mostly due to a maturing technology—optics, for example, costs less than before, and lasts longer. The second obstacle (part [ii]) is the more challenging concern, as it must be rigorously addressed if the use of the photons is to be optimized for the intended purpose. For example, if photons are wasted at the source, along the delivery path, or at the target, the economics of the enrichment process will become prohibitive, particularly if the photons are expensive to start with. Another way to express this criterion is in the specific energy consumption of the chemistry process (measured perhaps in moles of product dissociated per joule of laser energy consumed), including all the photon losses from the source to the point where the laser beam is finally discarded. As we will discuss, this economical parameter is closely connected to the control of the *shape* of the laser beam. For now it is enough to note that it is imperative that most of the laser photons be utilized in the separation process, and not lost on optics, or during propagation, or simply as heat in the interaction medium. This is what we refer to as *optimizing the photon utilization* of the process.

All of the interaction processes of laser beams and gas media in multiple-photon dissociation processes are based on nonlinear dependences of the laser beam intensity, with the result that the spatial and temporal distribution of the laser beam is of the utmost importance. Assume for the moment that we have applied our minds to the losses along the delivery path, and have come up with a suitably efficient delivery system, with optical components chosen to minimize the energy loss from the source to target. In our case, the target would be a reactor chamber where the photons interact with the molecules to be dissociated. Also assume that we have chosen the most efficient lasers possible for the task. That leaves only the reactor itself to consider. Invariably the reactor geometry is fixed (perhaps for gas handling reasons, or to satisfy gas flow constraints), and so very little leeway exists for reactor changes in order to optimize the use of the photons. Thus, good photon utilization must be achieved by making changes to the beam itself. One parameter to be addressed is the intensity distribution of the laser beam, which we will refer to as the *beam shape*. Without careful attention to the beam shape, the dissociation process becomes uneconomical. At first this would appear to be a relatively easy problem to solve. After all, several beam shaping options have matured to the extent that they could be used in a commercial laser plant. However, long reactor path lengths and multiple beams (often of differing wavelengths) often complicate the problem.

In this chapter, we will introduce some of the important variables in an isotope separation process and show how the beam shape influences the commercial success of the process. Rather than discuss the topic from a general perspective, we will use carbon isotope separation as a case study to illustrate the practical aspects of this application. Carbon isotope separation is topical because of the present and future benefits of monoisotopic carbon. In the medical industry, the ^{13}C isotope is already used as a *tracer* in special compounds, whereas ^{12}C has potentially huge benefits for the semiconductor industry as a solution to the heat removal problem in computer circuitry (isotopically pure diamond has a far greater heat conductivity than conventional diamond). Finally, we offer an up-to-date synopsis of progress in laser-based isotope separation.

7.2   MULTIPLE-PHOTON INFRA-RED EXCITATION

Multiple-photon infra-red (IR) excitation is the absorption of many IR photons by a single molecule; often these photons differ in frequency, requiring various sources to be used in the process. Observation of this phenomenon is only possible with the high-intensity sources typical of lasers. The physical process is well understood: a molecule absorbs light by interaction of its electrical dipole with the oscillating electric field of the source. This causes the molecule to start vibrating: radiant energy has been transformed into vibrational energy. If the molecule were an ideal oscillator, the successive vibrational levels would be equally spaced. However, all real molecules deviate from this, due to the fact that the restoring force drops to zero as the bonds stretch further toward breakage (dissociation). Therefore, all real molecules are anharmonic, which means that the spacing between adjacent vibrational levels decreases (is red shifted) with increasing energy.

The energy of an IR photon is very small, and is much less than the energy required to produce dissociation or even chemical reactions. Typically, between 15 and 65 photons are required to dissociate a stable molecule at room temperature. If the laser frequency is selected so that it matches the frequency difference of the first two vibrational levels of the molecule, then due to the anharmonicity, it would very soon be out of resonance as the molecule is excited to higher and higher vibrational levels. It therefore seemed improbable that a molecule could be excited via a multiple-photon process—how would the remaining photons be in resonance with the molecule? However, during the 1970s several experiments at the Institute of Spectroscopy in Troitzk [1], and at Los Alamos [2], clearly demonstrated that molecules could be excited via a multiple-photon process under certain conditions.

In multiple-photon processes, a molecule with a vibrational absorption band close to the frequency of the irradiating laser can absorb many photons and dissociate without the aid of collisions. The complete analysis of the multiple-photon effect is a complex theoretical problem. Effects that need to be included in such a model are, among others, anharmonicity, coriolis coupling, octahedral splitting, bandwidth considerations of the laser, absorption in the so-called quasi-continuum as well as the unimolecular dissociation of the excited molecule [1–10].

Multiple-photon excitation of a polyatomic molecule by an intense IR laser beam can be divided into three distinct regimes, as shown diagrammatically in Figure 7.1. The initial process is the resonant excitation of the low-lying vibrational transitions (discrete levels), the second stage consists of stimulated transitions in the vibrational quasi-continuum and, finally, photochemical transformation of the highly excited molecules (above the dissociation limit).

The excitation of the lower vibrational levels is of particular interest in isotope separation since this stage of the process determines the isotope selectivity of the multiple-photon process. A nonselective process would imply that both isotopes have been excited. This could happen if the bandwidth of the source is too large, overlapping with the absorption bands of both isotopes. The excitation of these lower levels is through multistep resonant multiple-photon excitations of successive vibrational levels. Anharmonicity causes a frequency detuning of successive vibrational levels,

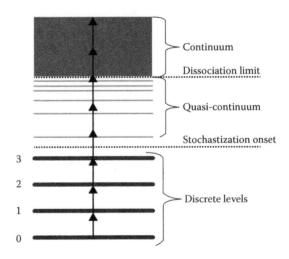

FIGURE 7.1   A simple model of multiple-photon dissociation of a molecule by an IR laser.

which would prevent resonant excitation beyond one or two vibrational levels. This is compensated for by the rotational levels of the molecule, and the anharmonic splitting of the higher vibrational levels. Thus, the same photon frequency can be used to excite the molecule.

The excitation of the lower vibrational levels is a coherent process, and a proper treatment of this subject requires a quantum mechanical model. As the energy of the excited molecule increases, so the density of vibrational states increases sharply. At some level of excitation, a coupling occurs between the overtones of the specific mode that is excited and the background of states.

The Hamiltonian for a molecule near the equilibrium position can be written as follows [3]:

$$H = \sum_{i}^{s} \omega_i \left[(p_i^2/2) + (q_i^2/2) \right] + (1/3) \sum_{i,j,k=1}^{s} \Phi_{ijk} q_i q_j q_k + \cdots \qquad (7.1)$$

Where q_i and p_i are the normal momenta and coordinates, respectively.

If the second term in Equation 7.1 is small, then one finds s independent harmonic oscillators, each with its own frequency. This simple picture is a good approximation at low vibrational energies but becomes increasingly inaccurate at higher vibrational energies. As the vibrational energy of the molecule increases, so the higher-order terms in the equation above increase faster than the quadratic ones. The result is that the normal modes start to intermix, giving rise to a change in the intramolecular dynamics. The motion changes from quasi-periodic to stochastic motion. The vibrational energy where this starts to occur is called the stochastization energy. The interested reader is referred to reference [3] for a more detailed explanation of this subject. The result of this stochastization is that the transition spectrum at high excitation energies forms a wide band, which is termed the vibrational quasi-continuum. If the energy of the excitation laser is high enough, then it is possible to excite the molecule to higher energy levels even though the absorption cross section of the transition is small. Owing to anharmonicity, there is a significant red shift with increasing energy. This would mean that as the molecule is excited to higher and higher energy levels, so the optimum excitation frequency would shift to the red. It is for this reason that more than one excitation laser is sometimes used in a multiple-photon excitation process.

Sometimes simple rate equations (also known as kinetic equations) can be used to model the excitation of a molecule. An excitation that can be described using rate equations is sometimes referred to as an *incoherent process*. Usually this approximation can be used when factors are present that would cause the broadening of the spectral lines. For example, if the line width of a transition is much larger than the power broadening introduced by the lasers, then the rate equations give a good approximation. This is the case in the quasi-continuum where the combined absorption band is much larger than the power broadening, and therefore rate equations can be used. Using the transitions in the quasi-continuum, a molecule can accumulate enough energy to exceed the dissociation energy. Owing to the fact that there are many combinations of vibrational overtones that can participate in the absorption in the quasi-continuum, the energy will be distributed over many vibrational modes.

If the vibrational energy of a molecule exceeds the dissociation threshold, then the molecule can decay spontaneously. The dissociation energy is the breaking energy of the weakest bond. However, the vibrational energy of the molecule will generally be distributed over all the vibrational degrees of freedom of the molecule. Therefore, a fluctuation is required to produce enough energy in the weakest bond for it to be larger than the dissociation energy. This is a statistical process and is analogous to water sloshing around in a bucket; at some stage enough of the energy will be concentrated in a specific degree of freedom (vibrational mode), and a drop of water will fall out of the bucket (the bond will break). The theory that is used to calculate the unimolecular decay is called the Rice–Ramsperger–Kassel–Marcus (RRKM) theory [11].

7.3   ISOTOPE SEPARATION BY LASER DISSOCIATION

Isotope shifts in atomic and molecular spectra are caused by mass differences of the isotopes, nuclear spin, and variations in nuclear volume. Laser isotope separation is based on these isotope shifts. In general, laser isotope separation is divided into the *atomic route* and the *molecular route*. In the *atomic route*, a low pressure metal vapor is generated, and then radiated by tuneable visible lasers. The lasers selectively excite the atoms until the atoms of a given isotope are ionized, thereby producing a selectively ionized species. The isotopes can then be separated electromagnetically.

In the *molecular route*, selective absorption in the infrared is used to excite the vibrational modes of the selected molecule. Since each isotope has a slightly different mode frequency, it is possible to excite selectively by careful choice of laser wavelength and bandwidth. An additional IR or UV laser can then either dissociate the excited molecules or facilitate their participation in a chemical reaction, whereas the unexcited molecule does not dissociate or react. After dissociating, the selected molecule with the required isotope is in a different chemical state to the other isotope(s), and can therefore be separated chemically. Multiple-photon excitation using infrared lasers usually refers to the molecular route, and the isotope separation technique based on this is called *Molecular Laser Isotope Separation* (MLIS).

For example, the enrichment of ^{12}C by the MLIS process involves photons from CO_2 laser systems irradiating Freon gas. Under ideal conditions, the resulting transformation is given by the following reaction:

$$^{13}CHClF_2(g) + {}^{12}CHClF_2(g) \xrightarrow{n \times h\nu} {}^{13}C_2F_4(s) + {}^{12}CHClF_2(g) + \text{by-products} \quad (7.2)$$

In Equation 7.2, the Freon gas ($CHClF_2$) is shown in its two stable forms—that where the carbon atom in the molecule is ^{13}C, and that where it is ^{12}C. The initial ratio of the two gases (right hand side of Equation 7.2) would match the natural isotopic ratio of ^{13}C to ^{12}C; thus, nearly 99% of all the carbon atoms are ^{12}C. This gas is then irradiated by photons ($n \times h\nu$). In this case, the laser photons ($h\nu$) are carefully chosen to interact with only a single vibrational frequency of the selected isotope (^{13}C). Once the selected bond in the ^{13}C Freon molecule is broken (at dissociation), the resulting radical reacts chemically with other radicals to form the solid C_2F_4. In an ideal scenario, the Freon molecule containing the ^{12}C isotope does not dissociate, and remains in the Freon form (gas phase). Since the ^{12}C isotope is now in the solid phase as C_2F_4, conventional chemical processes can be used to (easily) separate the two isotopes.

In the case of molecules, the isotope shift in the absorption spectrum is a result of the variation in isotope mass. Usually a molecule has a large number of molecular modes. However, only those in which the centre of symmetry of the molecule is disturbed show isotope shifts. This means that for certain normal vibrational modes, the absorption wavelength of two isotopes will be different; this is called the *isotope shift* of the absorption spectrum. The spectrum of a molecule is in reality made up of the superposition of rotational transitions, as well as from contributions arising from vibrational transitions of the thermally populated hot bands. This can be detrimental to isotope selectivity and for this reason the molecules are usually cooled, eliminating the contribution of the hot bands to the spectrum. A widely used method of cooling the gas while still keeping it in the gas phase is to use *flow cooling* through a supersonic de Laval type nozzle, as shown in Figure 7.2.

Lasers required for molecular laser isotope separation face severe technical challenges. Usually the lasers would be used in an industrial plant, which would require continuous operation (24 hours a day, seven days a week). In addition, the throughput in the plant is to a large extent determined by the pulse repetition rate of the lasers. For example, pulse repetition rates needed for a uranium enrichment plant should be approximately 16 kHz. Additional complications are the high pulse energies required for the multiple-photon process (typically 1 J per pulse), and the fact that the lasers sometimes need to be continuously tuneable (which in the case of pulsed CO_2 lasers means that the laser will be operated at a high pressure). Because the isotope shift in the molecule's absorption spectrum is usually rather small, frequency stable, narrow bandwidth operation of the lasers is often required. For the case of UF_6, the absorption band is in the 16 μm wavelength region, but no commercial lasers satisfying all the above requirements are available in this region. Therefore, the *stimulated Raman scattering* process is used to shift the wavelength of a TEA-CO_2 laser from the 10 μm wavelength region to the 16 μm wavelength region. This process usually

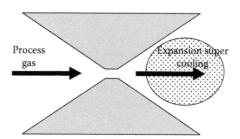

Process gas

Expansion super cooling

FIGURE 7.2 A de Laval type supersonic expansion nozzle can be used to create super cooled gas. The rapid expansion creates cooled monomers, having a *clean* vibrational spectrum; this is essential for many isotope separation processes. The shaded area indicates the region where the laser beam(s) pass through. The laser beam direction is usually perpendicular to the gas flow.

requires multiple passes through a Herriott type cell, filled with cooled hydrogen. The path length through the Herriott cell can be up to 40 m, implying that good beam quality (for example, Gaussian beams, or beams with beam quality factors of $M^2 \sim 1$) is required to ensure low losses through the system, and good focus ability.

When a techno-economic analysis is performed on the MLIS process, it becomes evident that the cost of photons is of critical importance. In order to obtain an economically feasible plant, it is critical that the photons are utilized efficiently—particularly in the dissociation zone, where the laser beam interacts with the cooled gas, and where the isotope separation actually occurs.

7.4 THE DISSOCIATION ZONE

The dissociation zone is that part of the reactor where the laser beam interacts with the cooled gas, finally causing selective dissociation of the molecule containing the isotope of interest. In the carbon separation process, IR laser beams interact with Freon gas (see Equation 7.2) to dissociate the Freon molecules with the ^{13}C isotope. The amount of dissociation that takes place is measured by the *yield* of the process, whereas the quality of the dissociation (i.e., how selective it is) is measured by the *alpha* of the process. In this section we discuss these concepts, and how beam shapes affect them.

7.4.1 ALPHA AND YIELD

Consider a molecule XY, existing in two states whose composition differs only in that one atom is found in two isotopes: let the molecule with isotope a be XY^a and the other with isotope b be XY^b. Assume that the gas is exposed to radiation that would dissociate the XY^a molecule preferentially. Therefore the concentration of XY^a in the dissociation zone will decrease after irradiation. However, due to the fact that the bandwidth of the laser often overlaps with the absorption bands of both molecules, a fraction of the XY^b molecules will also be dissociated. Assume that the dissociation yield for the two molecules is β_a and β_b, respectively. The selectivity α of the multiple-photon dissociation process is defined as

$$\alpha = \frac{\beta_a}{\beta_b} \tag{7.3}$$

This method of characterizing the selectivity of the process is equivalent to the conventional definition of selectivity [1]. Let $[XY^a]_0$ and $[XY^b]_0$ be the concentration of the molecules before irradiation and $[XY^a]_n$ and $[XY^b]_n$ the concentration of the molecules after irradiation. The conventional definition of selectivity is [1]

$$\alpha = \frac{\dfrac{\left[XY^a\right]_n}{\left[XY^b\right]_n}}{\dfrac{\left[XY^a\right]_0}{\left[XY^b\right]_0}} \tag{7.4}$$

As an example, consider the case of carbon enrichment; the ^{12}C isotope makes up roughly 98.89% of all stable carbon atoms, whereas ^{13}C makes up the remaining 1.11%. Thus initially the ratio of the two isotope fractions is

$$\left(\frac{^{12}C}{^{13}C}\right)_0 = \frac{0.9889}{0.011} = 89.09$$

After some irradiation, with selectivity α, the new ratio (from Equation 7.4) will be given by

$$\frac{^{12}C}{^{13}C} = 89.09\,\alpha.$$

Noting that we can write the percentage ^{12}C in the mix in terms of the fractions as

$$\%\,^{12}C = 100 \times \frac{^{12}C}{^{12}C + ^{13}C} = \frac{8909\,\alpha\,^{13}C}{89.09\,\alpha\,^{13}C + ^{13}C}$$

the purity of the ^{12}C isotope (in percentage terms) will be given by

$$\%\,^{12}C = \frac{8909\,\alpha}{1 + 89.09\,\alpha} \tag{7.5}$$

If there is no separation of the isotopes, then the final isotope ratios will be identical to the initial isotope ratios, since nothing would have changed (no dissociation). From Equations 7.3 and 7.4, we see that this would imply that the numerators and denominators are equal, giving a selectivity parameter of $\alpha = 1$. As the selectivity increases, so the enrichment increases. In the limit of $\alpha \to \infty$, the fraction of ^{12}C increases to 100%, from the initial 98.89% (Equation 7.5).

The two parameters mentioned above, that is, α and β, are of critical importance in the economic consideration of the separation process. Usually α and β follow an inverse relationship: a process with a high yield will have a low selectivity, whereas a process with a high selectivity will have a low yield. Why this is so will become clear later on, when we consider laser beam intensities. For now, it is sufficient to note that in order to achieve a high yield, the laser fluence needs to be high, the result of the high fluence is large power broadening, meaning that the laser spectrum increases, making selective absorption less likely, which in turn would result in a lower selectivity. In contrast, a high selectivity requires very small power broadening, which means low intensities, and consequently lower yields. Once the relationship between yield and alpha is known, a techno-economic analysis can be done to determine the most favorable operating point with regards to laser fluence.

At higher temperatures the higher vibrational modes are also populated, which means that the molecular vibrational spectrum will be scrambled by the overtone transitions. The overtone transitions usually have different isotope shifts as compared to the ground state transition. The result is a decrease in spectroscopic selectivity, which has a negative impact on the economics of the process. The result is that for many laser isotope separation processes, the process gas is flow cooled in a supersonic expansion nozzle (see Figure 7.2). It is obvious that it is important to irradiate as many of the molecules and to waste as few of the photons as possible. As can be seen from Figure 7.2, the irradiation zone has a very specific geometry. This geometry is determined from the flow conditions and is fixed. It is thus clear that an economical process will require a very specific beam shape in the irradiation zone, and also that this shape must remain constant over the length of the reactor zone.

7.4.2 Process Economics

It was mentioned earlier that there are two important criteria that must be met if an isotope separation process is to be economical: (1) The cost of laser photons or laser energy must be low and (2) the utilization of these photons on an industrial scale must be very good.

This can be written mathematically as follows:

$$\$/kg = \$/MJ \times MJ/kg \qquad (7.6)$$

This reads that the cost of producing a kilogram of final product (as measured in dollars per kilogram, $/kg) can be split into the cost per MJ of photons, multiplied by the amount of product (in kg) that those photons produce. The first term is determined by the choice of laser systems used, and relates directly to point (1) our criteria. For most lasers used for isotope enrichment, this value is approximately 100 $/MJ, and, rather surprisingly, does not vary that much from source to source. The second term is a measure of how efficient the entire process is in terms of how many photons are wasted and how many photons are used. Figures 7.3 a and b are simple schematics illustrating the interaction of the laser beam with the gas. In this case, the laser beam volume is described by a pipe, which represents propagation across the chamber used for enrichment. In the case of ^{13}C stripping, the *pipe* diameter might be roughly 6 mm, whereas the length could be in the order of 1 m. In such a volume, approximately 2 μg of product is generated, and only 10% of the photons are used. If the other 90% of the photons are discarded, the second term of the relation in Equation 7.6 can be as high as 500,000. This makes the process commercially unviable. However, if the photons can be used efficiently, which often means passing the beam through the chamber many times, then this cost of producing product in MJ/kg can be made orders of magnitude lower. This is thus the *make or break* step in the entire process; it is this term that plays a significant role in the commercial success of an isotope separation process, and is usually referred to as *photon utilization*.

7.5 BEAM SHAPES AND PHOTON UTILIZATION

All of the interaction processes of laser beams and gas media in multiple-photon dissociation processes are based on nonlinear dependences of the intensity of the laser beam, with the result that the spatial and temporal distribution of the laser beam is of the utmost importance. A common misconception is that the process has a threshold that must be overcome—implying that below a certain laser intensity, no dissociation

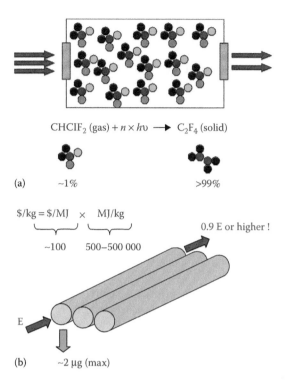

$CHCIF_2$ (gas) $+ n \times h\upsilon \longrightarrow C_2F_4$ (solid)

(a)   ~1%   >99%

$\$/kg = \$/MJ \times MJ/kg$

~100   500–500 000

0.9 E or higher !

E

(b)   ~2 µg (max)

FIGURE 7.3 (a) Energy enters a chamber with a mixture of isotopes. Molecules with one of the isotopes absorb energy, whereas the other molecules do not. After dissociation, the selected isotope is found in a different chemical compound, (b) A simple pipe version of the irradiation process, with each pipe representing the beam irradiation volume of a single pass.

takes place. The multiple-photon process does not have such a threshold; however, at a certain laser intensity the product will be negligible and below the sensitivity of the measurement technique. Thus, for all practical purposes, a *threshold* does exist. In this text, we will refer to threshold with this view in mind.

What happens if the intensity of the laser beam is above or below this threshold? Intensities far below this value will cause negligible dissociation, so the yield will be very small. Intensities far above this will cause power broadening of the molecular spectra, which adversely affects the selectivity of the process, but the yield will be high. In general then, as the intensity of the laser beam increases, so the yield of product increases, but at the expense of selectivity—the product will have a larger mix of isotopes than if the selectivity had been better. Likewise, as the intensity of the laser beam is decreased, so the selectivity increases until a threshold value is reached, below which no (measurable) enrichment takes place due to the accumulative photon energy being lower than the energy needed for dissociation. Thus while the selectivity increases, the yield decreases, and vice versa. One can liken the situation to trying to scoop your favorite sweets from a mixed box, but with different sized scoops. Using a high-intensity laser beam is like using a very large scoop—you maximize the total amount of sweets you get (yield), but at the expense

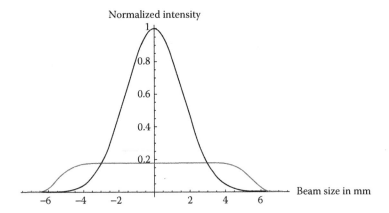

FIGURE 7.4   Example intensity profiles of super-Gaussian and Gaussian beams.

of ensuring that you get only your favorite (selectivity). Conversely, using a small scoop (low-intensity beams) allows you to select your favorites without the unwanted sweets, but the yield in this process is going to be low, that is, fewer of your selected sweets will be gathered in total. Thus, these two parameters need to be balanced, with some of this balancing done by changing the shape of the beam.

When we talk about the *shape* of a laser beam, we are invariably referring to the intensity distribution in the spatial domain. That is to say, if we took a cross section of the beam, the shape would tell us something about how the total energy is distributed across this area. In discussing beam shapes, we will restrict ourselves to two cases: the Gaussian beam and the super-Gaussian beam (see Figure 7.4 for example profiles of each) [12]. Super-Gaussian beams of intensity $I(r)$ and order p are defined as follows:

$$I(r) = I_o \exp\left(-2\left(\frac{r}{\omega}\right)^{2p}\right) \qquad (7.7)$$

where:
   ω is the beam half width
   r is the radial coordinate

The order p indicates the steepness of the intensity drop near $r = \omega$, with increasing values of p indicating flatter intensities with sharper edges. When $p = 1$, the beam is called Gaussian, and follows the standard Gaussian propagation equations. Because the distribution of energy in the two cases (Gaussian and super-Gaussian) is not the same, the peak intensity values will differ, and can be shown to be related by

$$I_s = \frac{p 2^{1/p} \omega_g^2 I_g}{2\omega_s^2 \Gamma(1/p)} \qquad (7.8)$$

where the subscripts s and g refer to the super-Gaussian and Gaussian parameters, respectively. This is an important relationship, because it shows that one requires more energy to get the same peak value in a super-Gaussian beam as in a Gaussian beam.

Gaussian beams are easily generated in resonators that use optical elements with spherical curvatures or even by resonators with diffractive optical elements [13]. Their propagation is well understood, and easily predicted [12]. However, by definition they have an intensity distribution that has a peak value of twice the average (average intensities are calculated by dividing the total power by the beam area, whereas the average fluence is calculated by dividing the total energy by the beam area). This variation in intensity across the beam means that, although some part of the beam will be above the dissociation threshold, other parts will be below the threshold. It also means that the selectivity and yield of the process will vary across the beam area, as well as along the propagation axis—in other words, the selectivity of the process will vary everywhere in the propagation volume (that part of the reactor volume filled by the beam).

Consider the example of enriching carbon using IR beams of both Gaussian and super-Gaussian shape. By ignoring the propagation effects (which will be discussed later), we consider the influence of the intensity distribution only, that is, we assume that the shape remains the same over a certain propagation length. In theory this can be achieved by making the beam size infinitely large, so that the divergence of the beam becomes infinitely small. Before calculating the influence of two different shapes on the final ratio of ^{12}C or ^{13}C that can be generated, we first need to know how the enrichment factor α varies with intensity. Figure 7.5 shows the relationship between the laser beam fluence (energy density) and the resulting enrichment factor α for a given photon wavelength. The measure is given in fluence and not intensity because the time envelop of the pulse was held constant over the experiment, thus only the fluence is needed to characterize the beam's power distribution.

As can be seen from Figure 7.5, as the fluence increases, the selectivity α decreases. At fluences below 800 mJ/cm^2 there is no enrichment due to the fact that the laser beam merely heats up the gas, without actually breaking any bonds. Recall that the Gaussian beam has very low enrichment near its centre, where the fluence is at a peak, and increases toward the edge of the beam area, where the fluence decreases. For this reason, experiments with Gaussian beams are difficult,

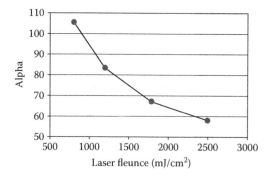

FIGURE 7.5 Alpha versus laser fluence in the isotope separation of carbon.

because the enrichment is an averaging process. By contrast, super-Gaussian beams have the property that the average intensity and the peak intensity have the same value. This means that the entire beam area can be used for dissociation, and that the selectivity and yield will be constant across the beam area. Since all the photons are used equally with this beam shape, very little *averaging* takes place. This makes super-Gaussian beams very attractive for isotope enrichment. However, the propagation effects that were neglected in this discussion play an important role in isotope separation processes. Thus while Gaussian beams can be easily generated and propagated, super-Gaussian beams need to be generated by specially designed optics, whether inside or outside the laser cavity, and have more complicated propagation characteristics as compared to Gaussian beams. These generation and propagation challenges need to be addressed when considering reactor designs.

7.6 REACTOR CONCEPTS

As was just highlighted, propagation effects are very important when considering reactors for isotope separation, and since the propagation of a laser beam is in part determined by the intensity profile of the beam [12], the chosen shape makes a considerable impact on the reactor design. The challenge is to find a beam shape that is optimal, and then design a reactor to hold this shape over an extended distance. Propagation distances in isotope separation processes tend to be long (more than several meters) due to the fact that most media are optically thin (low absorption).

7.6.1 Propagation Issues

The propagation of Gaussian beams can be derived analytically, and is well understood. The divergence of the beam is inversely proportional to the beam size, and together with the wavelength, determines how quickly a given sized Gaussian beam will diverge. The quoted indicator for this is the Rayleigh range, defined as

$$z_r = \frac{\pi \omega_0^2}{M^2 \lambda} \qquad (7.9)$$

where:
ω_0 is the beam size at the waist (minimum) position
λ is the laser wavelength
M^2 is the beam quality factor

For a perfect Gaussian beam, $M^2 = 1$, although in practice, values of less than 1.2 cannot be easily differentiated from a perfect Gaussian beam. This parameter is a measure of beam quality and increases in value for all shapes other than Gaussian beams (super-Gaussian beams will have $M^2 > 2$).

The propagation of super-Gaussian beams was traditionally achieved by using a suitable input field to the Fresnel diffraction integral, but has since been made easier by the flattened Gaussian beam (FGB) approximation. The propagation of such beams of various classes can now be easily predicted [12].

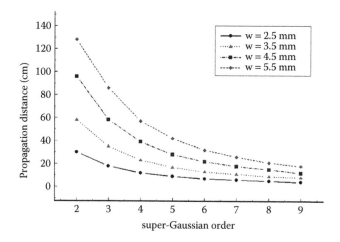

FIGURE 7.6 Maximum propagation distance as a function of super-Gaussian order and starting beam size (w), for a 10% rms change in intensity distribution. The starting field was taken as having a flat wavefront, so that the beam size w is equivalent to the w_0 of Equation 7.9.

The problem in isotope separation is to keep the fluence as constant as possible over the reactor length, while also ensuring that the fluence is above the dissociation limit. Equation 7.9 indicates that beams that are a combination of multiple modes (as a flat-top beam can be seen to be), will diverge faster than single, low order modes, since the M^2 parameter for multimode beams is considerably higher than that of Gaussian beams. Smaller beams will also diverge faster, but the size is limited by the available energy, since the average fluence is calculated by dividing the total beam energy by the beam area. Figure 7.6 shows the distance over which a super-Gaussian beam can be propagated, as a function of both beam size and super-Gaussian order, so that the intensity distribution changes by less than 10% rms. A typical reactor pass length (there will be many passes due to the low absorption) would be in the order of 1 m.

Figure 7.6 then gives an indication of the energy required for a given reactor length and threshold fluence. In order to propagate farther, a larger beam size is required, which in turn requires more energy if the beam is to have the same fluence as the small beam. In most delivery systems, amplification is needed to increase the beam size (while keeping the fluence above the threshold) to a point where there are no major changes in the beam shape over the reactor length. Often it is necessary to have multiple passes through the reactor, due to the fact that the absorption is relatively low through the system. This then complicates the shape-holding criteria—it is simply not good enough to create a given shape, one also must be able to hold (i.e., keep the intensity profile as constant as possible) this shape over an extended distance, which is usually achieved through multiple use of beam shaping optics.

7.6.2 PROPAGATION WITHOUT ABSORPTION

To consider the influence of propagation effects alone, consider the case of carbon enrichment with a laser system that generates 400 mJ of energy in a Gaussian mode, and a threshold fluence for suitable dissociation of 1.4 J/cm². This means that a Gaussian beam would have to have a beam radius of not more than 3 mm in order for the average fluence of the Gaussian beam to be above the dissociation threshold. For an order 5 super-Gaussian beam, because of the lower peak and average fluences (see Equation 7.8), the beam radius must not increase above 2.39 mm. The shape and size of the beam (with the wavelength) determines the distance that such a field can propagate before the intensity distribution changes (the initial phase of the beam also plays an important role, but for this calculation the phase is taken as flat). To determine how the beam shape changes over a 1 m distance, we propagate the two fields under the scenario of no absorption changes to the beam, that is, we are looking only at intensity changes due to propagation effects. The results are shown in Figures 7.7 and 7.8.

Clearly the benefits of starting with the super-Gaussian shape are diminished in this propagation scenario: the shape does not hold (stay constant to within 10% rms) for long enough to realize the benefits of constant fluence, and therefore constant enrichment and yield. The Gaussian beam on the other hand keeps its shape extremely well over 1 m; Figure 7.7 includes propagation up to 5 m to illustrate the change in shape.

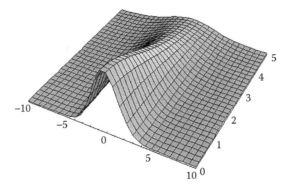

FIGURE 7.7   Intensity changes in a Gaussian beam ($\omega_0 = 3$ mm, $M^2 = 1$, $\lambda = 10.6$ μm) over 5 m.

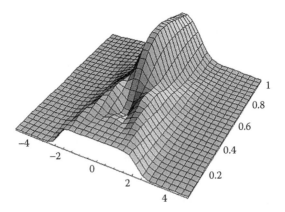

FIGURE 7.8   Intensity changes in a super-Gaussian beam ($\omega_0 = 2.39$ mm, $p = 5$, $\lambda = 10.6$ μm) over 1 m.

The fact that the Gaussian beam holds its shape for longer can be directly related to the fact that this shape has a low divergence for a given field size.

7.6.3 Propagation with Nonlinear Absorption

The important thing to remember about isotope enrichment is that the absorption of the field is nonlinearly dependent on intensity. We now propagate the same fields as in Section 3.2, through the same reactor distance, but take into account the absorption of the gas [14].

The influence of this absorption is determined as follows:

1. The beam, of an initial shape, is entered as a starting field into the Fresnel diffraction integral. An output field is calculated a short distance away.
2. The influence of absorption over this distance is approximated by applying the absorption influences only to the final field. When the distance is very small, this is accurate.
3. The new field is entered into the Fresnel diffraction integral, and the process is repeated.

Because the absorption is a function of intensity, both the absorption and the propagation influence the shape of the beam. The results for the Gaussian and super-Gaussian cases are shown in Figures 7.9 and 7.10, respectively.

The advantage of the super-Gaussian is clear—because of the near constant intensity over most of the beam area, the absorption is the same across this area, thus helping to maintain its shape. Furthermore, because the absorption is larger for high intensities, the formation of peaked shapes, which is characteristic of super-Gaussian propagation, is suppressed. The Gaussian beam clearly shows a flattening of the peak, due to the fact that the absorbance is more in this region than near the edges. At the end of the chamber, the beam is no longer Gaussian, and starts to approximate a low-order super-Gaussian beam. This is a surprising result, because it indicates that by introducing nonlinear

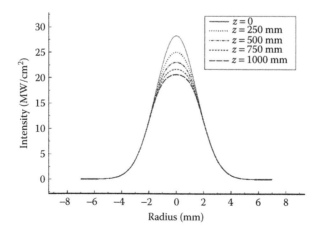

FIGURE 7.9 The Gaussian beam flattens as it propagates through the medium.

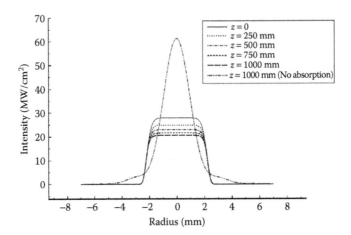

FIGURE 7.10 The super-Gaussian beam holds its shape for longer due to the suppression of intensity spikes.

absorption, the propagation problems mentioned earlier are reduced, as opposed to the expected result that further complexity would simply add to the problem.

The example just given considers a relatively strong absorbing medium, where we have used the absorption of the ^{12}C isotope. If we were instead to dissociate the ^{13}C isotope, the absorption would be correspondingly smaller. In this case, the Gaussian fields would maintain their shape for longer, while the super-Gaussian beam would not. *The choice of beam shape is therefore also influenced by the type of isotope that one chooses to separate.* This result has not been hinted at before in the literature. The choice of isotope, of course, also influences the wavelength that one must irradiate with. Figure 7.11 shows the absorption of ^{12}C in a Freon compound. The absorption is a function of wavelength, as is the propagation.

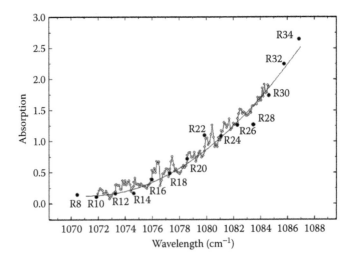

FIGURE 7.11 Multiple-photon absorption spectra of ^{12}C in a Freon compound.

In many cases, the complicating issue in propagation through reactors is the need to propagate several wavelengths together, with very careful spatial and temporal overlap. The major complication is often the beam shapers themselves, which can be wavelength dependent.

7.7 INFLUENCE OF PROPAGATION ON SELECTIVITY

So far we have shown how the intensity of the laser beam affects selectivity and yield. We have also shown how this intensity will change over the length of the reactor by considering propagation of two example fields through the absorbing medium. In the sections that follow, we combine this information to determine the enrichment one can expect for various beam shapes, when taking into account the entire interaction volume.

7.7.1 EXPERIMENT

To illustrate the influence of the wavelength and initial beam shape on isotope selectivity, consider the following carbon example:

Dissociation measurements were done on a closed loop system equipped with a 1 m irradiation length nozzle (the total volume of the loop was 16 liters), using a Gaussian beam. The loop was connected to a quadrupole mass spectrometer, allowing online measurement of the enrichment. Batch quantities of process gas were loaded and irradiated, and samples were taken at regular intervals. The laser wavelength was chosen to be in resonance with the ^{13}C isotope. The irradiation therefore was directed at stripping the ^{13}C content from the feed gas. Recall earlier (Equation 7.2) that after dissociation, the ^{13}C isotope is contained in the solid C_2F_4, so *stripping* this from the feed gas simply implies removing the solid product from the flowing gas. The higher the fluence of the laser beam throughout the reactor volume, the quicker the enrichment target of ^{12}C concentration is reached (in this experiment, the target was 99.9%). However, recall that higher fluences resulted in lower selectivity. Once the ^{12}C enrichment of 99.9% was reached, the average bulk ^{13}C enrichment of the dissociated product varied from 20%–60% depending on the specific fluence and wavelength used. Figure 7.12 a shows two different ^{12}C enrichments obtained on the dissociation product by changing the laser fluence, but keeping the wavelength constant. The plot shows how the total enrichment increases with the number of laser pulses used. In an industrial plant, as few pulses as possible would be optimal. Figure 7.12 b shows a plot of absorption peaks obtained by keeping the fluence constant, but varying the laser wavelength from being in resonance with ^{12}C, to between ^{12}C and ^{13}C resonances, to in resonance with ^{13}C. Single-wavelength ^{13}C enrichment of more than 90% has been measured, albeit at low yield.

7.7.2 NUMERICAL EXPERIMENT

To consider a slightly different case, we take a 6.5 mm size beam, with 1 J of energy and a pulse width of 80 ns. The beam interacts with the medium, absorbing ^{12}C. Two cross sections are shown for the Gaussian beam, and two for the super-Gaussian

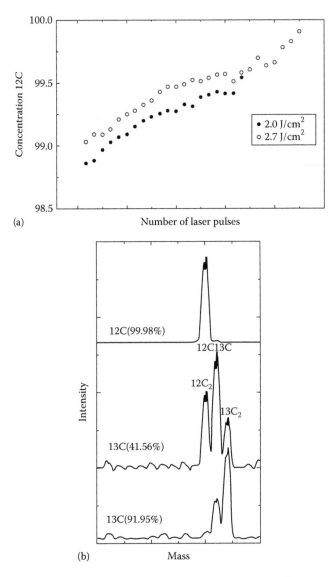

FIGURE 7.12 (a) Enrichment of ^{12}C for two laser fluences, as measured over a period of multiple pulses. The pulse count required for a given enrichment is classified, and so is not shown on the graph and (b) Enrichment of carbon when the laser wavelength is in resonance with ^{12}C (top panel), between resonances (middle panel) and in resonance with ^{13}C (bottom panel). The amplitude of the peaks is a direct indication of concentration of the products formed.

beam. Figure 7.13 shows cross sections that are taken along the propagation axis, in the centre of the beam ($r = 0$ for all z). This monitors the enrichment due to the central part of the beam, which in the case of a Gaussian beam is the peak value of the intensity. Figure 7.14 shows the cross sections in the middle of the reactor, across the beam area ($z/L = 0.5$, for all r).

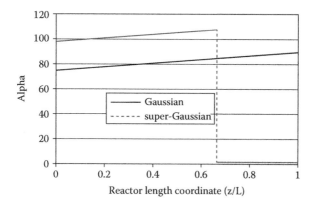

FIGURE 7.13 Alpha along the axis for both a Gaussian and a super-Gaussian beam. The reactor length is normalized to one.

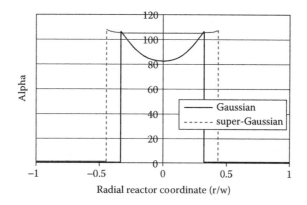

FIGURE 7.14 Alpha across the reactor area, in the centre of the reactor length, for both a Gaussian and a super-Gaussian beam. The reactor radial coordinate is plotted as a normalized value, that is, multiples of the beam size.

In the super-Gaussian case, the fluence along the propagation axis remains nearly constant, as the nonlinear absorption *holds* the shape of the beam. At some point, the intensity of the field is below the threshold, and so no further enrichment takes place. Because the peak intensity is the same as the average, when the intensity falls below the threshold, no further enrichment takes place across the entire field. In contrast, the Gaussian beam shows a steadily increasing selectivity in its central peak. This is because as the absorption decreases, so the alpha increases (recall that lower intensity or fluence gives a better enrichment). Because the peak value is twice the average, at certain stages during the propagation, parts of the beam will not be used.

Figure 7.14 shows the alpha across the beam area in the center of the reactor. Here, the entire super-Gaussian beam area is used, and the alpha is nearly constant across the beam. The Gaussian case is very different: The central peak intensity of the Gaussian beam results in a lower alpha. As the intensity drops toward the edges of the Gaussian beam, so the alpha increases, until the threshold is reached. Any part of the beam outside this area will not contribute to the enrichment. Thus the area of the reactor used in the Gaussian beam is smaller than that of the super-Gaussian beam.

The influence of the two beam shapes can be summarized as follows: the Gaussian beam results in some of the beam being used all of the time, whereas the super-Gaussian beam results in all of the beam being used some of the time. After adding the contributions across the entire reactor, the Gaussian beam results in an average enrichment factor of $\alpha = 25.3$, whereas the super-Gaussian beam results in an average value of $\alpha = 52.5$. Usually this process would be repeated in multiple stages, with each stage improving the enrichment by this factor. In multiple-stage systems, the final enrichment factor is the product of all the previous enrichment factors. Thus, with an increase by a factor of 2 in the alpha of each stage, after n stages, the enrichment will be higher by approximately a factor of 2^n. This is why the choice of beam shape is so important in isotope separation processes. The number of stages that are required for the same final product purity will be much lower using a super-Gaussian beam than using a Gaussian beam, resulting in huge cost savings.

7.8   BEAM SHAPING CONSIDERATIONS

Traditionally, beam shaping for isotope separation has been achieved with Gaussian beams, using multiple pass cells of the Herriot design [15–17]. This technique has proven inefficient in maintaining a beam with intensity above the dissociation threshold. In recent years there has been much development in both resonator beam shaping [13,18–26] and external cavity beam shaping [27–30] for the creation of flat-top beams. For a detailed discussion, the reader is referred to the references given, or to more general texts on the subject [31].

In making the intracavity versus external shaping choice, careful consideration must be given to the complete beam delivery system. Factors that influence the choice will include the following: the total delivery distance to the irradiation chamber; whether one or multiple wavelengths are used; whether or not amplification systems are used to increase the laser energy; and finally the energy stability of the output beam from the resonator. For example, if a super-Gaussian beam is shaped externally, then the shaping elements can be placed very close to the irradiation chamber. This will make the propagation to the irradiation chamber very easy (Gaussian beam propagation up to the shaping elements), but will be at the expense of beam energy, due to losses on the element, and lower extraction from the resonator and amplifiers. The opposite is true for intracavity shaping–the propagation will be more complex, consisting of relay imaging elements, but the energy extraction from the resonator and amplifiers will be better. As pointed out earlier, the reactor cross section may also have a specific shape due to gas flow considerations. Thus the laser beam area (circular, square, rectangular, etc.) cannot be chosen arbitrarily. To illustrate a typical reactor geometry, consider Figure 7.15.

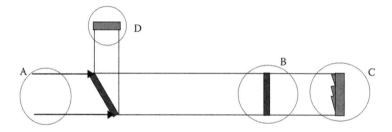

FIGURE 7.15 Example of a reactor model.

In Figure 7.15, the elements A, B, C, and D are as follows:

A. An incoming beam, as formed by one of the *beam shaping* examples. It is assumed to have a super-Gaussian intensity profile, and constant phase. The polarization of the beam is vertical.
B. A quarter wave plate, to induce a λ/4 change in electric field vector. After the second pass, in the return direction, the polarization is changed from vertical to horizontal.
C. Phase conjugate mirror. This diffractive optic takes the phase conjugate of the super-Gaussian field at this point, so that the *reflected* beam de-ripples itself during the return propagation.
D. A flat mirror, which returns the beam. Because the polarization is horizontal, the beam is reflected off the thin film polarizer window immediately before this mirror. The beam at point D is an exact duplicate of the beam at point A.

In the geometry shown in Figure 7.15, a super-Gaussian beam enters the reactor at point A. The polarization allows the beam to pass through the thin film polarizer (TFP) without reflection loss. At point C, diffraction has resulted in intensity variations across the beam, as is evident from the slightly rippled effect seen at some points on the intensity profile of Figure 7.16. In order to minimize this during the rest of the propagation, the beam is reflected off a phase conjugate mirror. After passing through the quarter wave plate (B) for a second time, the polarization ensures that the beam is now reflected off the TFP. The distances are chosen so that after reaching point D, in the absence of absorption, the beam is identical to the starting beam at point A. The return propagation to C is then the same as the initial propagation from A to C. Again the beam passes through the quarter wave plate twice, so that on the final path it is passed back out of the reactor.

The total distance covered is then

$$2 \times (A \rightarrow C) + 2 \times (D \rightarrow C), \text{ but note that } A \rightarrow C = D \rightarrow C$$

If the reactor length is 1 m, the beam will traverse a distance of 4 m inside the reactor. Figure 7.16 shows the propagation of a super-Gaussian beam through this system. It is important to propagate over as long a distance as possible, since the medium is

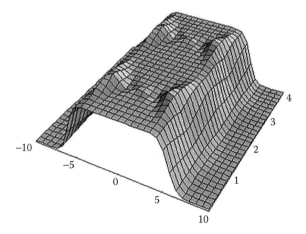

FIGURE 7.16 The intensity plot of a super-Gaussian beam while propagating through the reactor. The total propagation distance is 4 m. At each point during the propagation, the intensity cross section is taken. The reactor has a radius of 10 mm, so cross sections are taken from $r = -10$ to 10.

in general optically thin. If the path length inside the reactor was short, only a small fraction of the total beam energy would be used, and the rest wasted. The photon utilization in this case would be very poor, making the process uneconomical.

It is also obvious from this discussion that the chosen beam shape, and the method to shape it, both have a direct influence on how efficiently the reactor works.

7.9   OUTLOOK AND PROGRESS

In the past decade there has been renewed interest in this field due to advances in beam shaping, laser development, and the economics of certain isotopes [32]. Lithium isotopes have been enriched using a two-step photoionization process with a dye laser [33–35], Gaussian beams of dual wavelength have been used for Barium isotope enrichment [36], all using atomic vapor as the starting phase of the material for enrichment [37]. A major advance in this field has been at the isotope separator on line device (ISOLDE) facility [38]. In this facility, state-of-the-art lasers are used to isotopically enrich a wide variety of elements. The emphasis has been on spatial mode purity and temporal shaping to achieve better enrichment factors. A more recent development has been the approach of coherent-control by optical pulse shaping, usually at time scales faster than the decoherence time of the state under control. This typically makes use of shaped femtosecond pulses [39]. There has also been renewed interest in molecular isotope separation, including carbon [40] and uranium [41], where in the case of the latter a full production plant is presently underway at the General Electric Nuclear Fuel Fabrication Facility (North Carolina, the United States).

Recent advances in laser beam shaping tools are likely to impact on this field. First, the ability to extract maximum energy from the laser cavity and deliver it as a Gaussian mode makes downstream conversion of this mode possible [42].

Second, the multiple wavelengths required for most isotope separation processes can now be shaped with the same element with small perturbations in the size and location of the shaped light, but not in the intensity profile [43]. When these are combined with the new field of temporally shaping ultrafast light, new opportunities arise for highly efficient laser isotope separation.

7.10 CONCLUSIONS

For stable isotope enrichment, good quality beams that hold their shape over an extended distance are required, to ensure effective photon utilization. This is the fundamental difference between beam shaping for isotope separation, and beam shaping for many other applications—in most applications the shape of the beam is important in *one plane only*. For example, in some materials-processing applications, the beam shape at the material surface is very important, but the shape during propagation is not. In isotope separation, however, the beam shape over an extended distance is of importance. Furthermore, propagation through the irradiation chamber shows that the beam shape will change due to nonlinear absorption effects, as well as due to diffraction. In some cases the two work together to hold the beam over longer distances. Because of the intensity dependence of many of the separation parameters, the shape of the beam has a very large influence on the economy of the process, influencing both selectivity and yield as well as overall efficiency.

Today, with the advances in many beam shaping techniques, it is possible to efficiently generate laser beams with intensity profiles of any shape and size. It is this that gives renewed hope to isotope separation by laser-based systems.

ACKNOWLEDGMENTS

I thank posthumously Dr. Lourens Botha for his contribution to this chapter. He was always an enthusiastic supporter of laser isotope enrichment and would be delighted to see it realized one day. I also acknowledge useful data provided by my erstwhile colleagues at Klydon (Pty) Ltd., where isotope separation research continues.

REFERENCES

1. Bagratashvili, V. N., Letokhov, V. S., and Makarov, A. A., Ryabov, E. A., *Multiple Photon Infrared Laser Photophysics and Photochemistry*, Harwood Academic Publishers, Switzerland, 1984.
2. Jensen, R. J., Judd, O. P., and Sullivan, J. A., Separating isotopes with lasers, *Los Alamos Sci.*, 3(1), 2–33, 1982.
3. Letokhov, V. S., *Laser Spectroscopy of Highly Vibrationally Excited Molecules*, Adam Hilger, Bristol, 1989.
4. Shen, Y. R., *The Principles of Non-linear Optics*, John Wiley & Son, New York, 1984.
5. Loudon, R., *The Quantum Theory of Light*, 2nd edn., Oxford Science Publications, Oxford, 1983.
6. Shore, B. W., *The Theory of Coherent Atomic Excitation*, John Wiley & Sons, New York, 1989.
7. Suzuki, M. et al., Selective excitation of branched vibrational ladder in uranium hexa-fluoride laser isotope separation, *J. Nucl. Sci. Technology*, 31, 293, 1994.

8. Larsen, D., Frequency dependence of the dissociation of polyatomic molecules by radiation, *Opt. Comm.*, 19(3), 404, 1976.

9. Bloembergen, N., Comments on the dissociation of polyatomic molecules by intense 10.6 μm radiation, *Opt. Comm.*, 15(3), 416–418, 1975.

10. Botha, L. R. et al., Continuously tuneable CO_2 MOPA chain for a molecular laser isotope separation pilot plant, *Opt. Eng.*, 33(9), 2687–2691, 1994.

11. Frost, W., *Theory of Unimolecular Reactions*, Academic Press, New York, 1973.

12. Forbes, A., *Laser Beam Propagation: Generation and Propagation of Customised Light*, CRC Press Taylor and Francis Group, Boca Raton, FL, 2014.

13. Litvin, I. A., and Forbes, A., Gaussian mode selection with intracavity diffractive optics, *Opt. Lett.* 34, 2991–2993, 2009.

14. Forbes, A., Laser beam propagation in a non-linearly absorbing media, *Proc. SPIE 6290*, 629003, 2006.

15. Fub, W. et al., IR Multiphoton absorption and isotopically selective dissociation of $CHClF_2$ in a Herriot multipass cell, *Z. Phys.*, D29, 291–298, 1994.

16. Kojima, H., Uchida, K., Takagi, Y., A Waveguide reactor for IR multiphoton dissociation and its application to ^{13}C laser isotope separation, *Appl. Phys.*, B41, 43–48, 1986.

17. Gothel, J., Ivanenko, M., Hering, P., Fub, W., and Kompa, K. L., Macroscopic enrichment of ^{12}C by a high-power mechanically Q-switched CO_2 laser, *Appl. Phys.*, B62, 329–332, 1996.

18. Belanger, P. A., and Pare, C., Optical resonators using graded-phase mirrors, *Opt. Lett.*, 16, 1057–1059, 1991.

19. Pare, C., and Belanger, P. A., Custom laser resonators using graded-phase mirrors, *IEEE J. Quant. Electron.*, 28, 355–362, 1992.

20. Naidoo, D. et al., Transverse mode selection in a monolithic microchip laser, *Opt. Commun.*, 284, 5475–5479, 2011.

21. Ngcobo, S., Litvin, I. A., Burger, L., and Forbes, A., A digital laser for on-demand laser modes, *Nat. Commun.*, 4, 2289, 2013.

22. Ngcobo, S., Ait-Ameur, K., Litvin, I. A., Hasnaoui, A., and Forbes, A., Tuneable Gaussian to flat-top resonator by amplitude beam shaping, *Opt. Express*, 21, 21113–21118, 2013.

23. Hocke, R., CO_2 Slab Laser with a Variable Reflectivity Grating (VRG) as a Lineselective Element, *Proc. SPIE*, 4184, 427–430, 2000.

24. Leger, J. R., Chen, D., and Wang, Z., Diffractive optical element for mode shaping of a Nd:YAG laser, *Opt. Lett.*, 19, 108–110, 1994.

25. Zeitner, U. D., Wyrowski, F., and Zellmer, H., External design freedom for optimization of resonator originated beam shaping, *IEEE J. Quant. Electron.*, 36, 1105–1109, 2000.

26. Zeitner, U. D., Aagedal, H., and Wyrowski, F., Comparison of resonator-originated and external beam shaping, *Appl. Opt.*, 38, 980–986, 1999.

27. Hoffnagle, J. A., and Jefferson, C. M., Design and performance of a refractive optical system that converts a Gaussian to a flattop beam, *Appl. Opt.*, 39, 5488–5499, 2000.

28. Dickey, F. M., and Holswade, S. C., Gaussian laser beam profile shaping, *Opt. Eng.*, 35, 3285–3295, 1996.

29. Arrizon, V., Optimum on-axis computer generated holograms encoded into low-resolution phase-modulation devices, *Opt. Lett.*, 28, 2521–2523, 2003.

30. Hendriks, A., Naidoo, D., Roux, F. S., Lopez-Mariscal, C., and Forbes, A., The generation of flat-top beams by complex amplitude modulation with a phase-only spatial light modulator, *Proc. SPIE*, 8490, 849006, 2012.

31. Dickey, F. M., and Holswade, S. C., in *Gaussian Beam Shaping: Diffraction Theory and Design, Laser Beam Shaping—Theory and Techniques*, ed. Dickey, F. M., pp. 151–194, Taylor and Francis Group, Boca Raton, FL, 2014.

32. Forbes, A., Strydom, H. J., Botha, L. R., and Ronander, E., Beam delivery for stable isotope separation, *Proc. SPIE*, 4770, 13–27, 2002.
33. Saleem, M., Hussian, S., Rafiq, M., and Baig, M. A., Laser isotope separation of lithium by two-step photonionization, *J. Appl. Phys.*, 100, 053111, 2006.
34. Saleem, M., Hussian, S., Zia, M. A., and Baig, M. A., An efficient pathway for Li6 isotope enrichment, *Appl. Phys. B*, 87, 723–726, 2007.
35. Olivares, I. E., Selective laser excitation in litium, *Opt. J.*, 1, 7–12, 2007.
36. Jana, B., Majumder, A., Kathar, P. T., Das, A. K., and Mago, V. K., Ionization yield of two-step photoionization process in an optically thick atomic medium of barium, *Appl. Phys. B*, 102, 841–849, 2011.
37. Bokhan, P. A. et al., *Laser Isotope Separation in Atomic Vapor*, Wliey-VCH, Wienheim, Germany, 2006.
38. Fedosseev, V. N. et al., Upgrade of the resonance ionization laser ion source at ISOLDE on-line isotope separation facility: new lasers and new ion beams, *Rev. Sci. Instrum.*, 83, 02A903, 2012.
39. Goswami, D., Optical pulse shaping approaches to coherent control, *Phys. Rep.*, 374, 385–481, 2003.
40. Dong, M., Mao, X., Gonzalez, J. J., Lu, J., and Russo, R. E., Carbon isotope separation and molecular formation in laser-induced plasmas by laser ablation molecular isotope spectrometry, *Anal. Chem.*, 85, 2899–2906, 2013.
41. Weinberger, S., Laser plant offers cheap way to make nuclear fuel, *Nature*, 487, 16, 2012.
42. Litvin, I. A., and Forbes, A., Resonator with intracavity transformation of a Gaussian into a top-hat beam, patent 13/266,078, 2011.
43. Forbes, A., Dickey, F., DeGama, M. and du Plessis, A., Wavelength tunable laser beam shaping, *Opt. Lett.*, 37, 49–51, 2012.

8 Applications of Diffractive Optics Elements in Optical Trapping

Ulises Ruiz, Victor Arrizón and Rubén Ramos-García

CONTENTS

8.1 INTRODUCTION

Optical tweezers have become a powerful tool for contactless manipulation of matter at the nano and microscope scale since the pioneering work of Ashkin and coworkers.[1,2] Experimentally, it is straightforward to implement an optical tweezers setup (in its most simple version): just need to focus a laser beam using a high aperture lens at whose focus a particle can be trapped. Conceptually, the physical mechanism of trapping involves subtle properties of light to exert forces and torques on matter: transfer of linear and angular momentum. The physical mechanism of optical tweezers and applications has been nicely reviewed in many works and for such reason they will not be discussed here. Readers interested in the subject may check out the following references.[3-6] Optical tweezers have found many applications like particle sorting,[7,8] microrheology,[9,10] study of colloidal hydrodynamics,[11,12] compliance of bacterial tail,[13] stretching of single deoxyribonucleic acid (DNA) molecules,[14,15] among others. Moreover, optical tweezers can be used as quantitative tools to exert and measure forces on microscopic and nanoscopic particles such as those exerted by single motor proteins.[16,17]

Certain applications may require the use of two or more beams to study, for example, the viscoelastic properties of red blood cells.[18,19] This can be achieved by just adding more beams, but this may complicate the setup and increase its cost. One elegant solution is the time-sharing scheme trapping, where a single beam is scanned along predesigned positions, producing in this way a discrete array of spots[20,21] or even structured beams.[22] Although time-sharing trapping can be challenging to implement, it has already been commercialized by Elliot Scientific.[23] Since time-sharing involves no beam shaping, it will not be discussed further but readers interested in the subject are referred to.[24–26]

A more interesting approach for multiple trapping has emerged with the use of diffractive optical elements (DOEs). DOEs significantly expanded the versatility and applications of optical tweezers making possible parallel manipulation of particles and arrangements in complex spatial patterns. For example, Dufresne et al. proved that DOE lithographically fabricated in glass could be used to produce an array of up to 400 trapping spots.[27] Although in this case the trapping sites are static, a fast moving laser beam through the DOE can produce limited dynamic holograms. One serious disadvantage of DOE etched in glass, besides its permanent nature, is that it requires the use of photolithographic techniques that may be demanding and costly.

Some applications may require real-time reconfiguration and optimization of DOEs. For that purpose, computer-controlled spatial light modulators (SLMs) fulfill the task very well. An SLM is a device capable of modifying the amplitude, phase, or polarization of an optical wavefront in response to information-bearing control signals. SLMs are very versatile devices since complex fields can be encoded, but at the same time, they may correct for aberrations acquired in the optical path.[28–31]

8.2 OPTICAL SETUP FOR OPTICAL TWEEZERS

Before discussing how to implement DOEs for optical tweezers applications, let us analyze how they are implemented. The simplest optical tweezers system consists in a high numerical aperture (NA) microscope objective that focuses an incoming laser beam to a tight spot in order to optimize the gradient force.[32] Typical numerical apertures for microscope objective are 1.25 oil immersion (or 1.2 for immersion water). A recent study found that water immersion objective is more suitable for optical trapping due to reduced aberrations; however, they are much more expensive than oil immersion objectives.[33] If the beam is not well collimated, that is, its wavefront posses some curvature, it may focus before (for converging beam) or after (for diverging beam) the focal point as shown in Figure 8.1a. This means that if the beam is not well collimated, the trapping and image plane may not coincide. By means of DOEs implemented on SLMs, the wavefront can be fully controlled to continuously displace the particle along the propagation axis by encoding lenses.[34]

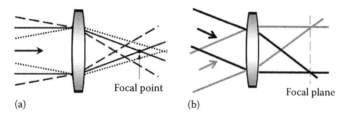

(a)   (b)

FIGURE 8.1 (a) A collimated beam (continuous line) focused at the focal plane but a converging (broken line) focused before the focal plane while a diverging beam (dotted line) focused after and (b) Collimated beam at some angle respect to the optical axis.

A highly desirable feature of the system is beam steering to manipulate particles on the focal plane of the lens. The real challenge of implementing beam steering in optical tweezers is to produce a collimated laser beam, which is always centered on the back aperture of the microscope objective despite angular deviations of the beam with respect to the optical axis (Figure 8.1b). A nice solution was provided by Dufresne et al.[27] using afocal telescope (see Figure 8.2). An afocal telescope can be created with a pair of lenses separated by a distance d equal to the sum of each lens focal length ($d = f_1 + f_2$). With a careful selection of lenses, the telescope does not alter significantly the divergence of a collimated beam but only changes the beam width by a factor given by the telescope magnification, $M = -f_2/f_1$. This allows filling the back pupil of the high aperture objective to optimize the gradient force and minimize losses.

The beam steering mirror is positioned at one conjugate plane and imaged onto the entrance aperture of the objective lens (the other conjugated plane). A change to the wavefront curvature of the trapping beam in the plane of the beam steering mirror would cause an axial shift of the trapping beam focus (see Figure 8.1a). On the other hand, an angular deviation in the beam at the mirror produces a lateral displacement of the focus at the focal plane as shown in Figure 8.2.

Figure 8.3 shows a schematic representation of a typical optical tweezers setup using a DOE, implemented in an LC phase SLM, which is illuminated by an expanded and collimated laser beam. The beam must be linearly polarized with the polarization along the SLMs' director vector. In this way, the phase modulation is effectively achieved. In general, the SLM could be a transmissive or reflective device. Here, we consider only a reflective SLM (Figure 8.3) to minimize losses. Typical size of

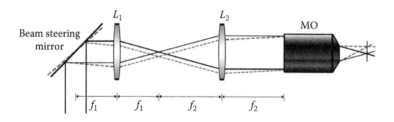

FIGURE 8.2 The beam steering mirror is placed at a conjugated plane of an afocal telescope and imaged into the back aperture of a microscope objective (MO).

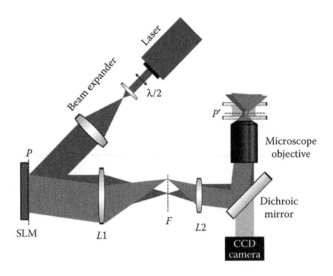

FIGURE 8.3   A typical OT setup using an SLM to generate DOE.

LC-SLM is ~8 × 8 mm², which means that the expanded beam must illuminate most of the LC active area to take full advantage of its spatial resolution. In addition, an expanded beam is required to avoid damage to the SLM. Typical intensity damage threshold lies on the range of 1 W/cm² but depends on the employed wavelength. For example, damage is more likely to occur in NIR part of the spectrum than in the visible range due to higher absorption by the ITO semitransparent electrodes. Reflective SLMs (see discussion below) can be operated at normal incidence, but this implies the use of a cube beam splitter producing unacceptable high losses (~75%). To avoid that, the incident beam makes a small angle (<10°) with the normal of the SLM surface. In order to obtain the spatial separation between the incident and reflected beam, a long propagation distance is required. This in turn sets limitations to the size and focal distance of the lenses for the afocal telescope. Typical focal lenses lie in the range of 40–75 cm and >2″ diameter lenses to collect as much light as possible. Fortunately, large focal lenses introduce little spherical aberrations and facilitate spatial filtering of diffraction orders introduced at the DOEs' Fourier plane. The SLM is placed at the input focal plane of lens L1, while the output focal plane corresponds to the spatial filter plane F, where the Fourier transform of the encoded hologram is generated and spatially filtered. The second lens of the afocal telescope is used to reduce the size of the beam, and finally the microscope objective is used to project the image obtained at the back focal plane into the trapping plane. Note that the SLM plane P and the objective focal plane P′ are conjugated to each other and that the image at plane P′ is the Fourier transform of plane P. Finally, a dichroic filter is used to direct the laser light into the microscope objective and to allow white light illumination to image the trapped particle into a CCD camera. There are, of course, many variations of this optical setup, but all rely on this basic configuration.

8.3 THE SPATIAL LIGHT MODULATORS (SLMs)

An SLM is a real-time reconfigurable device capable of modifying the amplitude, phase, or polarization of an optical wavefront as a function of two spatial dimensions and time in response to information-bearing control signals (either optical or electrical). SLMs can be classified in two general types according to their addressing mechanism: electrical and optical. Electrically controlled SLMs consist of a two-dimensional array of pixel matrix of individually addressed liquid crystal cells[35] or micromirrors.[36] In optically addressed SLMs, one light beam is used to change the reorientation of the liquid crystals and thus the phase or amplitude of a probe light beam.[37] Although this type of SLMs are quite attractive due to its simplified construction, they may require a spatially complex control beam that must be created with an SLMs or image forming systems, limiting the range of its applications.

Currently, two SLM–based technologies are the dominant: microelectromechanical system MEMS,[36] and LCD technology.[38] MEMS consist of electrically controlled micromirrors producing only binary amplitude modulation. In this work we will focus our attention to LCD technology because its ability to modulate amplitude, phase or polarization. In addition, the high resolution and minimal pixel size are valuable features of 2D LCDs. Ferroelectric liquid crystals SLMs provide very fast response time (microseconds) but are limited by its binary phase modulation. In particular, we will center our attention to nematic LC-based devices that provide 256 phase levels in a modulation range of 2π radians. Nematic LC SLMs are fabricated in two modes; transmission and reflection. Reflection-mode SLM are preferred over transmission mode because of thinner device and faster response time (operating up to 100 Hz), but still much slower than ferroelectric ones. In addition, its fabrication technology is quite mature such that high pixel integration is possible allowing high fill factor.

Liquid crystals are birefringent materials whose molecules can be oriented by an external electric field. The liquid crystal molecules may be oriented parallel aligned (PA), vertical aligned (VA), or twisted. In PA cells, alignment layers are rubbed in a common direction in both top and bottom substrates. The liquid crystals align along the groves, and due to long large ordering of liquid crystals, they align along the whole cell volume. In a twisted cell, the orientation of the molecules differs by typically 90 between the top and the bottom of the LC cell and the liquid crystal molecules self-arrange in a helix-like structure in between. In VA aligned cells, a surfactant forces the liquid crystals cells to align perpendicular to the substrate. Twisted nematic liquid crystals can be used for amplitude and phase modulation, while PA and VA may be used as phase only modulators.

When an electrical field is applied to a PA liquid crystal cell (or pixel in the case of SLMs), a torque is exerted on the liquid crystal molecules and tilt an angle θ rotating toward the applied electric field. The phase retardance $\Delta\varphi$ introduced by a cell pixel into a beam incident perpendicular to the substrate is given by[39]

$$\Delta\varphi = \frac{2\pi}{\lambda}\int_0^d [n_e(\theta) - n_0]dz \qquad (8.1)$$

where $n_e(\theta)$ is given by

$$\frac{1}{n_e^2(\theta)} = \frac{\cos^2(\theta)}{n_e^2} + \frac{\sin^2(\theta)}{n_o^2} \tag{8.2}$$

The tilt angle is dependent on the applied voltage V and position along the propagation distance z, that is, $\theta(z, V)$. Liquid crystals molecules in contact with the surface (or very close to it) are hard to align and only those molecules far from the surface respond linearly to the applied voltage. Ideally, a linear relationship between phase retardance and voltage should exist, but in practice, this relationship is far from linear. SLMs makers usually provide software to operate them in a quasi-linear regime, still reaching 2π phase modulation but in a limited range of gray levels.

Typically, phase-only SLM operates in reflection mode in order to reduce the thickness of the cell, which in turn reduces the response time to a few milliseconds. Amplitude modulation, usually obtained with twisted nematic liquid crystal (TNLC), is more complicated to operate than phase modulation mode.[40] Indeed, true amplitude modulation is hard to achieve since phase and amplitude are coupled in TNLC SLMs. In addition, they cannot be used in applications where high power is required like in optical tweezers.

Use of LC SLMs for optical tweezing was first proposed by Hayasaki et al.[41] followed by a number of publications of Dufresne and Grier[27] and Reicherter et al.[34] Since then, SLM has been widespread used for many applications in different configurations.

8.4   DOEs DISPLAYED IN SLMs

The vast majority of DOEs for optical trapping applications have been optimized to produce discrete arrays of spots. This was an obvious extension to conventional single trap, allowing arrays of lines, circles, squares, etc. In addition, DOEs displayed in SLMs also added dynamic and custom-designed 2D structures. Further optimization methods lead to a dynamic 3D array of trapped particles in complex spatial structures.

In order to implement DOEs in SLMs, let us assume that an optical field can be described as $s_h(x, y) = a(x, y)\exp[i\phi(x, y)]$, where $a(x, y)$ is the field amplitude and $\phi(x, y)$ is its phase. If the beam is directed toward a phase-only SLM, then the output field can be written as $s_h = a\exp(i\phi + i\phi_h)$, where ϕ_h is the phase delay introduced by the SLM. For simplicity, the spatial coordinates have been omitted. The main task in all holographic methods consist in finding the adequate phase that, after far-field propagation, will reproduce the desired intensity pattern. This task may be simplified if a lens is used, so the field just after the SLM and far-field are related by a Fourier transform:

$$U_i(x_i, y_i) = \Im\left[s_h(x_h, y_h)\exp(i\phi + i\phi_h)\right] \tag{8.3}$$

where subindex i and h indicate coordinates in the image and hologram plane, respectively (see Figure 8.4). This equation has important consequences for holographic methods. For example, if the hologram phase is known, then the field at the

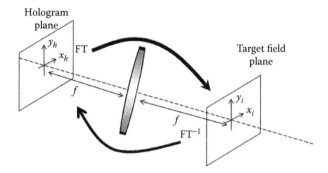

FIGURE 8.4   The relationship between the field at the hologram plane and image plane is given by Fourier and inverse transformation.

image plane is easily calculated using Equation (8.3). However, it is common that the intensity distribution $I_i = U_i U_i^*$ at the image plane is known, then the task is to find the phase distribution at the hologram plane that produce this intensity distribution. Algorithms such as Gerchberg–Saxton,[42] Additive-Adaptive methods,[27] direct binary search,[43,44] or its variations are commonly used. One drawback of these methods is that the obtained intensity distribution is valid only at the image plane and the phase at the image plane is arbitrary, which means that trapping in 3D is not possible and aberrations may be present. Fortunately, the Gerchberg–Saxton algorithm was modified to correct the phase and thus produce 3D trapping.[45–49] These Fourier transform-based methods can produce not only arrays of spots but also any arbitrary intensity distribution with pretty good results. There are other algorithms of optimization of DOEs based on interactive control of the pixel's phase on the hologram plane to maximize the intensity distribution at the image plane but also control its shape and uniformity.[50] These Fourier methods will not be discussed here since there is plenty of information in the literature as just pointed out above.

In order to understand how DOE elements form arrays of spots implemented in SLMs, one shall remember that spatial displacements (Δx, Δy) of the beam, along orthogonal transverse axes in the image plane, can be achieved by encoding a blazed grating, or equivalently a prism, with the linear phase:

$$\phi_{\text{prism}}(x, y) = \frac{2\pi}{\lambda f}\left(\Delta x x_h + \Delta y y_h\right) \tag{8.4}$$

where:
   λ is the light wavelength
   f is the focal length of the transforming lens

On the other hand, for displacements along the propagation axis, one can employ a Fresnel lens with the quadratic phase:

$$\phi_{\text{lens}}(x, y) = \frac{\pi}{\lambda f}\left(x_h^2 + y_h^2\right) \tag{8.5}$$

With these basic elements, a beam can be displaced sideways and also in and out of the focal plane if a hologram with phase modulation of the type

$$\phi_h(x,y) = \left[\phi_{\text{prism}}(x,y) + \phi_{\text{lens}}(x,y)\right] \bmod 2\pi \tag{8.6}$$

is encoded. An array of spots can be easily achieved by superposition of n holograms displayed at positions (x_n, y_n). The phase modulation of the composite DOE can be expressed as $\sum_n \phi_h(x - x_n, y - y_n)$ if the different phases are displayed in nonoverlapping regions of the SLM. The Fourier transform of the hologram is computationally efficient such that real-time modification can be achieved. In addition, each spot can be individually controlled or even corrected for aberrations.[28,31,45,50–52]

Besides discrete arrays of traps, the emergence of continuous potential optical landscapes is of great interest in optical trapping and manipulation. Spatially structured beams allow trapping not only in 2D but also in 3D. In addition, some structured beams may carry orbital angular momentum that can be transferred to matter and induce rotation of particles. Optical fields such as Bessel,[53,54] Laguerre,[55,56] Mathieu,[57] Ince,[58] Hermite,[59] and Airy[60] beams are of interest for trapping particles in complex spatial distributions. A general technique for the generation of arbitrary complex fields, such as the just mentioned beams as particular cases, is discussed in the next section.

8.5   SYNTHETIC PHASE HOLOGRAMS (SPHs)

As discussed above, an important task in optical manipulation is the generation of high-quality and high-efficiency complex fields. There is plenty of information for the generation of arrays of trapping sites but not much information on the generation of complex structured beams. Thus, here we intend to fill that gap by doing a thorough discussion for the generation, performance, and quality of three types of SPHs, already reported,[58,59] and two novel approaches.

Let us consider an arbitrary complex optical field whose amplitude and phase modulations are independently specified. This complex field can be expressed as

$$s(x,y) = a(x,y)\exp[i\phi(x,y)] \tag{8.7}$$

where the amplitude $a(x,y)$ and the phase $\phi(x,y)$ take values in the intervals $[0,1]$ and $[-\pi, \pi]$, respectively. The complex amplitude values of the function $s(x,y)$ belong to the set of complex numbers with modulus equal to or smaller than one, which is denoted as Ω_s. The aim is to encode both the amplitude and the phase of a complex field by means of a properly designed SPH. In general, an SPH that encodes the arbitrary complex modulation $s(x,y)$ has a constrained complex transmittance, on a subset of Ω_s, formed by the complex points of unity modulus. The transmittance of the SPH, expressed explicitly as a function dependent on both the amplitude and the phase of the encoded field, is given by

$$h(x,y) = \exp\left(i\psi(a,\phi)\right) \tag{8.8}$$

where $\psi(a,\phi)$ is the phase modulation. The explicit dependence of the amplitude a and phase ϕ on the spatial coordinates (x, y) has been omitted in Equation 8.8. The goal is to establish phase functions with the form of Equation 8.8 that provide the appropriate encoding of the complex field $s(x, y)$. A fruitful method to determine appropriate forms of the hologram phase modulation $\psi(a,\phi)$ is based on the representation of $h(x, y)$ in Fourier series in the domain of ϕ. Developing this Fourier series, the SPH transmittance can be expressed as

$$h(x, y) = \sum_{n=-\infty}^{\infty} c_n^a \exp(in\phi) \tag{8.9}$$

with

$$c_n^a = \frac{1}{2\pi} \int_{-\pi}^{\pi} \exp\left[i\psi(a,\phi)\right] \exp(-in\phi) d\phi \tag{8.10}$$

In Equation 8.10, it is noted that after integration in the variable ϕ, the resulting coefficients c_n^a remain explicitly dependent on the amplitude a and, therefore, implicitly dependent on the coordinates (x, y). Since the integrand in Equation 8.10 is a phase function, it is straightforward to obtain the inequality

$$\left|c_n^a\right| \leq 1 \tag{8.11}$$

The signal $s(x, y)$ is recovered from the first-order term in the series of Equation 8.11 if the identity

$$c_1^a = Aa \tag{8.12}$$

is fulfilled for a positive constant A. This identity is referred to as the signal encoding condition. Considering the restriction in Equation 8.11 and recalling that the amplitude $a(x, y)$ has been normalized, we note that the constant A must be smaller than or equal to one. This result provides a limit to the efficiency of the SPHs that belong to this class. Sufficient and necessary conditions to fulfill Equation 8.12 and recover the signal in the first order of the hologram are given by the following relations[61]:

$$\int_{-\pi}^{\pi} \sin\left[\psi(a,\phi) - \phi\right] d\phi = 0 \tag{8.13}$$

$$\int_{-\pi}^{\pi} \cos\left[\psi(a,\phi) - \phi\right] d\phi = 2\pi Aa \tag{8.14}$$

Equations 8.13 and 8.14 provide a useful basis for the determination of appropriate SPHs. The phase functions $\psi(a,\phi)$ that obey these equations define a specific class of phase SPHs. It is noted that the maximum of the integral in Equation 8.14 is 2π. Thus, the maximum possible value of the constant A in the encoding condition (Equation 8.12) is one. This result provides a limit to the efficiency of the SPHs that belong to this class. In the remaining discussion, we focus our attention on functions $\psi(a,\phi)$ with odd symmetry in the variable ϕ; such functions ensures the fulfillment of Equation 8.13.

We discuss three types of SPHs, which fulfill Equations 8.13 and 8.14. The first one is called, for the sake of simplicity, *type 1 SPH*.[62] Its phase modulation can be expressed as

$$\psi(a,\phi) = f(a)\phi \qquad (8.15)$$

where the factor $f(a)$ remains undetermined for the moment. The *n*th-order Fourier series coefficient for this SPH computed with Equation 8.8 is

$$c_n^a = \mathrm{sinc}(n - f(a)) \qquad (8.16)$$

where $\mathrm{sinc}(\eta) = \sin(\pi\eta)/(\pi\eta)$. Then, the encoding condition if $A = 1$ reduces to

$$\mathrm{sinc}(1 - f(a)) = a \qquad (8.17)$$

For the complete definition of the SPH, the function $f(a)$ is numerically inverted as shown in Figure 8.5 (see dotted line).

The second case corresponds to the *type 2 SPH*, whose phase modulation is given by

$$\psi(a,\phi) = \phi + f(a)\sin(\phi) \qquad (8.18)$$

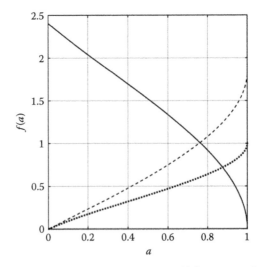

FIGURE 8.5 Numerical computation of the function $f(a)$ versus a, for the three types of SPHs. Type 1 (dotted line), type 2 (dashed line), and type 3 (solid line).

The phase SPH transmittance in this case is $h(x, y) = \exp(i\phi)\exp\left(if(a)\sin(\phi)\right)$. The Fourier series in the variable ϕ for this phase transmittance can be directly found by using the Jacobi–Anger identity.[63] According to this identity, the second phase factor in the SPH transmittance $h(x, y)$ is expressed by

$$\exp\left(if(a)\sin(\phi)\right) = \sum_{m=-\infty}^{\infty} J_m\left[f(a)\right]\exp(im\phi) \tag{8.19}$$

where J_m denotes the integer-order Bessel functions. It is readily proved that the SPH transmittance $h(x, y)$ is expressed by the Fourier series defined in Equation 8.9 with coefficients

$$c_n^a = J_{n-1}\left[f(a)\right] \tag{8.20}$$

Then, the encoding condition, valid with $A = 1$, is obtained from the relation

$$J_0\left[f(a)\right] = a \tag{8.21}$$

Equation 8.21 can be fulfilled for every value of a in the interval $[0,1]$ by taking the appropriate value of $f(a)$ in the domain $[0, x_0]$, where $x_0 \cong 2.4048$ is the first positive root of the Bessel function $J_0(x)$. The resulting function $f(a)$, numerically generated from Equation 8.21, is plotted in Figure 8.5 (solid line).

The *type 3 SPH* is specified by the phase modulation

$$\psi(a, \phi) = f(a)\sin(\phi) \tag{8.22}$$

It is noticed that this SPH phase modulation is a simplified version of the function in Equation 8.18. However, this SPH presents special features that justify its detailed discussion. To obtain the Fourier series and corresponding coefficients for this SPH, we employ again the Jacobi–Anger identity [Equation 8.19]. The resulting nth-order coefficient in this Fourier series is

$$c_n^a = J_n\left[f(a)\right] \tag{8.23}$$

and the encoding condition is fulfilled if $f(a)$ is inverted from the relation

$$J_1\left[f(a)\right] = Aa \tag{8.24}$$

The maximum value of A for which Equation 8.24 can be fulfilled is $A \cong 0.5819$, which corresponds to the maximum value of the first-order Bessel function $J_1(x)$, which occurs at $x = x_1 \cong 1.84$.

The function $f(a)$ obtained by numerical inversion from Equation 8.24 adopts values in the interval $[0, x_1]$ [see Figure 8.5 (dashed line)]. It is interesting to note, considering the limit values of $f(a)$ and Equation 8.22, that this SPH can be implemented with phase modulation in a reduced domain $[-f_0\pi, f_0\pi]$ with $f_0 \cong 0.586$. The phase range of the required modulator in this case is $\Delta\phi = 2f_0\pi \cong 1.17\pi$.

A further reduction in this phase range can be attained by adopting a smaller value of A in Equation 8.24. An advantage of this reduced phase domain is that it can be easily obtained with conventional SLMs employing relatively long wavelengths, for example, in the near-infrared domain. This type of illumination is appropriate for manipulation of living cells with optical tweezers.[64–66]

8.5.1 Modification of the SPHs by a Phase Linear Carrier

We assume that the reconstruction of the encoded field is performed by spatial filtering in the hologram Fourier spectrum plane. The optical setup for reconstruction is schematically represented in Figure 8.6. It is assumed that the SPH is placed at the back focal plane of the first lens (L_1). The Fourier transform of the field transmitted by the SPH is formed at the spatial filter plane. In general, the SPH spectrum is formed by the signal term (as those discussed above) and nonsignal or high-order spectrum field contributions. However, for reconstruction of the encoded field with high signal-to-noise ratio (SNR), a minimal overlapping is desired between the signal and the high-order terms in the SPH spectrum. Under this condition, a spatial filter pupil can be employed to transmit only the light corresponding to the Fourier spectrum of the encoded field. The encoded field itself is generated by the second Fourier transforming lens (L_2) at the output plane of the setup.

Next, we discuss a modification of the SPH defined in Equation (8.8) that enables the isolation of signal from noise in the SPH Fourier spectrum domain. The Fourier spectrum of the encoded field $s(x, y)$ is denoted by $S(u, v)$, where (u, v) represent the spatial frequency coordinates associated with the spatial coordinates (x, y). If we assume that the Fourier spectrum $S(u, v)$ is centered on the Fourier plane axis $(u, v) = (0,0)$, then the spectra for the different terms in the hologram Fourier expansion

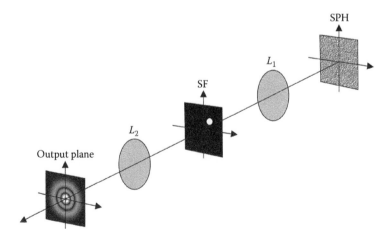

FIGURE 8.6 Four-f optical system for the generation of scalar complex fields by means of an SPH.

[Equation 8.9] are also centered on this axis. Thus, the encoded field cannot be recovered by spatial filtering from the SPH defined in Equation 8.8. To achieve the spatial isolation of the encoded field, the above definition of the SPH is modified by adding the carrier phase modulation $\phi_p(x,y) = 2\pi(u_0 x + v_0 y)$ with spatial frequencies (u_0, v_0) to the phase of the encoded field. The modified PSH transmittance $h_c(x,y) = \exp\left[i\psi(a,\phi+\phi_p)\right]$ can be expressed by the Fourier series

$$h_c(x,y) = \sum_{n=-\infty}^{\infty} h_n(x,y)\exp\left[i2\pi(nu_0 x + nv_0 y)\right] \qquad (8.25)$$

where $h_n(x,y) = c_n^a \exp(in\phi)$. The Fourier spectrum of this modified PSH is given by

$$H_c(u,v) = \sum_{n=-\infty}^{\infty} H_n(u - nu_0, v - nv_0) \qquad (8.26)$$

where $H_n(u,v)$ is the Fourier transform of $h_n(x,y)$. The structure of the SPH Fourier transform formed by laterally shifted copies of the Fourier spectra $H_n(u,v)$ allows the spatial isolation of the encoded field, whose Fourier spectrum appears as $H_1(u - u_0, v - v_0)$. The distribution of the SPH spectra terms $H_n(u - nu_0, v - nv_0)$ when $u_0 = v_0$ is schematically represented in Figure 8.7. The zeroth-order $H_0(u,v)$ appears at the center of this figure. The Fourier spectrum in Equation 8.26 corresponds to an SPH implemented with an element free of spatial quantization. Accurate implementation of SPHs without pixelation allows an efficient isolation of the encoded field from nonsignal diffraction orders of the SPH, enabling reconstruction with a

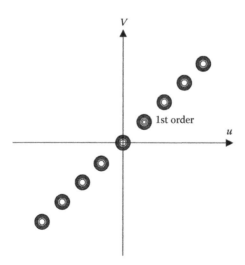

FIGURE 8.7 Distribution of the diffraction terms of the SPHs, by assuming the frequencies $u_0 = v_0$. The signal spectrum is on the first diffraction order, which allows spatial filtering and in principle high SNR.

high SNR. A requirement (common to all the SPH types) for obtaining an acceptable SNR is that at least one of the hologram carrier frequencies (u_0, v_0) be larger than the bandwidth of the encoded field $s(x, y)$, that is, the spatial extension of the signal spectrum. On the other hand, if an SPH is implemented with a low-resolution pixelated element, the reconstructed signal term can be significantly affected by high-order diffraction contributions.[67] In general, the noise level in the reconstructed field introduced by high-order diffraction terms is highly dependent on the SPH type.

So far, each one of the SPHs, $\psi(a, \phi)$, discussed above has been employed to construct the SPH transmittance $h_c(x, y) = \exp[i\psi(a, \phi + \phi_p)]$, which are modified by the linear carrier $\phi_p(x, y) = 2\pi(u_0 x + v_0 y)$, in order to isolate the encoded field from high-order terms of the SPH transmittance. In the particular case of the type 2 SPH, the phase modulation modified with the carrier is given by $\psi(a, \phi + \phi_p) = \phi + \phi_p + f(a)\sin(\phi + \phi_p)$. However, this procedure to obtain the SPH transmittance is not the only one. Interesting options appear by introducing the carrier function in different ways. An interesting example, in the case of the Type 2 phase modulation $\psi(a, \phi) = \phi + f(a)\sin(\phi)$, occurs when the carrier function is only added to the phase ϕ that appears as argument of the sin function. Following this procedure, we obtain the *type 4* SPH transmittance

$$h_c(x, y) = \exp(i\phi) + \exp[if(a)\sin(\phi + \phi_p)] \tag{8.27}$$

In this case, the Fourier series are expressed as

$$h_c(x, y) = \sum_{n=-\infty}^{\infty} J_n\big[f(a)\big]\exp\big[i(n+1)\phi\big]\exp\big[in\phi_p\big] \tag{8.28}$$

A second interesting modification of the type 2 SPH occurs when the phase carrier simply replaces the phase ϕ within the argument of the sin function. In this case, we obtain the *type 5* SPH transmittance

$$h_c(x, y) = \exp(i\phi) + \exp[if(a)\sin(\phi_p)] \tag{8.29}$$

whose Fourier series is

$$h_c(x, y) = \sum_{n=-\infty}^{\infty} J_n\big[f(a)\big]\exp\big[i\phi\big]\exp\big[in\phi_p\big] \tag{8.30}$$

It is interesting to note that the type 5 SPH coincides with one of the forms (in our opinion the most useful of the possible forms) of the SPHs proposed by Kirk et al.[68] For types 4 and 5 SPHs, the encoding condition that allows to obtain the desired complex field at the first order of the SPH Fourier series is given by Equation 8.21, which was established for the type 2 SPH.

An alternate method of attempting the separation of signal from noise is based on the addition of a quadratic phase carrier[69] to the phase of the encoded field. If the resulting SPH is illuminated by a plane wave, the Fourier transforms of the different terms $h_n(x, y)$ of the SPH Fourier series are generated at different planes in the

Fresnel domain of the PSH. A serious drawback of this method is that part of the high-order noise field contributions appear in a diluted form at the plane where the Fourier transform of the encoded signal is obtained; therefore, the SNR is strongly reduced.

8.6 SIGNAL-TO-NOISE RATIO (SNR) OF THE PROPOSED SPHs

Let us assume that the SPHs are implemented on an SLM. For simplicity, we assume that the pixel pitch δx is the same in the horizontal and the vertical axes and that the pixels are squares of side b. Thus, the Fourier spectrum of the pixelated SPH is given by

$$H_{\text{pix}}(u,v) = E(u,v) \sum_{n=-\infty}^{\infty} \sum_{m=-\infty}^{\infty} H_c(u-n\Delta u, v-m\Delta u) \qquad (8.31)$$

where:

$\Delta u = 1/\delta x$ is the largest possible spatial frequency or bandwidth

$E(u, v) = b^2 \, \text{sinc}(bu)\text{sinc}(bv)$ is the Fourier transform of the square pixel

$H_c(u,v)$ is the Fourier spectrum of the continuous SPH given by Equation 8.26

According to Equation 8.31, the Fourier spectrum of the pixelated SPH is formed by the superposition of laterally shifted replicas of the spectrum $H_c(u,v)$ modulated by the pixel Fourier transform $E(u, v)$. The spectrum function $H_{c,1}(u-u_0,v-v_0)$ that appears at the first term of the series in Equation 8.31 is equivalent to the signal spectrum $S(u-u_0,v-v_0)$. The distortion of the signal spectrum by the factor $E(u, v)$ can be avoided by an appropriate shaping of the encoded field.[70] This process consists in replacing the original encoded field $s(x, y)$ by a modified field $s'(x, y)$ that is defined by its Fourier transform $S'(u, v)$ obtained from the relation $S'(u-u_0,v-v_0) = S(u-u_0,v-v_0)E^{-1}(u,v)$. The function $E^{-1}(u, v)$ is appropriately defined in the domain of $S(u-u_0,v-v_0)$ since the carrier frequencies (u_0, v_0) are always chosen in such a way that $S(u-u_0,v-v_0)$ is enveloped by a nonzero sector of $E(u, v)$. This shaping is applied to the complex fields that are holographically encoded below either by numerical simulations or experimentally.

An inconvenient consequence of the pixelated structure of the SPH is that the domain of the signal spectrum term $H_1(u-u_0,v-v_0)$ that is centered at the spatial frequency coordinates (u_0, v_0) may also contain high-order spectrum contributions $H_n(u,v)$ of replicas of the spectrum $H_c(u-n\Delta u, v-m\Delta u)$. To analyze this, let us assume that the carrier spatial frequencies are $u_0 = v_0 = (P/Q)\Delta u$ with relative prime integers P and Q. This choice of P and Q values ensures that (u_0, v_0) is not an integer multiple of the bandwidth Δu. In this case, it is not difficult to prove that the spectra contributions that appear centered at the signal spatial frequencies (u_0, v_0) are $H_{QR+1}(u-u_0,v-v_0)$ for any arbitrary integer number R. The signal term in this set of spectra contributions corresponds to $R = 0$. To prove this result, we first note that for the assumed carrier frequencies $u_0 = v_0 = (P/Q)\Delta u$, the spectra terms $H_n(u,v)$ in Equation 8.26 take the form $H_n(u-un_0, v-nv_0)$. This means that all the terms $H_n(u,v)$ in $H_c(u,v)$ are centered at the axis $u = v$ in the SPH Fourier spectrum domain. As a particular case, the signal spectrum term $H_1(u-u_0,v-u_0)$ is also centered at the coordinates (u_0,u_0). Another important observation is that the only term $H_c(u-n\Delta u,v-m\Delta u)$ in Equation 8.31 that contains spectra functions $H_n(u,v)$ placed at the axis $u = v$ is those with indices $n = m$. Considering the relation $u_0 = v_0 = (P/Q)\Delta u$, the term $H_c(u-n\Delta u,v-n\Delta u)$ can be expressed as

$$H_c(u-n\Delta u,v-n\Delta u) = \sum_{q=-\infty}^{\infty} H_n[u-(nQ/P+q)u_0,v-(nQ/P+q)u_0] \quad (8.32)$$

It can be directly verified that the functions $H_n(u,v)$ in Equation 8.32 that appear centered at coordinates (u_0,u_0) are those corresponding to the combination of indices $q = -RP$ and $n = QR + 1$ for any integer R. This proves that the Fourier spectrum of the pixelated SPH contains the terms $H_{QR+1}(u-u_0,v-u_0)$ for any integer R that are centered at the coordinates (u_0,u_0). It must be emphasized that in this collection of spectra terms, the spectrum of the encoded field corresponds to $R = 0$.

To obtain a high SNR,[71] the high-order contributions $H_{QR+1}(u-u_0,v-u_0)$ (with $R \neq 0$) must be negligible compared with the signal spectrum $H_1(u-u_0,v-u_0)$. Considering Equation 8.10, it is clear that the power of the spectrum term $H_{QR+1}(u,v)$ is proportional to the squared modulus of the coefficient c_{QR+1}^a. It is interesting to note that coefficients c_n^a for the SPHs of types 2, 3, 4, and 5 defined in Equations 8.18, 8.22, 8.27, and 8.29 in terms of Bessel functions tend rapidly to zero when $|n|$ is increased. Thus, it is expected that (for a moderately large Q) these SPHs will enable reconstruction of the encoded field with relatively high SNR. The significance of noise contribution $H_{QR+1}(u-u_0,v-u_0)$ in relation to the signal term $H_1(u-u_0,v-u_0)$ can be measured by the ratio $\sigma = |c_{QR+1}^a|^2/|c_1^a|^2$. The parameter $\sigma = |c_{QR+1}^a|^2/|c_1^a|^2$ is useful to evaluate the quality of SPHs without requiring knowledge of the encoded complex function $s(x,y)$. Lower values of σ correspond to SPHs with higher immunity to the high-order noise terms that are transmitted in the SPH plane. The SNR provides another measurement of the SPH performance. Contrasting with the parameter σ, the SNR, which is employed below to evaluate the SPHs, is explicitly dependent on the form of the encoded field $s(x,y)$ and it is given by[70]

$$\text{SNR} = \frac{\iint_\Omega |s(x,y)|^2\, dxdy}{\iint_\Omega |s(x,y)-\beta s_r(x,y)|^2\, dxdy} \quad (8.33)$$

where $s_r(x,y)$ is the reconstructed signal that includes noise distortions due to high-order diffraction terms, and β is a constant that is determined to minimize the error power, which is given by

$$\beta = \frac{\iint_\Omega \text{Re}\left\{s(x,y)s_r^*(x,y)\right\}dxdy}{\iint_\Omega \left|s_r(x,y)\right|^2 dxdy} \tag{8.34}$$

It must be emphasized that low-order nonsignal spectrum field contributions (of order $n \neq QR + 1$) also present replicas due to the SLM pixelation. However, none of these spectrum replicas appears centered on the signal frequencies (u_0, u_0). On the other hand, the low-order nonsignal spectrum field contributions (e.g., of orders 0 or 2) can partially overlap the signal spectrum. However, this influence can be minimized either by increasing the carrier frequencies (u_0, u_0) or by reducing the bandwidth of the encoded signal. This control is possible, because, in general, a reduction in the encoded signal bandwidth also reduces the bandwidth of the high-order SPH terms.

8.7 SPHs' NUMERICAL PERFORMANCE

Bessel Gauss and Laguerre Gauss beams represent two broad interesting families of beams for optical trapping applications: nondiffracting and self-similar beams. In non-diffracting beams, its transverse spatial structure is propagation-invariant, which allows manipulation of particles on large distances along the beam axis as compared to Gaussian beams.[72] In addition, they are self-reconstructing, that is, their transverse field distribution is reconstructed after a portion of the beam is obstructed; this imply that multiple particles can be trapped along the beam, as it has been demonstrated. Finally, Bessel beams can also transfer optical orbital angular momentum to matter, so microparticles can rotate.[73] On the other hand, self-similar beams maintain their transverse shape during propagation but scale during free-space propagation or after passing through optical elements like lenses. In particular, helical Laguerre–Gaussian beams are known to carry optical orbital angular momentum.[55,56] To illustrate the performance of all five types of pixelated SPHs, we encode the same field using Laguerre–Gauss beams and Bessel beams.

It is worth mentioning that any arbitrary field can in principle be encoded with these SPHs. Laguerre beams are expressed in polar coordinates (r, θ) as

$$s(r,\theta) = C(\sqrt{2}r/w_0)^{|l|}L_p^{|l|}(2r^2/w_0^2)\exp(-r^2/w_0^2)\exp(il\theta) \tag{8.35}$$

Here, $L_p^{|l|}$ denotes an associated Laguerre polynomial, w_0 is the beam waist radius, p is the radial mode index, l is the phase singularity charge, and C is a normalization constant. For the numerical simulation, the following parameters where chosen: waist radius $w_0 = 75\,\delta x$ (recall that δx is the pixel pitch), the beam indices (p, l) = (2, 1), and the beam support is a circle of radius $R = 256\,\delta x$. Figure 8.8 shows the amplitude and phase modulation of the beam. The phase carrier added to the phase of the encoded field has spatial frequencies $u_0 = v_0 = \Delta u/6$.

The phase distributions obtained using all types of SPHs are displayed in Figure 8.9. On the other hand, Figure 8.10a shows the spectrum modulus of the encoded beam defined by Equation 8.35 and Figures 8.10b–f show the moduli of the signal spectra obtained with the pixelated SPHs. The last were normalized respect to the peak amplitude of the type 5 SPH; this means that the type 5 SPH shows the greatest intensity efficiency.

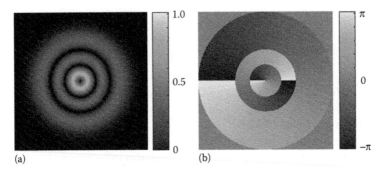

FIGURE 8.8 Amplitude (a) and phase modulation and (b) of the (2, 1) Laguerre–Gauss field.

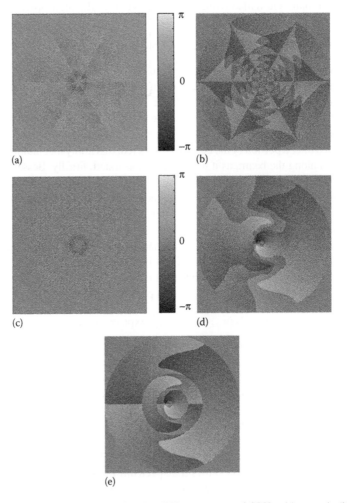

FIGURE 8.9 Phase distribution for the different types of SPHs: (a) type 1, (b) type 2, (c) type 3, (d) type 4, and (e) type 5. As can be noted, for the type 3 SPH, only the central part is adequately resolved because it has a reduced phase modulation ($\approx 1.2\pi$).

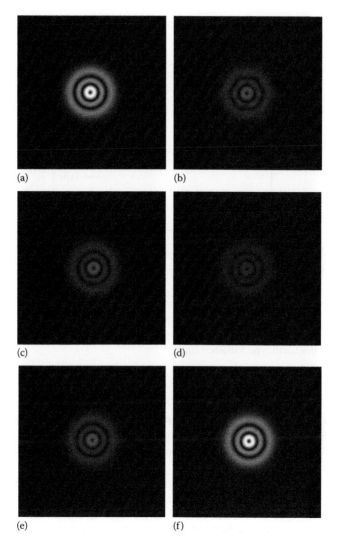

(a)　　　　　　　　(b)

(c)　　　　　　　　(d)

(e)　　　　　　　　(f)

FIGURE 8.10 Normalized spectra moduli for the signal (a) and for the SPHs, (b)–(f) types 1–5, respectively. In this case, the spectra were normalized with respect to the peak amplitude of the type 5 SPH.

We computed the SNR of the reconstructed beams employing Equation 8.35. Figure 8.11 shows a semilogarithmic plot of the SNR versus the radius of the spatial filter on the frequency space in terms of the inverse of the beam waist. According to Figure 8.11, we choice the radius of the spatial filter $\approx 1.45/w_0$, which corresponds to the maximum value of SNR. From the figure, one can see that the cleanest fields are obtained with type 2 and 3, while type 4 and 5 follow close. Type 1 SPH is the least clean and less efficient hologram.

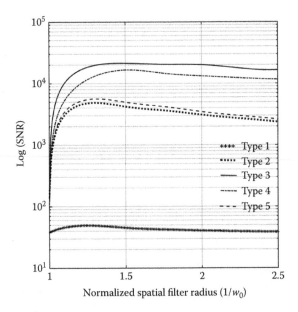

FIGURE 8.11 Semilogarithmic plot of the SNR vs the spatial filter radius in terms of the beam waist. SPH: type 1 (plus symbol line), type 2 (dotted line), type 3 (solid line), type 4 (dashed-dotted line), and type 5 (dashed line).

Figure 8.12 shows the SNR versus different frequency values of the carrier (u_0, v_0) on the range $[\Delta u/15, \Delta u/2]$, for the type 1 SPH. Meanwhile, Figure 8.12b, c shows similar results for the type 2 and 3 SPHs, respectively. Results for type 4 and 5 SPH are quite similar to the type 2.

As can be seen in Figure 8.12a, c the highest values for SNR are around the diagonal of the carrier frequencies for type 1 and 3 SPH, that is, when the value $u_0 \approx v_0$. In particular, the type 3 SPH shows the highest SNR values for some determined carrier frequencies ($u_0 = v_0 \approx \Delta u/3$). For the case of the type 2 SPH (Figure 8.12b), it presents a more extended range of carrier frequencies where the SNR achieves high values. These results give us a general idea to choose the optimal carrier frequencies to recover a high quality field.

Next, numerical reconstruction of the (2, 1) Laguerre–Gauss beam (Equation 8.35) is presented with the parameter described above; $w_0 = 75 \, \delta x$ and the beam support is a circle of radius $R = 256 \, \delta x$. Figure 8.13 shows the amplitude (column [a]–[e]) and phase modulation (column [f]–[j]) of the reconstructed beam for the types 1–5 SPHs, respectively. The phase carrier frequencies added to the phase of the encoded field are $u_0 = v_0 = \Delta u/6$. In Figure 8.13a and f for the type 1 SPH, it can be seen that both the amplitude and phase present evident distortions. Meanwhile, for the other types SPHs, the recovered amplitudes are very similar, so that the main differences are evident on the recovered phases.

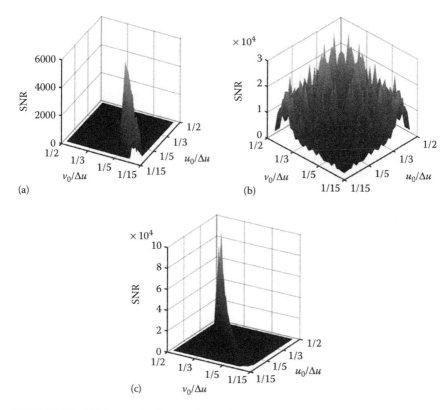

FIGURE 8.12 SNR vs carrier frequencies, on the range $[\Delta u/15, \Delta u/2]$, for the SPHs, (a) type 1, (b) type 2, and (c) type 3.

As a second example, we present the codification of a Bessel beam, which is expressed by

$$s(r,\theta) = J_n(2\pi r/r_0)\exp(in\theta) \tag{8.36}$$

where:
J_n is the nth-order Bessel function
r_0 is the radial period

Figure 8.14a, b shows the amplitude and phase modulation of a Bessel functions of order 1 and $r_0 = 121\,\delta x$, respectively. For this particular example, the spectrum distribution for the different SPHs is similar to the previous example. We will only show the results for the SNR and the recovered beams. As it is well known, the spectrum of the Bessel fields has an annular amplitude distribution with radius $1/r_0$; hence, we consider a spatial filter radius on the range $[1.5/r_0, 4.5r_0]$ to compute the SNR.

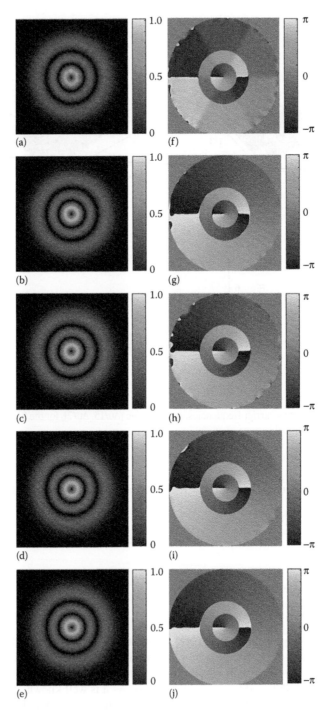

FIGURE 8.13 Recovered amplitude (a)–(e) and phase (f)–(j) corresponding to the types 1–5 SPHs, respectively.

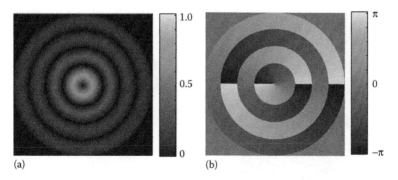

(a)   (b)

FIGURE 8.14   Amplitude (a) and phase modulation (b) of the first-order Bessel field.

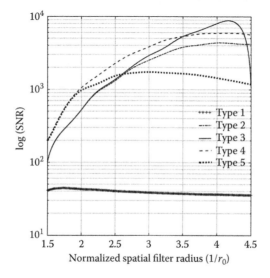

FIGURE 8.15   Semilogarithmic plot of the SNR vs the spatial filter radius in terms of the radial period r_0. SPH: type 1 (plus symbol line), type 2 (dashed-dotted line), type 3 (solid line), type 4 (dashed line), and type 5 (dotted line).

Figure 8.15 shows the semilogarithmic plot of the SNR vs the spatial filter radius. As can be noted, type 4 SPH is greater than the other SPHs, except in the range $[3/r_0, 4.4\ r_0]$ where the type 3 SPH achieves the maximum values.

Since all types of SPH give the best SNR for a filter radius around $r_{sf} = 4.25/r_0$ with $r_0 = 121\ \delta x$, the numerical reconstruction of the first-order Bessel beam, given by Equation 8.36, is presented using a circle of radius $R = 256\ \delta x$ as support. Figure 8.16 shows the amplitude (column [a]–[e]) and phase modulation (column [f]–[j]) of the reconstructed beam for all types of SPHs. The phase carrier added to the phase of the encoded field has spatial frequencies $u_0 = v_0 = \Delta u/6$. Evident distortions of both amplitude and phase can be seen (Figure 8.16a, f) for the type 1 SPH; meanwhile, as in the previous case, for the other types of SPHs, the recovered amplitude are very similar, so that the main differences are evident on the recovered phases.

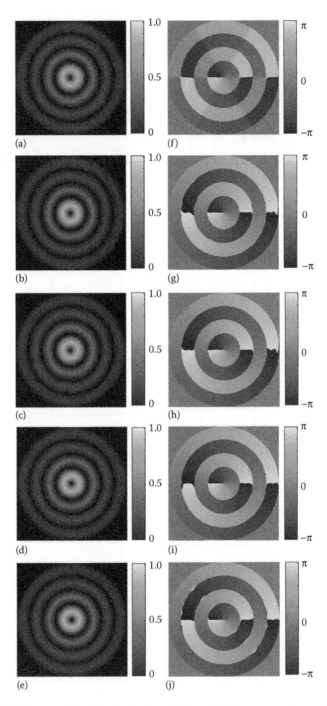

FIGURE 8.16 Recovered amplitude (a)–(e) and phase (f)–(j) corresponding to the types 1–5 SPHs, respectively.

8.8   EXPERIMENTAL GENERATION OF COMPLEX BEAMS

Here we show the experimental implementation of SPHs for the synthesis of nondiffracting Bessel beams and Laguerre–Gauss beams employing a phase-only liquid crystal spatial light modulator (SLM) (Pluto, Holoeye GmbH). We employed a circular central zone of 512×512 pixels in this device (pixel pitch $\delta x = 8$ μm). The voltage applied to the SLM pixels are related to the gray levels of images supplied to the SLM by a PC video card. The phase modulation (versus the gray level) provided by the SLM illuminated with a He–Ne laser (633 nm) is depicted in Figure 8.17.

The encoded Laguerre–Gauss and Bessel beams are analytically expressed by Equations 8.35 and 8.36, respectively. For both beams, we employed carrier spatial frequencies $u_0 = v_0 = \Delta u/6 = 1/(6 \ \delta x)$. For the case of Laguerre–Gauss beam, which was encoded by the type 2 SPH, the waist radius was adopted as $w_0 = 75 \ \delta x$. For the Bessel beam, which was encoded by the type 3 SPH, the asymptotic period was adopted as $r_0 = 121 \ \delta x$. A circular pupil in the Fourier domain of the SPHs employed as spatial filter for signal isolation was adjusted to optimize the quality of the generated fields. Because of the symmetry of the encoded field, the pupil employed as a spatial filter in the experiment was circular. The application of this pupil to the signal spectrum represents a low-pass filtering during reconstruction of the encoded field. To increase the fidelity of the reconstructed field, it is necessary to increase the pupil diameter. However, if the diameter is too large, higher amounts of noise contribution will be transmitted by the pupil. Thus, in practice, it is necessary to optimize this diameter for each particular SPH. The intensity distributions of the experimentally generated beam were recorded with a CCD camera. The intensity of the generated Laguerre–Gauss beam with indices (p, l) of $(2, 1)$ and Bessel beam with $n = 1$ appears in Figure 8.18a, b respectively.

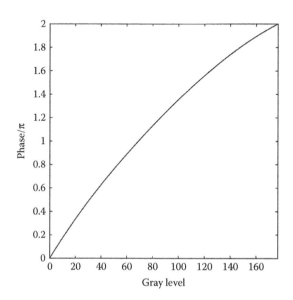

FIGURE 8.17   Phase modulation provided by the LC SLM.

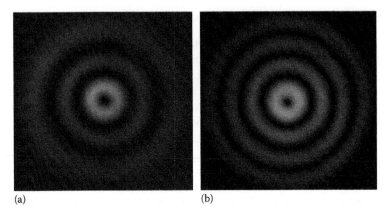

(a)   (b)

FIGURE 8.18 Recorded intensity of the (a) recovered (2, 1) Laguerre–Gauss and (b) first-order Bessel beams, codified with the types 2 and 3 SPH and displayed on an LC SLM, respectively.

8.9 SUMMARY

In this chapter, the use of DOEs in optical tweezers was analyzed and discussed in detail. A special case of DOEs, where the amplitude and phase are encoded in a phase holograms, are introduced. This DOEs are called synthetic phase holograms (SPHs). In particular, five classes of SPHs, which allow encoding arbitrary scalar complex beams, have been discussed. An important characteristic of this SPH is that the phase modulation $\psi(a, \phi)$ of the SPHs discussed must have odd symmetry in the variable ϕ. Numerical and experimental evaluation for energy performance and quality (SNR) were evaluated for all proposed SPHs.

ACKNOWLEDGMENTS

Rubén Ramos-García acknowledge financial support from CONACyT under grant CB-2010–153463.

REFERENCES

1. Ashkin, A., Acceleration and trapping of particles by radiation pressure, *Phys. Rev. Lett.*, 24(4), 156–159, 1970.
2. Ashkin, A., Dziedzic, J. M., Bjorkholm, J. E., and Chu, S., Observation of a single-beam gradient force optical trap for dielectric particles, *Opt. Lett.*, 11(5), 288, 1986.
3. Ashkin, A., Optical trapping and manipulation of neutral particles using lasers, *Opt. Photonics News*, 10(5), 41, 1999.
4. Molloy, J. E., and Padgett, M. J., Lights, action: Optical tweezers, *Contemp. Phys.*, 43(4), 241–258, 2002.
5. Block, S. M., Making light work with optical tweezers., *Nature*, 360(6403), 493–495, 1992.
6. Moffitt, J. R., Chemla, Y. R., Smith, S. B., and Bustamante, C., Recent advances in optical tweezers, *Annu. Rev. Biochem.*, 77, 205–228, 2008.

7. MacDonald, M. P., Spalding, G. C., and Dholakia, K., Microfluidic sorting in an optical lattice, *Nature*, 426(6965), 421–424, 2003.

8. Ricárdez-Vargas, I., Rodríguez-Montero, P., Ramos-García, R., and Volke-Sepúlveda, K., Modulated optical sieve for sorting of polydisperse microparticles, *Appl. Phys. Lett.*, 88(12), 121116, 2006.

9. Bishop, A. I., Nieminen, T. A., Heckenberg, N. R., and Rubinsztein-Dunlop, H., Optical microrheology using rotating Laser-Trapped particles, *Phys. Rev. Lett.*, 92(19), 198104, 2004.

10. Furst, E. M., Applications of laser tweezers in complex fluid rheology, *Curr. Opin. Colloid Interface Sci.*, 10(1–2), 79–86, 2005.

11. Henderson, S., Mitchell, S., and Bartlett, P., Direct measurements of colloidal friction coefficients, *Phys. Rev. E*, 64(6), 061403, 2001.

12. Meiners, J.-C., Quake, S. R., Direct measurement of hydrodynamic cross correlations between two particles in an external potential, *Phys. Rev. Lett.*, 82(10), 2211–2214, 1999.

13. Block, S. M., Blair, D. F., and Berg, H. C., Compliance of bacterial flagella measured with optical tweezers, *Nature*, 338(6215), 514–518, 1989.

14. Smith, S. B., Cui, Y., and Bustamante, C., Overstretching B-DNA: The elastic response of individual Double-Stranded and Single-Stranded DNA molecules, *Science*, 271(5250), 795–799, 1996.

15. Wang, M. D., Yin, H., Landick, R., Gelles, J., and Block, S. M., Stretching DNA with optical tweezers, *Biophys. J.*, 72(3), 1335–1346, 1997.

16. Finer, J. T., Simmons, R. M., and Spudich, J. A., Single myosin molecule mechanics: piconewton forces and nanometre steps, *Nature*, 368(6467), 113–119, 1994.

17. Neuman, K. C., and Nagy, A., Single-molecule force spectroscopy: Optical tweezers, magnetic tweezers and atomic force microscopy, *Nat. Methods*, 5(6), 491–505, 2008.

18. Dao, M., Lim, C. T., and Suresh, S., Mechanics of the human red blood cell deformed by optical tweezers, *J. Mech. Phys. Solids*, 51(11–12), 2259–2280, 2003.

19. Mills, J. P., Qie, L., Dao, M., Lim, C. T., and Suresh, S., Nonlinear elastic and visco-elastic deformation of the human red blood cell with optical tweezers., *Mech. Chem. Biosyst.*, 1(3), 169–180, 2004.

20. Visscher, K., Brakenhoff, G. J., and Krol, J. J., Micromanipulation by 'multiple' optical traps created by a single fast scanning trap integrated with the bilateral confocal scanning laser microscope, *Cytometry*, 14(2), 105–114, 1993.

21. Visscher, K., Gross, S. P., and Block, S. M., Construction of multiple-beam optical traps with nanometer-resolution position sensing, *IEEE J. Sel. Top. Quant. Electron.*, 2(4), 1066–1076, 1996.

22. Grinenko, A., MacDonald, M. P., Courtney, C. R. P., Wilcox, P. D., Demore, C. E. M., Cochran, S., and Drinkwater, B. W., Tunable beam shaping with a phased array acousto-optic modulator, *Opt. Express*, 23(1), 26, 2015.

23. www.elliotscientific.com., <http://www.elliotscientific.com/140–0/Optical-Tweezer-Systems/> (15 October 2015).

24. Sasaki, K., Koshioka, M., Misawa, H., Kitamura, N., and Masuhara, H., Pattern formation and flow control of fine particles by laser-scanning micromanipulation, *Opt. Lett.*, 16(19), 1463, 1991.

25. Yamamoto, J., Iwai, T., Spatial stability of particles trapped by Time-Division optical tweezers, *Int. J. Optomechatronics*, 3(4), 253–263, 2009.

26. Visscher, K., Gross, S. P., and Block, S. M., Construction of multiple-beam optical traps with nanometer-resolution position sensing, *IEEE J. Sel. Top. Quant. Electron.*, 2(4), 1066–1076, 1996.

27. Dufresne, E. R., Spalding, G. C., Dearing, M. T., Sheets, S. A., and Grier, D. G., Computer-generated holographic optical tweezer arrays, *Rev. Sci. Instrum.*, 72(3), 1810–1816, 2001.

28. Love, G. D., Wave-front correction and production of Zernike modes with a liquid-crystal spatial light modulator, *Appl. Opt.*, 36(7), 1517, 1997.

29. Roichman, Y., Waldron, A., Gardel, E., and Grier, D. G., Optical traps with geometric aberrations, *Appl. Opt.*, 45(15), 3425–3429, 2006.

30. Wulff, K. D., Cole, D. G., Clark, R. L., Dileonardo, R., Leach, J., Cooper, J., Gibson, G., and Padgett, M. J., Aberration correction in holographic optical tweezers, *Opt. Express*, 14, 4170–4175, 2006.

31. Jesacher, A., Schwaighofer, A., Fürhapter, S., Maurer, C., Bernet, S., and Ritsch-Marte, M., Wavefront correction of spatial light modulators using an optical vortex image, *Opt. Express*, 15(9), 5801, 2007.

32. Ashkin, A., Dziedzic, J. M., Bjorkholm, J. E., and Chu, S., Observation of a single-beam gradient force optical trap for dielectric particles, *Opt. Lett.*, 11(5), 288, 1986.

33. Alexeev, I., Quentin, U., Leitz, K.-H., and Schmidt, M., Optical trap kits: Issues to be aware of, *Eur. J. Phys.*, 33(2), 427–437, 2012.

34. Liesener, J., Reicherter, M., Haist, T., and Tiziani, H. J., Multi-functional optical tweezers using computer-generated holograms, *Opt. Commun.*, 185(1–3), 77–82, 2000.

35. Neff, J. A., Athale, R. A., and Lee, S. H., Two-dimensional spatial light modulators: A tutorial, *Proc. IEEE*, 78(5), 826–855, 1990.

36. Solgaard, O., *Photonic Microsystems: Micro and Nanotechnology Applied to Optical Devices and Systems*, Springer, New York, 2010.

37. Li, F., Mukohzaka, N., Yoshida, N., Igasaki, Y., Toyoda, H., Inoue, T., Kobayashi, Y., and Hara, T., Phase modulation characteristics analysis of Optically-Addressed Parallel-Aligned nematic liquid crystal Phase-Only spatial light modulator combined with a liquid crystal display, *Opt. Rev.*, 5(3), 174–178, 1998.

38. Armitage, D., Underwood, I., and Wu, S. T., *Introduction to Microdisplays*, John Wiley & Sons, Chichester, 2006.

39. Yeh, P., Gu, C., *Optics of Liquid Crystal Displays*, John Wiley & Sons, New York, 1999.

40. Marquez, A., Iemmi, C., Moreno, I., Davis, J. A., Campos, J., and Yzuel, M. J., Quantitative prediction of the modulation behavior of twisted nematic liquid crystal displays based on a simple physical model, *Opt. Eng.*, 40(11), 2558–2564, 2001.

41. Hayasaki, Y., Itoh, M., Yatagai, T., and Nishida, N., Nonmechanical optical manipulation of microparticle using spatial light modulator, *Opt. Rev.*, 6(1), 24–27, 1999.

42. Gerchberg, R. W., and Saxton, W. O., A practical algorithm for the determination of phase from image and diffraction plane pictures, *Optik (Stuttg).*, 35(2), 237–246, 1972.

43. Seldowitz, M. A., Allebach, J. P., and Sweeney, D. W., Synthesis of digital holograms by direct binary search, *Appl. Opt.*, 26(14), 2788, 1987.

44. Laczik, Z. J., 3D beam shaping using diffractive optical elements, in *Laser Beam Shap. III*, Dickey, F. M., Holswade, S. C., and Shealy, D. L., eds., pp. 104–111, SPIE, 2002.

45. Curtis, J. E., Koss, B. A., and Grier, D. G., Dynamic holographic optical tweezers, *Opt. Commun.*, 207(1–6), 169–175, 2002.

46. Piestun, R., and Shamir, J., Generalized propagation-invariant wave fields, *J. Opt. Soc. Am.*, A 15(12), 3039, 1998.

47. Leach, J., Wulff, K., Sinclair, G., Jordan, P., Courtial, J., Thomson, L., Gibson, G., Karunwi, K., and Cooper, J. et al., Interactive approach to optical tweezers control., *Appl. Opt.*, 45(5), 897–903, 2006.

48. Leach, J., Sinclair, G., Jordan, P., Courtial, J., Padgett, M., Cooper, J., and Laczik, Z., 3D manipulation of particles into crystal structures using holographic optical tweezers., *Opt. Express.*, 12(1), 220–226, 2004.

49. Whyte, G., and Courtial, J., Experimental demonstration of holographic three-dimensional light shaping using a Gerchberg–Saxton algorithm, *New J. Phys.*, 7, 117–117, 2005.

50. Di Leonardo, R., Ianni, F., and Ruocco, G., Computer generation of optimal holograms for optical trap arrays., *Opt. Express*, 15(4), 1913–1922, 2007.

51. Roichman, Y., Waldron, A., Gardel, E., and Grier, D. G., Optical traps with geometric aberrations, *Appl. Opt.*, 45(15), 3425, 2006.

52. Wulff, K. D., Cole, D. G., Clark, R. L., DiLeonardo, R., Leach, J., Cooper, J., Gibson, G., and Padgett, M. J., Aberration correction in holographic optical tweezers, *Opt. Express*, 14(9), 4170, 2006.

53. Volke-Sepulveda, K., Garcés-Chávez, V., Chávez-Cerda, S., Arlt, J., and Dholakia, K., Orbital angular momentum of a high-order Bessel light beam, *J. Opt. B Quantum Semiclassical Opt.*, 4(2), S82–S89, Opt. Soc. America, 2002.

54. Volke-Sepulveda, K., Chavez-Cerda, S., Garces-Chavez, V., and Dholakia, K., Three-dimensional optical forces and transfer of orbital angular momentum from multiringed light beams to spherical microparticles, *J. Opt. Soc. Am.*, B 21(10), 1749, 2004.

55. Simpson, N. B., Allen, L., and Padgett, M. J., Optical tweezers and optical spanners with Laguerre–Gaussian modes, *J. Mod. Opt.*, 43(12), 2485–2491, 1996.

56. Simpson, N. B., McGloin, D., Dholakia, K., Allen, L., and Padgett, M. J., Optical tweezers with increased axial trapping efficiency, *J. Mod. Opt.*, 45(9), 1943–1949, 1998.

57. López-Mariscal, C., Gutiérrez-Vega, J. C., Milne, G., and Dholakia, K., Orbital angular momentum transfer in helical Mathieu beams, *Opt. Express*, 14(9), 4183, 2006.

58. Woerdemann, M., Alpmann, C., and Denz, C., Optical assembly of microparticles into highly ordered structures using Ince–Gaussian beams, *Appl. Phys. Lett.*, 98(11), 111101, 2011.

59. Sato, S., Ishigure, M., and Inaba, H., Optical trapping and rotational manipulation of microscopic particles and biological cells using higher-order mode Nd:YAG laser beams, *Electron. Lett.*, 27(20), 1831, 1991.

60. Baumgartl, J., Mazilu, M., and Dholakia, K., Optically mediated particle clearing using Airy wavepackets, *Nat. Photonics*, 2(11), 675–678, 2008.

61. Arrizón, V., Ruiz, U., Carrada, R., and González, L. A., Pixelated phase computer holograms for the accurate encoding of scalar complex fields, *J. Opt. Soc. Am.*, A 24(11), 3500, 2007.

62. Davis, J. A., Cottrell, D. M., Campos, J., Yzuel, M. J., and Moreno, I., Encoding Amplitude Information onto Phase-Only Filters, *Appl. Opt.* 38(23), 5004–5013 1999.

63. G. N. Watson., *A Treatise on the Theory of Bessel Functions (Cambridge Mathematical Library)*, Cambridge University Press, Cambridge, 1995.

64. Ashkin, A., Dziedzic, J. M., and Yamane, T., Optical trapping and manipulation of single cells using infrared laser beams, *Nature*, 330(6150), 769–771, 1987.

65. Ashkin, A., Schütze, K., Dziedzic, J. M., Euteneuer, U., and Schliwa, M., Force generation of organelle transport measured in vivo by an infrared laser trap, *Nature*, 348(6299), 346–348, 1990.

66. Wright, W. H., Sonek, G. J., Tadir, Y., and Berns, M. W., Laser trapping in cell biology, *IEEE J. Quant. Electron.*, 26(12), 2148–2157, 1990.

67. Arrizón, V., Optimum on-axis computer-generated hologram encoded into low-resolution phase-modul ation devices, *Opt. Lett.*, 28(24), 2521, 2003.

68. Kirk, J. P., and Jones, A. L., Phase-Only Complex-Valued Spatial Filter, *J. Opt. Soc. Am.*, 61(8), 1023, 1971.

69. Moreno, I., Campos, J., Gorecki, C., and Yzuel, M. J., Effects of Amplitude and Phase Mismatching Errors in the Generation of a Kinoform for Pattern Recognition, *Jpn. J. Appl. Phys.*, 34(Part 1, No. 12A), 6423–6432, 1995.

70. Arrizón, V., Méndez, G., and Sánchez-de-La-Llave, D., Accurate encoding of arbitrary complex fields with amplitude-only liquid crystal spatial light modulators, *Opt. Express*, 13(20), 7913, 2005.

71. Arrizón, V., Méndez, G., and Sánchez-de-La-Llave, D., Accurate encoding of arbitrary complex fields with amplitude-only liquid crystal spatial light modulators, *Opt. Express*, 13(20), 7913, 2005.

72. Garcés-Chávez, V., McGloin, D., Melville, H., Sibbett, W., and Dholakia, K., Simultaneous micromanipulation in multiple planes using a self-reconstructing light beam, *Nature*, 419(6903), 145–147, 2002.

73. Volke-Sepulveda, K., Chavez-Cerda, S., Garces-Chavez, V., and Dholakia, K., Three-dimensional optical forces and transfer of orbital angular momentum from multiringed light beams to spherical microparticles, *J. Opt. Soc. Am.*, B21(10), 1749, 2004.

9 Laser Beam Shaping through Fiber Optic Beam Delivery

Todd E. Lizotte and Orest Ohar

CONTENTS

9.1   INTRODUCTION

Since its inception, laser innovation has led to great technological achievements within various markets, including medical and industrial. Some of the most notable advancements from an early industrial perspective were the work conducted by individuals at major corporations such as General Electric (GE).[1] Over the period between 1964 and 1985, the application of fibers for the delivery of laser light to a surface accelerated. Looking back over history, the growth and diversity of fiber beam delivery and its application to industrial material processing, medical therapy, surgery, and communication technology are staggering. After half a century, fiber optic beam delivery has become a mainstay for providing flexibility to the delivery of laser light by reducing the limits inherent with standard geometric optical-based mirror beam delivery systems. Some of the most influential research and development of fiber optic-based laser beam delivery for industrial applications were conducted by corporations such as GE.[2]

Many of the early patents covering fiber optic beam delivery during the 1970s and 1980s emphasized the need for increased power delivery and general flexibility, either through a handheld device or using a robotic system for manufacturing processes such as drilling, cutting, and welding.[3] As industry adopted lasers into their manufacturing facilities at greater rates, the emphasis shifted toward better methods or techniques to reduce losses, increase power transmission, extend fiber lifetimes, and improve beam stability delivered through the fiber.[4] Increased control of the laser at the work surface improved the overall quality of the results.

From the earliest point in its history to the present day, fiber optic laser beam delivery has incrementally expanded into most traditional laser materials processing markets within heavy industry. Both fiber innovations and new laser technologies continued evolving on separate paths. As both found greater acceptance within industry, these technologies were applied to more precise laser fabrication techniques such as thin-film patterning, micromachining, micropercussion drilling, and 3D laser-assisted manufacturing.

Demands for improved quality within the aerospace industry have seen a shift from brute force laser processes toward higher finesse laser processes, which offer greater resolution in optical settings and sensor feedback adjustments, effectively fine-tuning the quality produced. This shift toward more control has created the realization that with increased quality comes economic benefits. For instance, within traditional markets such as drilling and milling, the demand for consistency and quality is being realized through the addition of laser beam shaping, both temporal and spatial. A good example of this within the aerospace industry is the shift to laser percussion drilling. The aerospace industry laser drills billions of precision holes in turbine components for effusion cooling. With ever-evolving technologies, the aerospace industry continues developing and refining their laser processes to improve precision and control. By adding innovations such as laser beam shaping, including pulse shaping (temporal) and laser beam shaping (spatial), turbine parts have evolved to handle higher combustion temperatures netting greater engine efficiencies.[5]

Consumer product global demand has spurred product innovators to continually develop and create new technologies, even to adapt old technologies in new ways. Engineers and technologists who seek to drive innovation have shifted from the normal mindset of creating entirely new innovation to a method of leveraging the concept of combinatorial innovation—*the melding of old ideas and new ideas.* Within industry, the adoption of wholly new technologies can be disruptive and potentially disastrous to established market leaders. Since applied research is that which is imperative to the long-term strength of a company, the following sections describe the interest of its authors in exploring innovative approaches to enhance the stability of fiber optic laser beam delivery, with an application example of micromachining technology integrated onto a robotic laser thin-film patterning system.

9.2   INDUSTRIAL LASER ROBOTIC FIBER BEAM DELIVERY

At the present time, there are thousands of laser systems working on production factory floors worldwide—24 h a day, seven days a week. A large portion of these laser systems are involved in the mass production of automobiles, consumer electronics, and aerospace products. Many of these applications have fiber optical-based laser beam delivery systems, which are integrated within multiaxis robotic arms. These laser systems utilize some form of laser beam delivery configuration for transmitting a high-power laser beam onto a workpiece for drilling, milling, cutting, or welding the workpiece.

In the early days of industrial beam delivery, engineers would utilize an assembly of mirrors and prisms to transmit the laser beam to the workstation. The use of articulated joint beam delivery systems coupled to the robotic arm requires many mirror links to reflect the laser beam.[6,7] As power demands increased, fibers were still in their infancy and were limited to laser applications less than 5 W. On the other hand, mirror systems could handle 2-kW power levels. As laser power increased, so did the complexity of the systems, with the need for water cooling. The mirror-based systems began to get ungainly, limiting robotic arms' ability to articulate into tight areas or even limiting its total range of motion.[6]

Where fiber optical laser beam delivery systems began to show promise, then dominance was within the automotive and aerospace markets, where the application of high-power infrared (IR) sources, such as CO_2 (10.6 micron Far IR) and Nd:YAG (1.064 micron Near IR) were used for high-speed spot welding, drilling cooling holes, and laser trimming of steel frame structures.[8] Figure 9.1 shows a modern laser robotic system using a fiber optic laser beam delivery system. When these lasers first emerged into the industrial market, no matter how precise the motion of the robotics or the laser beam quality being produced by the laser resonator, a high-quality and highly efficient process could not coalesce without the right optical configuration for delivering the laser beam to the workpiece.[9] Since those early days, incremental improvements have played a decisive role in the performance of fiber optic-based beam delivery within industrial robotic laser processes. Effective laser injection into the fiber optic became the weakest link.

FIGURE 9.1 Robotic laser welding system (courtesy of HighYAG II-VI.)

9.2.1 TRADITIONAL SYSTEMS OVERVIEW: FIBER DELIVERY

To better understand the general benefits and limitations of optical fiber laser beam delivery, its design and performance need to be discussed in detail. In its basic form, a fiber optic laser beam delivery system entails an injection housing, the fiber optical cable assembly, and output focusing/imaging housing.[6] The injection housing is where the laser is injected into an optical assembly to couple the laser energy into the fiber core. At the injection point, it is imperative that the power has to be coupled into the fiber core with high efficiency. Excessive thermal loading can occur if the coupling system is not designed appropriately, which leads to damage to the fiber core, its cladding, as well as the injection optics.

In most systems, the laser coupling optics are simply a convex lens which focuses the laser beam to illuminate the core of the fiber.[10] The fiber at this point should be positioned beyond the focal point of the convex lens such that the fiber core is not exposed to excessive power densities. The selection of the convex lens must meet some basic criteria. The first being an appropriate focal length to ensure that the compressed beam efficiently fills the fiber core as well as maintains an acceptance beam angle to match the fiber optics numerical aperture (NA); see Figure 9.2.

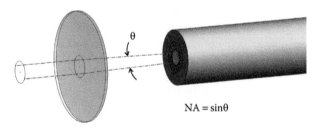

$$NA = \sin\theta$$

FIGURE 9.2 Fiber acceptance angle design requirements.

The lens injection system is typically attached to a multiaxis mount to allow either the lens or fiber to be adjusted in x, y, z as well as pitch and roll. Mechanical fine-tuning allows the compressed laser beam to adjust to the center of the fiber for greater injection efficiency. To decrease the sensitivity of alignment, the injection optic design ensures that the compressed beam spot diameter beyond the focal point is ~80% that of the diameter of the fiber core as shown in Figure 9.2, in effect; the laser spot size must be smaller than the core diameter.

Fibers used for laser beam delivery applications are commonly step-index fibers with a core of pure silica and an F-doped silica cladding encased in an armored stainless steel jacket and coated with a vinyl exterior for protection against impact, debris, and fluids. The standard cross-sectional core geometry of the fiber is typically circular, but over the last 10 years, a variety of geometric core shapes have emerged, including a square-formed fiber core cross section as well as hexagonal, octagonal, and rectangular cores. Figure 9.3 shows images of various core geometries.

These new geometric cross sections provide new opportunities for laser processes when considering the application of fiber optic beam delivery to more precision-based laser material processing. As for core dimensions, circular and geometric cores range in size from 75 μm × 75 μm to as large as 1.6 × 1.6 mm. The core size chosen is based on the power levels being transferred, as well as the feature or focal spot size being chosen for the laser process itself. Using square cores, for example, offer better edge finish of linking square shapes, reducing the scalloping effect consistent with round spots.

The output housing is where the beam emerges from the fiber end and gets delivered to the workpiece. The output can consist of a simple collimation lens or a complex optical assembly depending on the final laser beam spot characteristics required to perform the process. The injection optics should be of sufficient quality that the focused spot size is limited by the beam quality rather than the aberrations in the lens. Knowledge of the laser's beam quality such as the M^2 value allows designers to determine the desired performance of the injection optics, while taking into consideration laser beams' characteristics.[11] The beam quality becomes the limiting factor when the actual M^2 value of the beam deteriorates. Obviously, the laser plays a vital role, however, the problem of selecting the collimating lens needs to be given consideration if performance of the overall system across its operational range is to

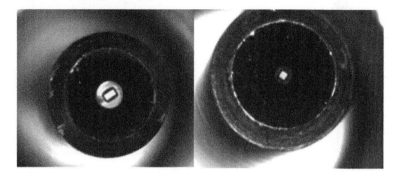

FIGURE 9.3 Geometric core fibers (rectangular and square.)

be understood. Performance degradations will appear somewhere in the system's operational range; understanding the limitations of each component will be helpful during the design process. In order to realize the best possible performance from the beam delivery system, well-corrected optics, such as aspheric, Gradient-index (GRIN), or multielement lenses are absolutely required.

Eliminating aberrations increases the energy available for useful work. Conventionally, well-corrected optic assemblies are produced by using doublets, triplets, or aspheres. Aspheres require specialized manufacturing processes and are substantially more costly than spherical optics.[11] Therefore, manufacturers of high-power laser systems have often used air-spaced or cemented doublets. Over the years, it has been recognized that a superior alternative to doublets or triplets would be the use of GRIN material.[12] GRIN optics refers to the branch of optics covering optical effects produced by a gradual variation of the refractive index of a material. Such variations can be used to produce physically thin lenses with flat surfaces, or lenses that do not have the aberrations typical of traditional spherical lenses. GRIN lenses can be designed to have refraction gradient that are spherical, axial, or radial.[13]

The refractive-index gradient, coupled with an appropriate surface curvature, allows well-corrected optics to be produced using planar or spherical surfaces. The effect is analogous to a single-term asphere. GRIN optics perform as well as comparable conventional doublets, yet often provide diffraction-limited performance. For example, a source of such GRIN lenses used for fiber optic laser beam delivery is manufactured by a company called LightPath Technologies, Inc., Fort Lauderdale, Florida, whose trademarked GRADIUM® lens uses its unique refractive index profile to bend laser rays while traveling through the lens resulting in improved focus with a smaller spot size. Figure 9.4 shows the difference between conventional and GRIN technology.

Large-sized GRIN lenses have been developed and are available today with diameter sizes ranging from 5 mm to 100 mm with various focal lengths offering an ample off-the-shelf selection for system designers. Not only can collimators based on GRIN optics handle more power, they also enhance beam quality, reduce maintenance, increase power delivered to the target area, and improve mode stability. The result is faster and more reliable material processing.

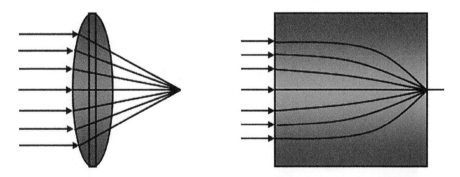

FIGURE 9.4 Shows a conventional lens versus a GRIN lens with varying indexes of refraction.

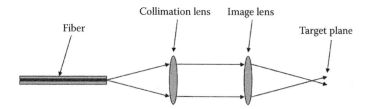

FIGURE 9.5 Shows the fiber output collimator and imaging lens configuration.

Finally, the output housing includes the final element collimation and the imaging optics. The purpose of the imaging optics is to reimage the fiber face to the material surface without distortion or variations that could impact the quality of the laser material interaction, be it drilling, milling, or welding. To ensure that these conditions are met, the design should take into consideration the various parameters that could affect the process itself. Since reimaging the fiber face involves imaging finite conjugates, the most efficient layout requires two lenses: a collimating and a focusing lens. While, for fixed finite conjugates, it is unlikely that the best optical performance can be realized when each lens works at infinite conjugates, forcing each lens to work at infinite conjugates allows the focal length of the focusing optic to be changed as needed without requiring any changes to the collimating optic. This technique simplifies the calculation of magnification.

As discussed previously, each lens should be at or near the best form shape to minimize spherical aberration. M^2 is an unnecessary parameter for calculating the performance of the system with large (200–1000 µm) fibers. Simple geometrical optics are sufficient.[11] As the diagram indicates in Figure 9.5, the collimator and imaging lens work together to define the image size on target. If the system needs to be changed to produce smaller or larger spot sizes, the only options available are to change the fiber core size or to change the collimation and imaging lens. In either case, there could be limitations or adverse conditions, such as a reduced working distance or less power-carrying capacity. Furthermore, consideration must be taken to cover process support subsystem, such as process nozzles for cover gas and protective windows, and in situ-process monitoring such as cameras or active sensors.

9.3 GEOMETRIC-SHAPED CORE FIBER OPTICS

10 years ago, the predominant core geometry of fiber optics for industrial laser-based fiber optic beam delivery was circular. Although the fiber material technology was well established at that time for many laser-based applications, the method of drawing the fiber core and cladding was evolving to increase the efficiency and to develop new geometric shapes. Another development during that period was research being conducted to develop complex preforms, which were the genesis of today's more complex microstructured optical fibers. These led to many advances in fiber lasers and fibers used in spectroscopy.

For industrial laser applications, it is easier to manufacture optical fibers having a circular cross section core, but there is no fundamental reason that fiber cores cannot have other shapes, including octagonal, rectangular, and square.[14] Today, a number of companies currently offer industrially qualified geometrically shaped cores for laser-based applications, including CeramOptec, Fujikura, Mitsubishi, and OptoSkand to name a few.

9.3.1 FORMATION AND MATERIALS

The basic configuration for geometrically shaped core is step-index multimodal fiber beginning with a pure silica material. It is shaped into specific geometries such as a square, then surrounded by a relatively thin region of silica-doped optical cladding. Additional pure silica outer clad is wrapped around the shaped core with a final protective polymeric buffer surrounding the whole to enhance the fiber's strength.[15] The typical steps in manufacturing optical fibers consist of preform fabrication, fiber drawing, cooling, and coating. The silica glass preform is drawn to a desired fiber geometry or diameter by radiation heating above its melting temperature under an applied axial tension. During the draw, the preform encounters a drastic change in temperature and geometric size. Upon exiting the drawing furnace, the temperature of the drawn fiber is lowered just below its melting temperature of approximately 1900 K (~1626°C) down to 500 K (~226°C) through convective and radiative cooling.[16] The neck-down process from the preform core geometry to its final fiber core geometry is critical for producing the desired final fiber quality.

Fibers with a geometric-shaped core, such as a square core, can have a circular cladding so that they are compatible with standard ferrules and mountings. Or they can have a uniform cladding which maintains the square shape. NAs between 0.12 and 0.37 are typical and with recent advances, larger NA fibers are popularizing the use of small core sizes, which have led to further applications within micromachining, where smaller feature sizes on target are desired. Depending on the operational target wavelength, the core silica material can be varied, with either ultra low OH silica for the visible to near IR, or high OH for UV to visible operation.[14] Silica fiber optics, as noted above are used primarily to deliver Nd:YAG (1064 nm) laser powers in excess of 2000 W, whereas lasers operating beyond 2 μm require nonsilica fiber optics. As more industrial applications begin to emerge for wavelengths from 2 μm to 4 μm, specifically IR transmissive fiber optics for Er:YAG (3 μm) or CO_2 (10.6 μm), these laser beam deliveries will require the use of fiber optic materials such as sapphire, fluoride, chalcogenide glass, or hollow waveguides, capable of transmitting these longer wavelengths.[17]

9.3.2 FIBER CORE GEOMETRY

With various core geometries available on the market today, including square, hexagonal, octagonal, and rectangular, fiber optics are now not only guiding or transporting the laser beam, but also actively forming the beam shape along the way. What is unique about geometrically shaped cores is that they perform similar to channel integrators or kaleidoscopes, which are discussed in detail in Chapters 2, 5, and 12 within this book. Figure 9.6 shows a diagram of cross section images of various fiber core geometries.

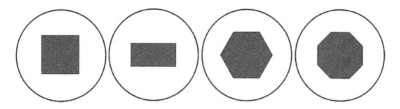

FIGURE 9.6   Square, rectangular, hexagonal, and octagonal cores.

Hexagonal-shaped step index fibers are advantageous in compact fiber bundle packaging for lighting applications, however the hexagonal output also provides an optimum on-target laser pattern used in large-area ablation where the beam can be overlapped to produce homogeneous intensity dosage or integrated ablative pulse distribution, creating a uniform depth during laser micromachining or micromilling operation as shown in Figure 9.7.

A square-shaped fiber core delivering a square formed near field beam is especially well suited for ablation-based thin-film removal. The unique behavior of the square core fiber to act as a homogenizer allows the final output beam to process thin films without creating thermal gradients which could lead to nonuniform delamination of the bulk coating. Rectangular-shaped fibers similar to square also enhance the efficiency in fiber laser delivery applications and in some cases, provide better beam utilization allowing for a rectangular pattern at the workpiece. The rectangular cross section minimizes energy density reduction at the output end of the fiber core, which occurs when a rectangular output of a solid state laser is coupled to a circular cross section fiber core.

Innovations in fiber manufacturing led to core shaping which optimized laser light transportation to the workpiece provided for cost-efficient system designs by reducing the need for complicated or expensive micro-optic-based beam shaping optics. These shapes alter the modal structure of the propagated laser light, improving the optical performance in varying ways for different system designs and application.[15]

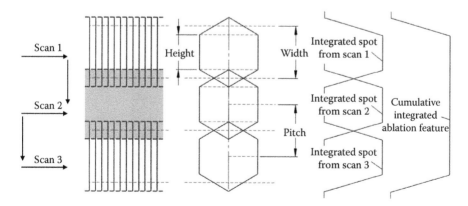

FIGURE 9.7   Hexagonal stitch-scanning sequence that can be used for micromachining.

9.3.3 FIBER DAMAGE THRESHOLDS

There are three discrete locations where damage and failure could occur within a fiber used for laser delivery. These include the front face, the rear face, and the internal face.[10] Depending on the design of the fiber ends, the damage threshold can vary, but in general, the fiber surface finish is the overriding factor in its durability, if properly illuminated by the laser injection optics. Knowledge of power limitations to light propagation through optical fibers is important in the development of fiber amplifiers and for the delivery of laser light to precise locations for materials processing.[18]

The most common approach to telecom fiber application is the formation of fiber ends by cleaving. Since the telecom boom in the late 1990s, fiber cleaver innovation has led to cleavers that produce high-quality fiber ends. Modern fiber cleavers are fairly good at producing high-quality cross-cut surfaces. The fiber is placed under slight tension and is then nicked, usually with a diamond blade. The tension causes the defect crack to propagate across the fiber. Some tools also allow for cleaving on an angle. Cleaved fibers are best used for telecom applications, where industrial laser applications are often compromised due to residual damage from the cleaving process. Figure 9.8 shows an example of a fiber that has been cleaved.

Note the small damage on the outer edge of the fiber as a result of the cleaving operation.[19] Anytime this method is used, deeper damage or induced stress can be created and go undetected using normal optical inspection. Such damage, when subjected to the thermal load of high-powered lasers, reduces the normal damage threshold. It should be noted that research has shown that the cleaving process can work, however for larger fiber core sizes, the dynamics of cleaving can create unacceptable surface variations that once again lead to reduced damage thresholds.

From an industrial laser perspective, a fiber used in a laser beam delivery system needs to be field replaceable. This means the use of different types of fiber

FIGURE 9.8 Mechanically cleaved fibers retain the notch used to form the cleave.

connections such as bayonet style connectors or the SMA (subminiature) connector that bonds a fiber optic into a 1/8-inch diameter male ferrule, which can be connected into a female-based injector or output connector. For large-scale industrial laser applications, armored cables and a bayonet style connection are used for increased stress reduction over the traditional SMA. The robustness designed into bayonet assemblies allows the fiber to handle greater stress and strain induced by multiaxis movements made by robots. This means that fiber endfaces must be prepared in a manner that allows durability and fast disconnection.

For high-powered industrial laser applications, the fiber cleave needs to be polished. The most common approach to polishing the fiber end of ferrule-based systems employs mechanical incremental polishing steps. The fiber is epoxied into the ferrule assembly and cut by cleaving, then a polishing regimen is used to improve the endface optical quality. A coarse polishing grit is used to remove excess epoxy and bring the fiber endface close to the top of the ferrule surface. Following this method, all scratches and other damage that will affect the optical throughput are removed by incremental polishing.[18] Industry research has shown that improper polishing can generate further subsurface defects, which often cannot be corrected using finer polishing compounds. Therefore, even polishing can limit the damage threshold values of the fiber endface. For high-power applications, the mechanical polish starts with a fairly small initial grit sizes and polishes over a longer time period to achieve the optimum surface finish. This procedure requires more time, but it is the preferred method for producing high-power fiber endface finishes.[18]

Modern industrial fiber assemblies have recently shifted to a new method for improving potential endface damage using laser-heated polishing. The wavelength of the heating laser needs to be chosen such that it is readily absorbed on the fiber surface but does not transmit down the fiber, causing deeper structural damage. CO_2 lasers at 10.6 µm are an excellent choice for fused silica fiber.[18] For example, research using a low-power longitudinal 10 W, continuous wave (CW) CO_2 laser demonstrated a decrease of surface roughness on a fiber endface form about 5 microns to only a few hundreds of a nanometer.[20]

Various studies have compared the three methods of preparing fiber endfaces described above. When comparing the optical damage threshold of cleaved fibers, mechanically polished fibers, and laser-polished fibers, the general outcome demonstrates that fibers polished using mechanical-polishing routines exhibit long lifetimes at relatively low laser powers. Cleaved fibers and those polished using the slower high-power techniques show substantially improved damage thresholds, but fibers polished using a final CO_2 exposure show the best optical damage threshold and the longest lifetimes.[18,21,22]

As mentioned earlier, another location for potential failure is within the length of the fiber itself. Typically within the first 3 mm from the endface where the laser is injected into the fiber, a phenomenon can occur called self-focusing. A fiber optic-based delivery carrying high power is susceptible to self-focusing which occurs when an intense laser pulse causes damage to the coupling end surface and refocuses the energy deeper inside the fiber. This leads to fiber material damage due to nonlinear effect of the fiber's refractive index. Simply stated, self-focusing results

FIGURE 9.9 Perfect square core fiber, internal damage, and external AR-coating damage.

around an intensity peak on the fiber face, focusing the light deeper inside the fiber. Figure 9.9 shows damage inside the bulk fiber used in an industrial laser application. The failure was due to the laser failing to maintain its mode and a resultant hot spot was formed. Other types of damage are also represented such as coating damage due to contamination.

9.3.4 OUTPUT UNIFORMITY IN A DYNAMIC SYSTEM

When launching laser light into a step-index multimode optical fiber, the optical pattern at the fiber output endface when imaged to the target plane and analyzed will be found to be split into many irregular light spots, called speckle. The degree of speckle and its uniformity is a function of the optical coherence of the laser. The method of injection into the fiber must take into account how well the fiber's optical NA is filled. Robotic systems continuously shift and bend or contort as they process a workpiece. Any shifting of intensity or hot-spot formation will be present at the fiber output as the fiber is flexed during that sequence of robotic movements. By dynamically measuring the fiber beam delivery's imaged output, it is possible to determine how stable the laser beam uniformity remains maintained at the process image plane. Where this becomes critical depends on the application and the tolerance of the process to withstand such transient hot-spot formations. For instance, to ensure the laser process is capable of etching thin films or scribing various coating materials, it is important to stabilize the uniformity of the beam on-target so that output energy density and uniformity remain near constant. Acceptable uniformity ranges within ±5% of beam quality across the imaged laser output. Laser processes destined for the solar market require the utmost in precision and reliability, so uniformity and stability become critical. In terms of automotive applications, variations around ±8% might be acceptable. Figure 9.10 shows the intensity distribution from an imaged square core fiber. The images show how the intensity uniformity and speckle distribution change as the fiber is flexed in the simulated scan over a 3m length. By simulating the movement of a fiber optical beam delivery in a robotic track, it is possible to determine how well the system will perform overall and provide valuable feedback for finding ways to improve the process.

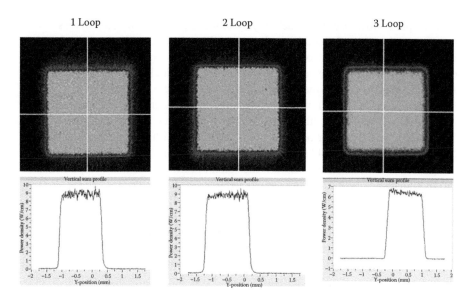

FIGURE 9.10   Example intensity variation of fiber output due to bending loops.

9.4   INJECTION OPTICS: EFFICIENCY, STABILITY, AND UNIFORMITY

When developing a laser system for industrial applications, technologists should start the design from the laser material interaction standpoint and work backward. It is common for people to overspecify a system based on past biases, discounting the need to understand the laser material interaction portion of the overall process during the system development. A failure to test a laser process end-to-end over its dynamic range, including the material interaction at the output end, will ultimately lead to missed opportunities for cost reduction and laser optimization. The final design generated might be too exotic or technically challenging in practical terms, leading to a lost opportunity for using beam shaper methods which can be relatively inexpensive to implement.

The fact is, beam shaping provides a more uniform target material removal rate and increases the overall quality of the work being processed. It is a well-known fact that customers prefer their parts to look good. This makes it critical to understand the design options based on real laser material interaction process results. This is important, because system requirements usually determine several important laser beam characteristics, such as wavelength, pulse energy, pulse width and shape, repetition rate, and in most cases a specific spatial beam shape. If these requirements are specified too tightly, the cost of both the laser and the subsequent system design could be out of reach for the particular market targeted. That is why it is beneficial to test the conditions that the laser process might encounter from the laser end all the way to the target surface. In terms of robotic-based fiber optic beam delivery, this means testing throughout the full range of motion of the fiber as it would be attached to the optical housing articulated by the robotic arm.

The optical fiber laser beam delivery system design, as discussed previously, will have its requirements defined by such testing. The parameters discovered after testing will dictate the optical power and the injected spot size; from this, the optical energy density being injected into the fiber may be calculated. Typically, the fiber size is selected to deliver the desired optical energy density, however, given these fixed requirements, several other aspects of the laser design will determine the laser beam mode structure. To filter injection for high-power lasers, a uniform spatial field is desired. If the laser beam exhibits small areas of higher intensity, this leads to a lower damage threshold for the fiber during injection.[11,15,18] The injection system design can be chosen to help smooth the spatial profile of the laser output, such as through the use of beam-shaping optical elements.

The main design criteria for high-power fiber injection are minimizing peak fluence in air before the fiber, minimizing peak fluence on the fiber endface, aligning the fiber axis to the incident beam axis, minimizing laser *hot spots*, preventing conditions that lead to focusing within the fiber, and broadening the initial-mode power distribution within the fiber.[10,23,24] The following sections will cover three approaches to injection of lasers into geometrically shaped core fiber optics used as optical fiber laser beam delivery systems for industrial applications. These particular techniques are explained in relation to injection into square core, step-index fiber optics. The square core fiber optic is the primary fiber core geometry used for high-end micromachining and milling applications. The methods described include traditional lens, diffuser, and integrator-based designs; however, each can be adapted to various other core geometries including hexagonal, octagonal, and rectangular shapes.

9.4.1 Traditional Lens Laser Injection

To understand the requirements for designing and building an injection system, an overview of the effects needs to be discussed. The information outlined here addresses the beam injection issues for a square core, step-index fiber. It has been observed by researchers in the laser beam shaping community that strong patterns can be observed in the output irradiance of optical fibers when the laser source is focused onto the entrance endface of the fiber. This is due to the fact that, although the mode volume of the fiber is uniformly filled, the phase is not suitably randomized. This is analogous to the fact that summing a large number of Fourier series terms with uniform amplitude and phase will approximate a delta function. If the phases are randomized, the result will be a uniform speckle pattern. The Fourier components correspond to the modes of the fiber.

Further, in the case of uniform phase input, the phase of the different modes will change as light propagates down the fiber due to the difference in mode propagation velocities. But the phase will not assume a random distribution if the fiber is not perturbed. This problem is frequently alleviated using mode scramblers, which usually employ bends or microperturbations designed into the fiber path. The problem can also exist if the mode volume of the fiber is not uniformly filled. This will happen if the source NA does not match the NA of the fiber.

When using a traditional lens-based injection system, it should be noted that during transmission through the fiber, the light rays take various paths, which results in homogenization. In general, homogenization is a benefit and often a primary goal for using fibers for laser beam delivery. In the case of square core fibers, the fiber acts similar to a refractive channel integrator, since the four faceted sides form a remarkably similar channel.

9.4.2   DIFFUSER BASED INJECTION OPTICS

The main criterion for fiber injection is to produce a beam that matches (or fills) the modes of the fiber as previously mentioned. It is assumed that the energy is uniformly distributed among the fiber modes. If the phase of the modes is random, the optical field at any plane along the fiber will resemble a speckle pattern. This is the pattern observed at the output end of the fiber when the modes are effectively filled. It is commonly observed, but not well understood, that when the modes of a step-index fiber are effectively filled, the pattern does not change (statistically) with subsequent bends in the fiber.

Any input beam shape that duplicates the radiation pattern of the fiber mode will be accepted by the fiber and will propagate down the fiber unchanged if the fiber is straight with no stress-induced birefringence. If there are bends or birefringence in the fiber, the excited mode will couple into other modes of the fiber. This is the basis for mode scramblers. Mode scramblers typically employ a section of tight loops in a fiber or a series of microstresses (microbends) to mix modes at output. Some dynamic high-frequency vibration devices, such as piezo electric actuators operating at high frequencies have also been used to create this type of scrambling effect.

The fiber injection technique discussed here uses an optional far-field diffuser that matches the SBP (also known as the space-bandwidth-product) of the optical fiber. The SBP is a measure of the number of modes (Eigen functions) needed to represent a beam. The fiber modes are the Eigen functions of the boundary problem representing a fiber waveguide.

9.4.2.1   The Space-Bandwidth-Product

It is well known that a space-limited function cannot be band-limited, or conversely a band-limited function cannot be space-limited. A space-band-limited function is a function that is identically zero outside some region, this is based on the theory of prolate spheroidal wave functions.[25-31] Such a function can be arbitrarily close to a function that is both space-limited and band-limited if the SBP is sufficiently large.

The number of modes required to represent an arbitrary space-limited and band-limited function is determined by the SBP of the function given by the following:

$$N = WX \tag{9.1}$$

In this equation, W is the bandwidth of the function, and X is the space-width of the function. The number of modes is just the number of orthogonal functions (basis functions) need to represent the function. N is sometimes referred to as the *number of degrees of freedom* of a function, and is closely related to β, the uncertainty principle, and M^2.

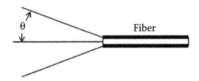

FIGURE 9.11   The fiber NA.

9.4.2.2   Fiber Space-Bandwidth-Product

A fiber is represented schematically in Figure 9.11. The angle θ is defined by the NA of the fiber as follows:

$$NA = \sin\theta \tag{9.2}$$

The NA of a multimode step-index fiber is measured by filling the modes of the fiber and measuring θ of the radiation pattern output of the fiber. The angle is measured at the 5% point of the radiation pattern. The far-field radiation pattern of the fiber is given by the Fraunhofer diffraction integral as follows:

$$U(x_0, y_0) = \frac{e^{ikz} e^{i\frac{k}{2z}(x_0^2 + y_0^2)}}{i\lambda z} \iint U(x_1, y_1) e^{-i\frac{2\pi}{\lambda z}(x_0 x_1 + y_0 y_1)} dx_1 dy_1 \tag{9.3}$$

where:

$U(x_1, x_2)$ is the field at the output face of the fiber.

Equation 9.3 is a Fourier transform where the frequencies are given as follows:

$$f_x = \frac{x_0}{\lambda z}, \quad f_y + \frac{y_0}{\lambda z} \tag{9.4}$$

If we let θ_x and θ_y to be the angles in the far-field radiation pattern of the fiber and note that

$$x_0 = z \tan\theta_x, \quad y_0 = z \tan\theta_y \tag{9.5}$$

Equation 9.4 simply becomes the following:

$$f_x = \frac{\tan\theta_x}{\lambda}, \quad f_y = \frac{\tan\theta_y}{\lambda} \tag{9.6}$$

Substituting θ for θ_x or θ_y in Equation 9.6 gives the spatial bandwidth W of the fiber radiation as follows:

$$W = \frac{2 \tan\theta}{\lambda} \tag{9.7}$$

The factor of 2 in Equation 9.7 is due to the fact that the angle θ is the half angle of the radiation pattern. If the fiber core dimension is d, the SBP for the fiber is given by the following:

$$N = \frac{2d \, \tan\theta}{\lambda} \tag{9.8}$$

Note that the development thus far is 1D. To get the number of modes for a fiber, we still need to use the quantity N^2 for a rectangular core fiber or the area for a circular core fiber. Also, N gives an upper bound for the number of modes in a fiber; the actual numbers will be generally less because the boundary conditions for the waveguide do not allow all the functions accounted for in the SBP calculation. For a 150 μm × 150 μm square fiber with an NA of 0.22, Equation 9.8 gives a value N of 65.7, making $N^2 = 4315$. This is considerably more than the number of modes (3225) calculated for this fiber based on fiber guide theory.[32] This might be explained partly by the fact that the formula from fiber guide theory is an approximation. However, the major difference is due to the fact that the fiber guide boundary conditions limit the number of modes; the above theory does not include such boundary conditions. The best explanation is that the above analysis assumes a flat spectrum for fiber radiation (a flat profile), but reality is closer to a cosine irradiance (over the angle defined by the NA). This might account for the differences in the mode numbers between the two calculations.

Equation 9.8 can also be applied to the diffuser. From the design for this fiber (See Section 4.1), we have an effective diffuser (input beam) diameter of 7.75 mm for a diffuser angle of 0.5° divergence angle, giving a value for N as 65.7. Thus, the diffuser gives the same value for N as that for the fiber, which is the basis of this approach.

9.4.2.3 Fiber Injection

Two diffuser approaches can be considered: image or Fourier transform-based. The first is interesting but does not give reliable solutions. The Fourier transform technique gives very good solutions and greatly simplifies the optical design, which is critical for reliability in applications of industrial fiber optic laser beam delivery system.

9.4.2.3.1 Imaged Diffuser Fiber Injection

The first concept that comes to mind is illustrated in Figure 9.12. The idea was to fabricate a random diffuser that had a phase autocorrelation approximating the

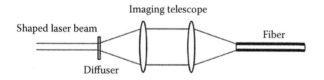

FIGURE 9.12 Imaged diffuser injection.

autocorrelation function of the field at the output face of a mode-filled fiber. The diffuser should also have the same radiation pattern given by the NA of the fiber. The diffuser needs to be illuminated by a collimated uniform beam that is the same size and shape as the fiber core. The diffuser output would then be imaged onto the fiber by a unit magnification telescope, preserving the NA of the beam.

This turns out to be impractical. When considering the fabrication of gratings with the large angle required, it is clear that the cost is prohibitive. This type of design also runs into issues involving iterative algorithms when attempting to produce quasi-random functions. On the other hand, the DC term might be a problem in this case, where it is anticipated that a DC term from the diffuser would not be a problem for this configuration if it is not too large, since it would not be focused.

Shaping a small beam to illuminate a diffuser is problematic, and turns out not to be required if taking a different approach. We have looked into the possibility of using a larger beam to illuminate and then imaging the diffuser output onto the fiber with demagnification, while preserving the beam NA. This would make the diffuser easier to fabricate, but it does not appear to work without extremely long distances between optical elements and large optics which can be an issue when dealing with industrial optical systems constraints.[33]

9.4.2.3.2 *Fourier Transform Diffuser Fiber Injection*

The Fourier transform approach illustrated in Figure 9.13 is based on diffusers available from RPC Photonics (Rochester, NY), SUSS Micro-Optics (Neuchatel, Switzerland), and Holo Or (Haifa, Israel). These diffusers are designed to produce a uniform far-field speckle (random) pattern. This means that the diffusers produce a pattern determined by Equation 9.3, the Fourier transform of the field produced by the diffuser. There is an implication that the diffuser output is independent of the shape of the illumination beam, however this is only approximately true. Assuming that the diffuser was designed using uniform illumination due to the Fourier transform property of Equation 9.3, the output of the diffuser would be a convolution of the Fourier transform of the illumination beam field and the Fourier transform of the diffuser phase transmission function. If the illumination beam was sufficiently large, the Fourier transform of the beam would be very narrow and thus would have little effect on the output pattern. RPC Photonics suggest a minimum illumination beam size of 3 mm, however the illumination beam should be significantly greater than the diffuser feature size.[34]

Assuming the diffuser has the above properties, it is placed before the transform (focusing) lens, and the fiber face is placed in the focal plane of the lens as shown

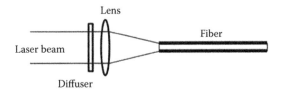

FIGURE 9.13 Fourier-transformed diffuser injection.

in Figure 9.11. The pattern produced by the diffuser at the focal length of the lens is given by Equation 9.3 with the substitution, $z = F$, where F is the focal length of the lens. Let ϕ be the full angle of the diffuser radiation pattern. We can use Equations 9.4 and 9.7 to determine the width w of the spot in the focal plane as follows:

$$w = 2F \tan\left(\phi/2\right) \tag{9.9}$$

Equation 9.9 can be approximated for small angles by the following:

$$w \cong F\phi \tag{9.10}$$

The diameter of the illumination beam is determined by the requirement to produce a cone angle matching the NA of the fiber, giving the following:

$$D = 2F \tan\theta \tag{9.11}$$

Equations 9.2, 9.9, and 9.11 are used to select a lens focal length and illumination beam size once a fiber NA and core size are specified. Using Equations 9.9 and 9.11, we can obtain the basic design equation as follows:

$$D = \frac{d \tan\theta}{\tan\left(\phi/2\right)} \cong \frac{2d \tan\theta}{\phi} \tag{9.12}$$

Note that $d = w$ was used to match the diffuser spot size to the core of the fiber. There is not a unique solution to these equations, iterating the diffuser angle is required to obtain an illumination beam diameter that meets a minimum size requirement (based on the diffuser feature size).

Finally, it is appropriate to address the randomness of the diffuser. Coherence theory[35] gives an approximation for the angular divergence of a random diffuser as follows:

$$\phi = \frac{2\lambda}{\Delta} \tag{9.13}$$

where:
Δ is the autocorrelation width of the diffuser phase function.

For example, an RPC Photonics diffuser with an angle of 0.5° divergence angle has a specified feature size of 0.500 mm. Equation 9.13 gives a Δ of 0.236. This is in reasonable agreement with the autocorrelation of a diffuser; using the specified diffuser size can be considerably less than the feature size due to the randomness generated by the array of the features making up the diffuser. Autocorrelation width of a square lenslet would be considerably less than the dimension of the lenslet. This confirms that the RPC Photonics and Holo Or designs have the degree of randomness they claim.

9.4.2.4 Designs

Designs are given below in Sections 4.2.7 and 4.2.8, respectively, for a 1,030 nm laser and 532 nm laser. RPC gives the wavelength dependence of the diffusers as "0.5 (1.0) deg divergence angle at 532 nm translates into 0.488 (0.977) deg divergence angle at 1030 nm." RPC also states that, with respect to the number of elements included in the illumination beam, "If fall-off is the primary requirement, the design has to make sure that the lens size is appropriate and that the illumination fills a minimum of 100–200 elements."

9.4.2.4.1   1030 nm Designs

The designs for the 1030 nm wavelength using diffusers are given in Table 9.1. The calculations are based on a step-index fiber with an NA of 0.22, utilized for evaluation and standards. The part numbers are listed in the RPC catalog for design evaluation units. Actual units are needed to be fabricated using fused silica with AR coatings on both sides. For the purpose of this discussion, the units represented here were selected from the manufacturer Holo Or and are referenced below.

A design for a SUSS MicroOptics diffuser is given in Table 9.2 and 9.3.

9.4.2.4.2   532 nm Designs

A design for two Holo Or diffusers is given in Table 9.4 of which the $\phi°$ 1.0 was utilized for the testing based on the best match for the specific system.

9.4.2.5   Diffuser Illumination Beam

The beam diameters given in the above designs do not address the shape of the illumination beam. The illumination beam requires expansion and collimation. Since the radiation pattern of the fibers measured in a plane is a cosine-like function,

TABLE 9.1
RPC Engineered Diffusers

Fiber	Core Dimension	$\phi°$ divergence	F	D
A	$\Phi = 150\,\mu m$	0.488	17.6 mm	7.94 mm
A	$\Phi = 150\,\mu m$	0.977	8.80 mm	3.97 mm
B	$150\,\mu m \times 150\,\mu m$	0.488	17.6 mm	7.94 mm
B	$150\,\mu m \times 150\,\mu m$	0.977	8.80 mm	3.97 mm
C	$75\,\mu m \times 75\,\mu m$	0.488	8.81 mm	3.97 mm
C	$75\,\mu m \times 75\,\mu m$	0.977	4.40 mm	1.98 mm

TABLE 9.2
Suss Micro Optics Engineered Diffusers

Fiber	Core Dimension	$\phi°$ divergence	F	D
A	$\Phi = 150\,\mu m$	0.8	10.74 mm	4.85 mm

TABLE 9.3

Suss Micro Optics Engineered Diffusers

Fiber	Core Dimension	$\phi°$ divergence	F	D
A	$\Phi = 150\ \mu m$	0.5	17.1 mm	7.75 mm
A	$\Phi = 150\ \mu m$	1.0	8.59 mm	3.88 mm
B	$150\ \mu m \times 150\ \mu m$	0.5	17.1 mm	7.75 mm
B	$150\ \mu m \times 150\ \mu m$	1.0	8.59 mm	3.88 mm
C	$75\ \mu m \times 75\ \mu m$	0.5	8.59 mm	3.88 mm
C	$75\ \mu m \times 75\ \mu m$	1.0	4.30 mm	1.94 mm

TABLE 9.4

Holo or Engineered Diffusers

Fiber	Core Dimension	$\phi°$ divergence	F	D
A	$200 \times 200\ \mu m$	0.5	22.92 mm	10.34 mm
A	$200 \times 200\ \mu m$	1.0	11.46 mm	5.17 mm

it seems like Gaussian beam approximations would be reasonable. Best practices suggest the use of the diameter defined by the following equation:

$$D = 2.14 r_0 \qquad (9.14)$$

where:

r_0 is the $1/e^2$ Gaussian beam radius.

Some consideration must be given to the polarization of the laser beam. The random nature of diffusers provides depolarization, so there is a slight benefit (if the laser output is polarized) to align the polarization along the diagonal of a square fiber. The grating pattern output would then effectively have two outputs, one for each polarization vector parallel to the sides of the fiber core.

9.5 INTEGRATOR-BASED INJECTION OPTICS

The fiber injection technique developed and discussed in this section covers the concept of utilizing a refractive optical integrator as a means to fill the modes of a square core step-index fiber. The system specification for the fiber injection integrator was constructed as follows:

$$\lambda = 1.06\ \mu m$$

Laser divergence: $\phi = 13$ mrad.
Beam diameter: $D = 5$ mm.
Fiber core: $800\ \mu m \times 800\ \mu m$.
Fiber length: $L = 2.5$ M.
Fiber NA $= 0.22$.

9.5.1 BEAM HOMOGENIZER FUNDAMENTALS

The basic *lenslet array beam integrator* configuration is shown in Figure 9.14. In designing a beam integrator, one needs to consider the system β and the size of the lenslets. β is the factor that determines how well one can approximate the flat-top beam profile. Low β means that the irradiance will not be uniform over the desired spot and, that considerable energy will fall outside the designed spot boundary. The size of the lenslets determines the amount of averaging. The lenslets should be small compared to bright spots in the input beam. There is a trade-off to be made between the lenslet size and β. The β greater than 40 is recommended for beam integrators.[36]

Beta for an integrator is given by the following equation:

$$\beta = \frac{\pi dS}{\lambda F} \tag{9.15}$$

The theoretical pattern for β = 40 and a coherent (single mode) source is shown in Figure 9.15. A multimode laser with decreased spatial coherence will generally provide a smoothing of the irradiance profile.

The output spot size of an integrator is given by the following equation:

$$S = \frac{F}{f/d} \tag{9.16}$$

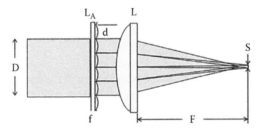

FIGURE 9.14   The basic form of a lenslet array beam integrator.[34]

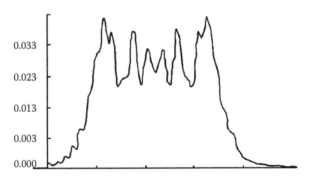

FIGURE 9.15   Single lenslet profile for β = 40.[37]

Since S is usually given, the problem is to select F and d using Equation 9.16 to determine f. In selecting d, one should consider the effects on the coarseness of the speckle; it should be noted that other techniques such as slightly tilting a lens array, sometimes called chirping, provides a means to affect the characteristics of the speckle pattern. The correlation width of the speckle pattern is approximated by the following equation:

$$W \leq \frac{\lambda F}{D} \tag{9.17}$$

The result is that larger d corresponds to a finer speckle pattern. It should be noted that when there is a regular array of spots, the spacing is given in an equation in Chapter 12 under MLAs. In addition to selecting the lenslet size and β, one needs to design the integrator to fill the NA of the fiber. The lens focal length is given by the following equation:

$$F = \frac{D}{2 \tan\theta} \tag{9.18}$$

where:
θ is given as follows:

$$\sin\theta = NA \tag{9.19}$$

For a 5-mm beam diameter, Equation 9.18 gives a focal length of 11.09 mm. The corresponding F-number is 2.22. To avoid complex lenses (due to aberrations), it is common to use an F-number of the primary lens that is equal to or greater than three.

9.5.2 First-Order Integrator Design

Table 9.5 is a summary of integrator designs for three input beam diameters and a focal length that fills the NA of the fiber. Notice that the minimum dimension for the output spot size was chosen to be slightly smaller than the desired input spot size. This is meant to counter the effects of the input beam divergence. The beam divergence spreads the pattern by producing a roll-off at the edges and results in coupling the energy with greater efficiency. To the first order, the output pattern of the beam integrator is obtained by convolving the pattern of the focused laser beam with the

Table 9.5

D mm	S mm × mm	F mm	f mm	d mm Minimum Dimension	β	ρ mm
5	0.7 × 0.7	11.1	7.9	0.5	93.6	0.14
10	0.7 × 0.7	22.2	31.7	1.0	93.6	0.14
20	0.7 × 0.7	44.3	94.9	1.5	70.2	0.14
20	0.7 × 0.7	44.3	63.3	1.0	46.8	0.14

ideal integrator pattern. The width ρ of the pattern spread is approximated by the following equation:

$$\rho = F\phi \qquad (9.20)$$

where:
    ϕ is the divergence of the input beam.

The spread of the pattern would be measured at the half-power points. The estimated spread of the shaped pattern in Table 9.5 is 20% of the design width. The spread increases to 29% for the designs in Table 9.5.

Any of the designs in the table have an adequate β, however larger values are preferred. In selecting a lens for the F-number of 0.22 lenses shown in the table, one might want to consider aberrations associated with the lens. Aberrations together with the primary integrator lens, further spread the output pattern. Smaller F-numbers increase the effects of the beam divergence. So for comparison, Table 9.6 gives the corresponding results for a system F-number of 3.

9.5.3 Optical Modeling

Based on theoretical design analysis, the next step is to create a prototype to test the concept. Using the first-order design as a guideline for specifications, the prototype can be constructed using off-the-shelf components purchased from Suss Micro-Optics. To test the performance of the prototype, component specifications were entered into Zemax to run a simulation. At a $\lambda = 1.06\,\mu m$, laser divergence: $\phi = 13$ mrad, beam diameter: $D = 5$ mm, fiber core: $800\,\mu m \times 800\,\mu m$, fiber length: $L = 2.5$ M, and a fiber $NA = .22$; the components consisted of two MLA lens arrays from Suss Micro-Optics, part#12-1531-500S, with a 1.2° engineered divergence angle and an imagining lens from LINOS HALO LENS, part #038901, which is a multielement lens with a focal length of 14 mm making it a compromise between 11 mm and 22 mm in Table 1. All the optics were coated for 1064-nm AR/AR. The results indicated that the uniformity and performance matched the requirements as outlined in the theoretical design. Figure 9.16 shows the spot diagram generated by Zemax and a ray trace modeling the injection.

With optical modeling showing successful corroboration to the theoretical calculations, a prototype was assembled for testing.

Table 9.6

D mm	S mm × mm	F mm	f mm	d mm Minimum Dimension	β	ρ mm
5	0.7 × 0.7	15	10.7	0.5	69.2	0.20
10	0.7 × 0.7	30	42.9	1.0	69.2	0.20
20	0.7 × 0.7	60	127.6	1.5	51.8	0.20
20	0.7 × 0.7	60	85.7	1.0	34.6	0.20

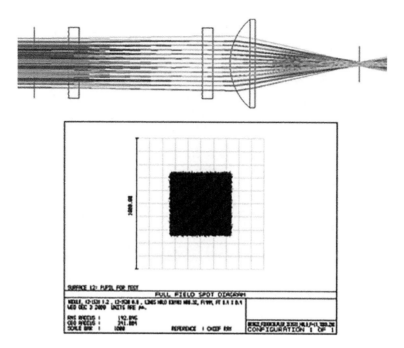

FIGURE 9.16 Spot diagram and ray tracing.

9.5.4 Prototype

An image of the prototype integrator-shaping unit mounted together with the fiber injection system is shown in Figure 9.17. Suss Micro-Optics supplied the components that formed the microlens array integrator, using the LINOS $F = 14$-mm focal length lens picked out during modeling to image the integrated beam onto an 800 μm × 800 μm square fiber. The fiber injection platform utilizes standard SMA connectors to couple the fiber to the optomechanics employing a LINOS rail assembly. The test laser chosen is an LDP-200MQG by Lee Laser of Orlando, Florida. It produces 1064 nm, 25 W of pulsed, multimode output energy. The beam was collimated to a size of ~Ø7.0 mm, then down collimated to ~Ø5.0 mm to illuminate the MLA integrator assembly.

9.5.5 Testing

The goal of assembling and testing the prototype is to determine baseline results of how well a simple lens injector compares to MLA integrator-based fiber injection system. A paper from a fiber beam delivery company called Optoskand, Sweden, discussed the use of square core fibers and traditional launch techniques; Blomster et al. compared a round core fiber with the square core, which both showed similar top hat uniformity when each fiber was tested static, for example, movement and/ or twisting of the fiber was not compared to simulate the articulation of the fiber in a robotic arm.[37] The results showed the beam output intensity with root-mean-square (RMS) of approximately 4% for both round and square cores, while in a static

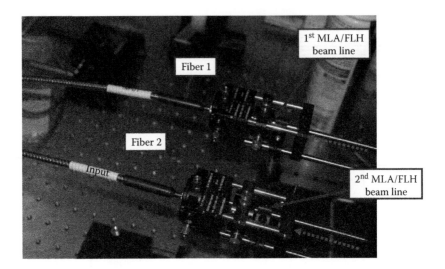

FIGURE 9.17   Prototype MLA integrator and fiber injection system assembly.

position.[35] The results in this work were compared and contrasted with the two methods of fiber injection: a traditional-focused beam approach and a lenslet array beam integrator–homogenizer approach. Each design offers similar and uniform benefits in a static position, however their dynamic behavior is what industrial applications should find most intriguing.

To simulate the original intent of designing a dynamic fiber optic beam delivery system for a moving robotic application, the two MLA/FLH assemblies were tested by inducing changes in the fiber position. The fiber position changes comprised of introducing loops such as 1 loop, 2 loops, and then 3 loops to see if perturbations changed the performance of the output from each fiber, using both techniques. As a baseline, the results shown in Figure 9.18 reveal the output measurements of a traditional lens-based fiber injection.

The simple lens-only design shows an excellent beam uniformity of 4% RMS, however as each loop is added, a preferential shift of intensity to one axis is detected and the shift correlates with the axis of curvature. This is similar to a traditional mode scrambler where perturbations are created in the fiber path. The results were consistent with previous research, however the speckle distribution outcome showed an interesting phenomenon. The second loop created a denser speckle pattern in comparison to the first loop and third loop (this will also be seen in the upcoming MLA results). Since illumination of the fiber with the simple lens only takes into consideration the beam diameter, while the MLA takes into consideration the illumination plus the lenslet diameter, it seems that the major influence becomes the number of turns or loops induced in the fiber. Figure 9.19 shows the results from testing the MLA integrator fiber injection prototypes.

Using an MLA integrator, the results show improvement with slightly better performance. As each additional loop was added, the intensity sloped preferentially to the axis of curvature of the fiber, similar to the traditional lens approach, however

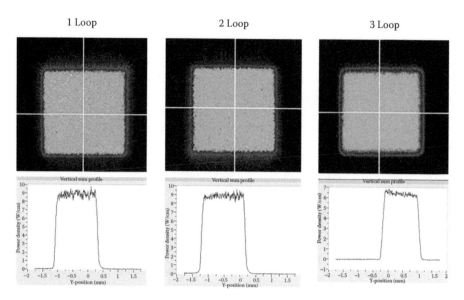

FIGURE 9.18   Traditional lens fiber injection results.

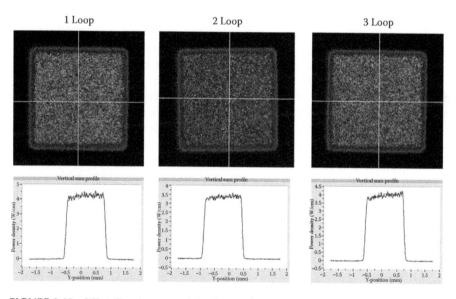

FIGURE 9.19   MLA fiber integrator injection results.

less speckle hot-spot formation, higher speckle uniformity, leading to greater intensity uniformity, measured with the introduction of one loop in comparison to the simple lens. The resulting uniformity RMS was improved between ~3.5% and 4%, while maintaining reduced hot-spot formation.

9.5.6 CONCLUSION: INTEGRATOR INJECTION DESIGNS

The designs in both Tables 9.5 and 9.6 give adequate β, except for the last row in each table. If aberrations are not problematic, the approaches found in the first two rows of Table 9.5 are recommended. For the grand scheme of industrial integration, it is not absolutely clear that the use of an MLA provides any substantial benefit in this simple static test with added loops. It is instead a unique approach that shows a reduction in speckle hot-spot formation with reduced speckle intensity spikes. However, the use of such designs within a dynamic system, such as robotic arm fiber beam deliveries, is likely never going to have an elegant solution in applications where the fiber will be dynamically perturbed in multiple simultaneous axes.

9.6 DEFINING SYSTEM PERFORMANCE

Chapter 12, by Fred Dickey and Scott Holswade, provides an excellent overview of how to approach general guidelines or characteristics for defining the performance of a beam-shaping system. The three general guidelines include uniformity of the intensity across the shaped profile, the steepness of the slope of the shaped beam (the steeper the better) and the efficiency or energy spread beyond the central zone (less loss is always desirable). In terms of an output from a square or other geometric core step-index fiber optics, the goal is to achieve a homogenized image output. The quality of that intensity pattern needs to be balanced against the actual process under development.

Not all materials can resolve small undulations present in a flat top beam intensity pattern, therefore specifying an optical system to achieve uniformity that is theoretical might lead to an expense that is unnecessary and which can become costly and detrimental to the success of commercializing the system. As a rule of thumb (as mentioned in the opening paragraph of Section 4.0), work backward from the material. Understand the material interaction and its dynamic range (which is referred to as its process window) to establish the true requirements for beam uniformity at the image plane of a fiber-based laser beam delivery system. Then continue working backward to determine the fiber, injection, and laser parameters required for design success.

9.7 LASER THIN-FILM PROCESSING

Methods of laser fiber injection into geometrically shaped fiber cores—specifically square core fiber optics, are gaining wide acceptance in the field of thin-film material laser processing for large format substrates such as glass. Key market segments have emerged within the solar and automotive markets. Laser systems under development are required to produce uniform thin-film material removal without damaging the surrounding thin film remaining at its edges of the substrate. Such precise work requires uniform beam intensity employing homogenized and imaged output.[37] Properly designed fiber optic laser beam delivery systems offer such capabilities, including the flexibility for dynamically scanning across large substrate areas, utilizing beam shaping to achieve the level of on-target precision required.

Tasked for patterning functional thin-film coatings deposited onto glass, such as antireflective (metal/dielectric stack for IR reflectivity), transparent heating elements (ITO—Indium Tin oxide, SnO2:F—fluorinated Tin oxide), and capacitive coatings for touch screen applications (ITO); these new fiber optic laser beam delivery systems are beginning to find niche market growth. The following sections describe one such application solution.

9.7.1 AUTOMOTIVE GLASS INDUSTRY APPLICATION

Automotive windshields are an application that demands multiple degrees of freedom if processing is required along the complex curvatures designed for various car and truck models. During the manufacture of modern car windshields, the application of multilayered oxide coatings can offer benefits to the automobile manufacturers which can be passed down to their customers. Solar reflective, better known as infra-red reflective (IRR), windshields have appeared in limited world-wide production over the last 15 years, originally found in selective high-end luxury sedans. Research has found that cars equipped with front windshields glazed with coatings are capable of reflecting greater than 95% of IR sunlight, yet continue to transmit visible light, deliver a more comfortable passenger cabin, increase the vehicle's fuel efficiency, and lead to reductions in both pollution and CO_2 greenhouse gas emissions.

National Renewable Energy Laboratory of the U.S. Department of Energy, the United States Environmental Protection Agency, and the Society of Automotive Engineers have conducted several joint studies and found that reducing 30% of the air conditioning (A/C) demand of a car can reduce fuel consumption by 26%. Since the A/C pump is the largest single-ancillary load added to the vehicle's motor, reducing its size delivers significant savings. As automobile regulatory demands increase and car buyers continue shopping for greener products, trends point to solar reflective automotive glass market share growth moving beyond the luxury sedans into lower cost automotive market sectors.

For over 40 years, solar reflective glass with its energy-saving benefits have enjoyed broad acceptance throughout the architectural market, from large sky scrapers to energy-efficient homes.[38] In the automotive industry, patterning and etching solar reflective coatings have traditionally followed the architectural method—being processed as flat glass. However, windshields are produced utilizing complex curves unique to each and every model of automobile and because of this, the coating must be processed once the windshield is formed into its curved shape. The laser process is tasked with removing the coating in specific patterns and around the windshield edge.

The coating along the windshield's edge is removed, in a process called edge deletion. Since the IRR coating is made from various stacks of metal and dielectric coatings, it must be isolated from the outer edge where it might be exposed to moisture and road salt causing corrosion. The windshield is manufactured through a lamination process so that the coating can be protected with a second cover glass and a vinyl interlayer. The edge-deleted area becomes the hermetic seal protecting the coating from corrosion infiltration. A second task that the laser process can perform is producing functional zones of deleted coatings within the central area of the

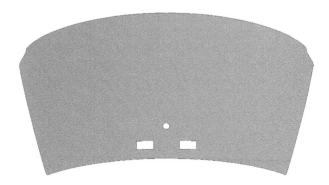

FIGURE 9.20   Schematic windshield with rain sensor and two RFID toll pass openings.

glass. Since the IRR coating is made up of stacked layers of thin metal designed to attenuate certain wavelength of light, the metallization effectively shields long wavelengths, such as cellular radio signals. Typical central coating deletion zones include removing material in geometric patterns designed to allow the operation of rain sensors, EZ-Pass™ RFID tags, night vision driver-assistance sensors, as well as many new technologies coming in the future to benefit passenger protection. Figure 9.20 shows a typical windshield with three features for functional elements.

Looking toward growing future demand for automotive solar reflective glass products and seeking to improve production yields by processing windshields in curved space, the industry had to adapt laser micromachining and etching technology knowhow from flat to curved surfaces. Figure 9.21 is the realization of moving from flat formats to complex curved surfaces by integrating the benefits of a multi-axis robotic system with a 100 W, 532 nm wavelength nanosecond pulse width laser employing a fiber optic beam delivery system.

9.7.1.1   Industrial Integration of Diffuser Injection System

Success was achieved by combining a configuration of a square core fiber optic beam delivery system with a diffuser-based laser injection system. Figure 9.22 shows the injection optical system schematic as well as the actual injection module modified with the diffuser element. The fiber coupling unit on the 532 nm Lee Laser LDP200-MQG model was dialed in to operate a 100 W output with a 280 ns pulse duration. As described in Section 4.2, a $\phi°1.0$ deg divergence angle, diffuser from Holo Or, Israel, utilized for testing during development, was adapted to the actual robotic beam delivery system installed in the automotive glass-manufacturing facility. The fiber chosen for installation is an Optoskand, 200 μm × 200 μm, QBH series for high-power laser applications and coated for 532 nm operation.

9.7.1.2   Industrial Integration Output Housing-Processing Head

The output housing or processing head, mounted to the end of a robotic arm, can take either simple or complex forms depending on the processing tasks required. Figure 9.23 is the processing head that was configured for automotive windshield applications. It is a unique configuration using traditional optical approaches with

FIGURE 9.21 Robotic system with an integrated fiber optic laser beam delivery and close-up of the integrated processing head with collimator, imaging lens, and scanner.

the integration of machine vision, height distance monitoring, and a secondary beam line for changing the spot size on-target. Starting at the top of the assembly, notice a rotary bearing integrated into the QBH fiber bayonet-coupling device. This bearing assembly allows the fiber to rotate, so that the processing head can freely rotate just over 360° of rotation about the wrist of the robotic arm. The robot is a FANUC M20iA model with six axes of freedom with a wrist.

By running the fiber optic through the existing cable pathway on the robotic arm and through the center of the wrist, the system was able to scan across the curved windshield surface continuously and smoothly.

9.7.1.3 Process Results: Diffuser Versus No Diffuser

This system was installed into an industrial production environment which provided an opportunity to perform dynamic testing. The test was conducted to determine if a noticeable process improvement was achieved based only on the quality of ablation

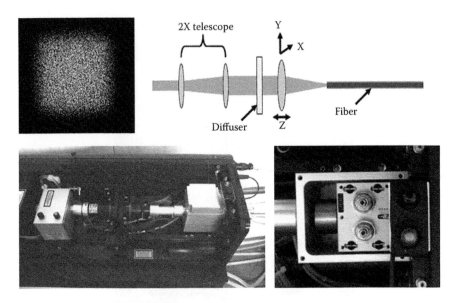

FIGURE 9.22 The diffuser pattern (top left), injection system schematic (top right), assembly of telescope and injection housing actual unit (bottom left), and cover-off showing diffuser-mounting position (bottom right.)

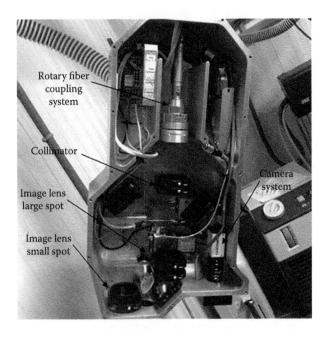

FIGURE 9.23 Output housing processing head assembly (uncovered.)

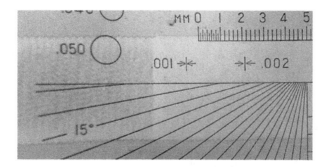

FIGURE 9.24   Test to dynamically show the quality shift (diffuser removed versus diffuser.)

over the robotic arm tool path. One zone was evaluated, which represents an area within the process tool path where the fiber is at its furthest extension and with the greatest number of fiber twists possible within the robotic arm cable ways. The process results are shown in Figure 9.24, pre- and post-diffuser insertion. As can be seen on the right-hand side of the image, there is a shifted hot spot in the beam. The output laser beam when imaged onto the surface creates a hot spot, that results in an incomplete removal of material, shown in Figure 9.24 as a texture to the thin film. When the diffuser is added, the uniformity is maintained over the travel of the robotic arm and the coating shows uniform removal. The use of the diffuser provides the process window needed to ensure quality edge deletion for this automotive windshield application.

9.8   CONCLUSIONS AND OPPORTUNITIES

Traditional fiber injection into a fiber optic laser beam delivery with simple lenses has limitations in terms of maintaining dynamic output uniformity. This means that the ability of the fiber optic laser beam delivery to maintain a uniform output when the fiber is bending in various directions represents problems at the output end. An improved uniformity of the laser beam at the image target plane as well as stable uniformity over the dynamic range of a multiaxis robotic arm can be achieved if the design incorporates the entire process end-to-end. Leveraging a diffuser-based injection technique provided the best overall performance while balancing tradeoffs, such as cost, complexity, and durability. Within industrial laser applications, various new possibilities have emerged for thin-film processing over large area substrates through the introduction of laser sources with higher powers and higher pulse repetition rates. The hybrid technologies brought together have created unique laser beam delivery options that can be configured to meet a wide variety of scanning and gantry-based system configurations that are capable of handling the new laser power and repetition rates. It is clear that further research and development of fiber optic injection represent a worthwhile investment. As beam-shaping technology continues to evolve and the costs reduce further, new techniques or hybrid optical configurations will provide further enhancements to laser injection into fiber optics and find opportunities to penetrate broader market sectors within industry.

REFERENCES

1. The HistoryMakers® Video Oral History Interview with Marshall Jones, August 4, 2012. *The HistoryMakers® African American Video Oral History Collection*, 1900 S. Michigan Avenue, Chicago, Illinois.

2. Jones, M., High power laser energy delivery system, Patent US 4681396 A, 1987.

3. Ortiz, A., Fiber optic delivered beam quality control system for power lasers, Patent US 5245682 A, 1993.

4. Roessler, D.M., and Michael G.C., Laser systems in the automotive industry, *Opt. Photonics News.*, 7(6), 16–21, 1996.

5. Low, D.K.Y., and Lin, Li., Effects of interpulse and intrapulse shaping during laser percussion drilling, *Proc. SPIE* 4426, Second International Symposium on Laser Precision Microfabrication, 191, February 19, 2002.

6. Marszalec, J.A., and Elzbieta, A., Marszalec. 4, *Integration of Lasers and Fiber Optics into Robotic Systems.* Bellingham, WA: SPIE Optical Engineering, 40–55, 1994.

7. Lizotte, T.E., and Dickey, F.M., Beam shaping diffuser based fiber injection for increasing stability of industrial robotic laser applications *Proc. SPIE* 8843, Laser Beam Shaping XIV, 88430H, September 28, 2013.

8. Pietro, F.A., Robotic Laser Welding Systems in Automotive operations, in *Laser Applications for Mechanical Industry*, Vol. 238, Martellucci, S., Chester, A.N., and Scheggi, A.M., eds., Springer Netherlands, pp. 271–275, 1993.

9. Chasse, E., *Laser Optics: Special Delivery, The Fabricator July 2008*, FMA, July 15, 2008. http://www.thefabricator.com/article/.

10. Greenaway, M.W., Proud, W.G., Field, J.E., and Goveas, S.G., The Development and study of a fiber delivery system for beam shaping, *Rev. Sci. Instrum.*, 73(5), 2185–2189, 2002.

11. Hunter, B.V., et al., Designing a Fiber Optic Beam Delivery System, *Lasers as Tools for Manufacturing II, Proc. SPIE* 2993, 1997.

12. GRADIUM Lenses, *GRADIUM Lenses.* Lightpath Technologies Inc., Oralando, FL, 21 Aug, 2012.

13. Gradient-index Optics, *Wikipedia.* Wikimedia Foundation, 12 May, 2005. https://en.wikipedia.org/wiki/Gradient-index_optics.

14. Schuberts, F., SQUARE Fibers Solve Multiple Application Challenges, *SQUARE Fibers Solve Multiple Application Challenges.* Photonics Spectra, 1 Feb, 2011. http://www.photonics.com/Article.aspx?AID=45913.

15. Shannon, J., Shaped Core Multimode Fibers for Improved Power Delivery. *Polymicro Fiber Optics White Papers.* Molex, 2005. http://www.molex.com/mx_upload/superfamily/polymicro/pdfs/Shaped_Core_Multimode_Fibers_for_Improved_Power_Delivery_Apr_2009.pdf.

16. Mawardi, A., and Pitchumani. R., Optical fiber drawing process model using an analytical Neck-Down profile., *IEEE Photonics J. IEEE Photonics J.*, 2(4), 620–29, 2010.

17. Harrington, J.A., Overview of power delivery and laser damage in fibers, *in Laser-Induced Damage in Optical Materials: 1996*, Bennett, H.E., Guenther, A.H., Kozlowski, M.R., Newnam, B.E., and Soileau, M.J. eds., Proc. SPIE 2966, 536, May 13, 1997.

18. Ronald, H.G., and Smith, A.V., Self-focusing in High-power optical fibers. *Integrated Optics: Devices, Materials, and Technologies XI, Proc. SPIE* 6475, February 8, 2007.

19. Joe Thomes, Jr. W., Ott, M.N., Chuska, R.F., Switzer, R.C., and Blair, D.E., Fiber optic cables for transmission of High-power laser pulses. *Nanophotonics and Macrophotonics for Space Environments V, Proc. SPIE* 8164, September 13, 2011.

20. Orun, H., Udrea, M.V., and Alacakir, A., Polishing of Optical Fibers Using a CO_2 Laser. *Proc. SPIE* 4068, *SIOEL '99: Sixth Symposium on Optoelectronics*, 570, February 23, 2000.

21. Poprawe, R., *Tailored Light 2: Laser Application Technology*. Vol. XVI. Heidelberg: Springer, 2011.

22. Nowak, K.M., Baker, H.J., and Hall, D.R., Analytical model for CO_2 laser ablation of fused quartz. *Appl. Opt. Applied Opt.*, 54(29), 8653–654, 2015.

23. Harrington, J.A., Infrared Fibers and Their Applications. SPIE, 2004. https://spie.org/Publications/Book/540899.

24. Harrington, J.A., Infrared fibers, in *Handbook of Optics: Fiber and Integrated Optics*, Vol. 4, Bass, M., Enoch, J.M., Van Stryland, E.W., and Wolfe, W.L., eds., McGraw-Hill, New York, 2000.

25. Slepian, D., and Pollak, H.O., Prolate spheroidal wave functions, fourier analysis and uncertainty—I, *Bell Syst. Tech J.*, 40, 43–63, 1961.

26. Landau, H.J., and Pollak, H.O., Prolate spheroidal wave functions, fourier analysis and uncertainty—II, *Bell Syst. Tech J.*, 40, 65–84, 1961.

27. Landau, H.J., and Pollak, H.O., Prolate spheroidal wave functions, fourier analysis and uncertainty—III: The dimension of the space of essentially time and Band-Limited signals, *Bell Syst. Tech J.*, 41, 1295–1336, 1962.

28. Lepian, D., Prolate spheroidal wave functions, fourier analysis and uncertainty—IV: Extensions to many dimensions; Generalized Prolate Spheroidal Functions, *Bell Syst. Tech J.*, 43, 3009–3057, 1964.

29. Landau, H.J., and Widom, H., Eigenvalue distribution of time and frequency limiting, *J Math Anal Appl.*, 77, 469–481, 1980.

30. Frieden, R., Evaluation, design and extrapolation methods for optical signals, based on use of the prolate functions, Chapter VIII, in *Progress In Optics, Vol. IX*, Wolf, E., ed., North-Holland, Amsterdam, 311–406, 1971.

31. Dickey, F.M., Romero, L.A., Delaurentis, J.M., and Doerry, A.W., Superresolution, degrees of freedom and synthetic aperture radar, IEE P-Radar Son Nav, 150(6), 419–429, 2003.

32. Wolf, H.F., *Handbook of Fiber Optics: Theory and Applications*, New York: Garland Stpm, 1979.

33. Lizotte, T.E., and Dickey, F., MLA Fiber Injection for A Square Core Fiber Optic Beam Delivery System: Design versus Prototype Results, in *OSA International Optical Design Conference and Optical Fabrication and Testing*, Optical Society of America, 2010.

34. Lizotte, T.E., Laser Beam Uniformity and Stability Using Homogenizer-based Fiber Optic Launch Method: Square Core Fiber Delivery. *Optical Fibers, Sensors, and Devices for Biomedical Diagnostics and Treatment XI, Proc. SPIE* 7894, February 16, 2011.

35. Goodman, J.W., *Stastistical Optics*, John Wiley & Sons, New York, 1985.

36. Dickey, F.M., and Holswade, S.C., *Laser Beam Shaping: Theory and Techniques*, 2nd edn., Taylor and Francis Group, CRC Press, Boca Raton, FL, 2014.

37. Blomqvist, M., Campbell, S., Latokartano, J., and Tuominen, J., Multi-kW Laser Cladding using Cylindrical Collimators and Square-formed Fibers, *High Power Laser Materials Processing: Lasers, Beam Delivery, Diagnostics, and Applications, Proc. SPIE* 8239, February 9, 2012.

38. Devonshire, J.M., Effects of IRR glazing on radiant heat and thermal comfort for on-road conditions. Report UMTRI-2004-40, Ann Arbor, MI: University of Michigan, 2004.

10 Laser Beam Shaping by Means of Flexible Mirrors

T. Yu. Cherezova and A. V. Kudryashov

CONTENTS

10.1 INTRODUCTION

The widespread use of lasers in technological processes highlights the problem of controlling the laser beam parameters. For example, metal-cutting processes typically demand the tightest beam focus possible.[1] On the other hand, metal-hardening processes require a laser beam with the most uniform transverse irradiance distribution possible.[2] For several laser applications in material processing and manufacturing, nonlinear conversion of a laser beam to a shape with uniform rectangular crosssection is often desirable.[1,2] Examples of such uniform beams include highly multimode laser beams,[3] flattened Gaussian beams,[4,5] and super-Gaussian beams.[6,7]

In general, the specified laser irradiance distribution can be formed in different ways—both extracavity[8] and intracavity.[9] The main advantages of intracavity shaping include not only the ability to form the desired irradiance structure, but also the ability to increase the laser output power. One of the most well-known intracavity approaches is to apply graded reflectivity mirrors.[7] However, such mirrors introduce large intrinsic power losses and thus are suitable only in lasers with high-gain active media and generally with unstable resonators. Another intracavity approach is to use graded-phase mirrors.[10,11] However, such mirrors can only serve in the specific applications for which they were designed; every change of laser parameters requires its own unique mirror. A single flexible controlled mirror, on the other hand, can form a number of desired laser outputs. Flexible controlled mirrors can also compensate for various phase distortions caused, for example, by thermal deformations of resonator mirrors or by aberrations of active media. Uncontrolled phase distortions can destroy the desired laser output distribution. Accurate prediction of phase distortions is not possible because they can depend, for example, on the laser pumping power, the inhomogeneity of the active medium, and so on. For general intracavity beam shaping tasks, it is thus easier and more universal to use flexible controlled mirrors.

As the key element of any adaptive optical system, the flexible corrector and its properties determine the performance of the whole system. Demands on the wavefront corrector element include the following:

- A wide range of surface deformation
- A small number of control actuators
- Efficiency of reproducing wavefront aberrations
- Temperature stability of the surface figure
- The ability, if necessary, to conjugate with a wavefront sensor (essential for closed-loop applications)
- Simplicity of fabrication and application
- Low cost

Bimorph mirrors are generally the most suitable correctors to satisfy these demands. It has been shown that semipassive bimorph mirrors with 13 actuators effectively reproduce low-order wavefront aberrations with large amplitudes.[12] For example, bimorph mirrors can theoretically reproduce aberrations with root-mean-square (RMS) amplitudes of the following: defocus, 0.3%; astigmatism, 0.7%; coma, 5%; and spherical aberration, 6%.

On the other hand, deformable bimorph mirrors are not standard optical elements. They are relatively thin and consist of several different material layers with different properties. As a result, there are no standard optical technologies to produce bimorph mirrors. Special methods of piezoceramic treatment, such as surface polishing and reflecting coating deposition, have to be developed to produce high-quality bimorph correctors. Some applications of bimorph correctors in lasers and imaging optical systems require an especially wide range of deformation and high stability of the mirror surface. Some of these problems are considered in the next section.

10.2   DIFFERENT TYPES OF BIMORPH MIRRORS

Pure bimorph mirrors consist of two comparatively thin piezoceramic plates polarized in opposite directions. Manufacturing issues with these pure bimorph flexible mirrors have led to the development of so-called semipassive bimorph correctors.[12,13]

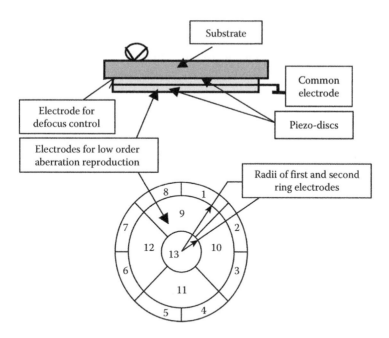

FIGURE 10.1 Design of a semipassive bimorph mirror.

A traditional semipassive bimorph mirror consists of two joined plates: a comparatively thick passive glass or metal substrate and a thin active piezoceramic plate (see Figure 10.1). The operational concept for the semipassive bimorph corrector is similar to that for the pure bimorph case, but its sensitivity is lower. Application of an electrical signal to the electrodes of the piezoceramic plate causes tension of the piezodisc due to the inverse piezoelectric effect. The piezodisc thus expands in the radial direction. The bonded substrate prevents free expansion, and this results in bending of the reflective surface. To reproduce different types of aberrations with the help of such a corrector, the outer electrode is divided among several controlling electrodes. The size as well as the number of controlling electrodes depends on the type of the aberrations to be corrected. Sometimes, it is useful to introduce into the mirror design an additional piezodisc with one round electrode for reproducing defocus (Figure 10.1).

Table 10.1 shows the main characteristics for piezoceramic materials that are currently commercially available. Piezoceramic material PKR-7M has the largest value of the piezomodule d_{31}. This material is thus the most sensitive one for the development of bimorph mirrors. The drawbacks of PKR-7M include a low Curie temperature and the fact that it is a soft segnetomaterial. Since the Curie temperature defines the stability of material properties at high temperatures, a low Curie temperature demands low-temperature technologies for ceramic treatment, bonding of the mirror components, and coating deposition.

Figure 10.2 represents a comparison of the sensitivity of three experimental bimorph mirrors with different types of piezoceramic material. All mirrors include a glass substrate of 2.5mm thickness and 40mm aperture, and PKR6, PIC151,

TABLE 10.1

The Main Characteristics of Piezoceramic Materials

Material	$d_{31} \times 10^{12}$ (C/N)	T_c (°C)	Hardness	α ($\times 10^{-6}$ K^{-1})	Firm Manufacturer
ZTS-19	170	290	hard	~3	ION, Russia
PKR-6	195	300	middle	1–3	Ultrasound Ltd, Russia
PKR-7M	350	175	soft	1–3	Ultrasound Ltd, Russia
PZT-5H	275	193	soft	1.5	Morgan Matroc, UK
PZT-5A	171	365	middle	1.5–2	Morgan Matroc, UK
PIC-151	210	250	middle	–	Physik Instrumente, Germany
P1 94	305	185	soft	–	Quartz&Silica, France
PCM-33A	262	–	soft	–	Matsushita Electric, Japan

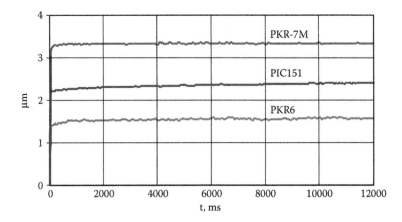

FIGURE 10.2 Deformation of bimorph mirrors made from different piezoceramic materials versus time.

or PKR-7M ceramic discs of 0.3mm thickness and 40mm aperture. The same +100 V potential was applied to the electrodes of the mirrors. Surface deformation dynamics were recorded by a Shack–Hartmann sensor every 50 ms. Sensitivity for the PKR-7M ceramic mirror was 1.4 and 2 times higher than that for mirrors with PIC151 and PKR6 ceramics, respectively. For PKR-7M, initial measurements upon application of the control voltage showed 3.2 μm of deformation. This increased to 3.33 μm in 2 s, then stayed constant for the remaining measurement. This indicates that the creep effect for PKP-7M does not exceed 4%.

Typical ceramic hysteresis curves exhibit a *butterfly* pattern (Figure 10.3). Depolarization or even reverse polarization phenomena occur at potentials opposite to the polarization direction voltage. We measured the threshold negative voltage at which depolarization begins for PKR-7M, which equals −500 V per mm of ceramic. To avoid unpredictable behavior of the mirror deformation, the range of negative control voltage should be restricted. For example, the range of control voltage for

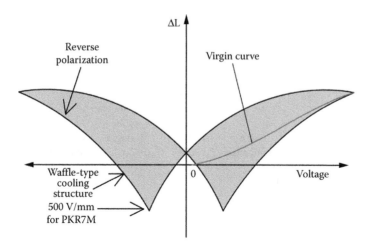

FIGURE 10.3 Response of a PZT actuator to a bipolar drive voltage.

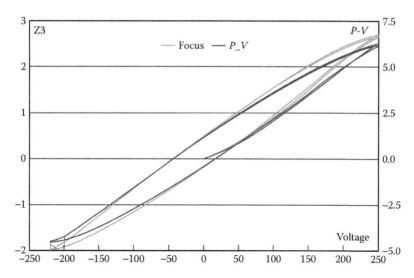

FIGURE 10.4 Hysteresis curve for a 40mm bimorph mirror. The y-axis represents the value of mirror deformation $P–V$ (peak-to-valley) measured in μm; the x-axis represents the voltage given in volts.

a 0.4mm thick ceramic could be varied from −200 to +300 V. The response of the bimorph actuator remains linear within this range (Figure 10.4).

One of the important characteristics of correctors is the temperature stability of the mirror reflecting surface. Surface instability is basically caused by the difference between the thermal expansion coefficients for substrate material α_1 and piezoceramic material α_2. In the simplest cases, the instability behavior manifests itself in additional defocus deformation of the mirror surface due to changes in ambient temperature.

TABLE 10.2

Parameters of Materials for Mirror Substrates

Material	Parameters					
	Density, g (kg/m³)	Index of Refraction	Index of a Thermal Expansion, α (×10⁻⁶ K⁻¹)	The Module of the Young, E (Pa × 10⁻⁹)	Thermal Conductivity, β (W/(m K))	Specific Thermal Capacity (J/(kg K))
Quartz glasses KU1, KU2, KU	2.2	1.46	0.55	98	1.36	733
Optical glass LK-5	2.27	1.4846	3.5	69	1.2	—
Piezoceramics PZT	7.4–7.7	—	1–3	46–90	1.4	400
Monocrystalline Si	2.33	3.3	2.54	126–131	150	700
Copper Cu	8.96	—	15.9	129.5	401	385

Application of an additional voltage to the electrodes of the piezodisc $V_{\Delta T} = (t/d_{31})(\alpha_1 - \alpha_2)\Delta T$ can compensate for thermal deformation. This limits the dynamic range of the control voltage. For example, for a copper substrate ($\alpha_1 = 15.9 \times 10^{-6}$ K⁻¹), almost 20% of the maximal range of the control voltage would be used to compensate for thermal deformation caused by a 5°C change in temperature.

The main criterion for the choice of a substrate material is the thermal expansion value α_1, which should be close to α_2 for the chosen piezoceramic material (~3 × 10⁻⁶ K⁻¹). Table 10.2 shows the parameters of optical materials, which are used in the manufacturing of experimental samples of bimorph mirrors. Piezoceramic lead–zinconate–titanate (PZT) is the most suitable material for a mirror substrate. In this case, we can ensure equality of thermal expansion coefficients, leading to an ideal mirror for thermal stability. However, there can also be problems with polishing ceramic materials that have a tendency to crumble. Ceramic material PKR-7M, which is produced under special heat baking technology, allows polishing of the surface to optical quality.

Another method for thermally stabilizing the mirror shape is water-cooling. Water-cooling is very important in high-power laser applications, where the mirror is heated by the laser beam. In this case, channels for cooling water circulation have to be formed inside the thin substrate of the mirror (see Figure 10.5). We are producing such cooled mirrors from copper, because copper is a good material for machining, diffusion welding, and optical polishing. Such cooled mirrors are used for formation and correction of continuous wave (CW) CO_2 laser beams with powers up to 5 kW.[13]

Figure 10.6 shows experimental samples of bimorph mirrors that we have developed for different applications. Mirror apertures vary from 30 to 100 mm. The number of control electrodes on the mirrors varies from 17 to 33. The mirror surfaces were

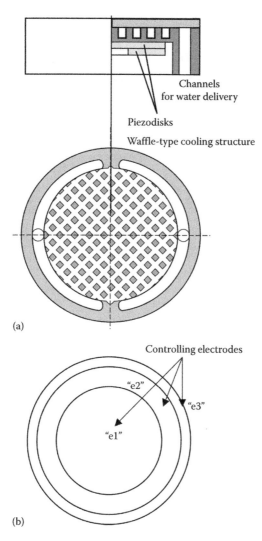

FIGURE 10.5 (a) Sample of cooling structure for a water-cooled bimorph mirror and (b) layout of mirror electrodes (e1 is underneath e2 and e3).

deposited with multilayer dielectric and protected silver coatings. These coated mirrors were subjected to beam irradiances up to ~10^{12} W/cm^2 without observable damage.

The 30mm aperture bimorph mirror shown in Figure 10.6 is used for the formation of the near-field laser beam distribution for the last amplifier of the ATLAS Ti:Sa laser at MPQ, Garching, Germany.[14] The homogenization of the irradiance distribution produced by this flexible mirror allowed pulse energy to increase from 0.5 to 1.5 J. The next mirror, shown in Figure 10.6, having an 80mm aperture and 33 control actuators, is used in the same laser for far-field wavefront correction. Closed-loop control of the mirror produced an almost diffraction-limited spot at the focal plane of the nonaxial parabolic mirror.

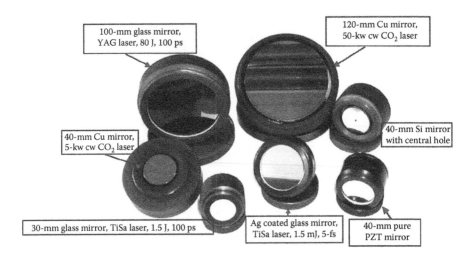

FIGURE 10.6 Photograph of experimental bimorph mirrors.

In addition to their utility for extracavity beam correction and formation, bimorph mirrors are very useful in intracavity applications. The following sections detail an investigation of bimorph correctors for intracavity laser beam formation. For intracavity applications, we suggest the use of the well-developed water-cooled flexible mirror designs (Figure 10.5) to avoid any undesirable surface thermal deformations. As Figure 10.5 illustrates, such mirror designs contain two piezodiscs, where the interface between the piezoceramic discs contains a continuous conducting *ground* electrode. Another continuous conducting electrode between the piezodisc and the copper plate e1 is used to control the overall curvature of the entire mirror. Two controlling electrodes, e2 and e3, having the form of concentric rings, were attached to the outer surface of the piezodisc. The response function for each mirror electrode (the deformation of the mirror surface while applying voltage to each mirror electrode) is shown in Figure 10.7 and was measured using a modified Fizeau interferometer.[15] Three-dimensional profiles of the response functions are shown in Figure 10.8. These response functions will be used in the intracavity procedures described in this chapter.

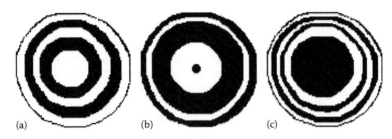

FIGURE 10.7 Level map of response functions of three electrodes: (a) common focus—defocus electrode e1 ($P–V = 0.81$ μ for applied voltage 20 V); (b) second ring electrode e3 ($P–V = 0.35$ μ for applied voltage 40 V); and (c) first ring electrode e2 ($P–V = 0.79$ μ for applied voltage 40 V).

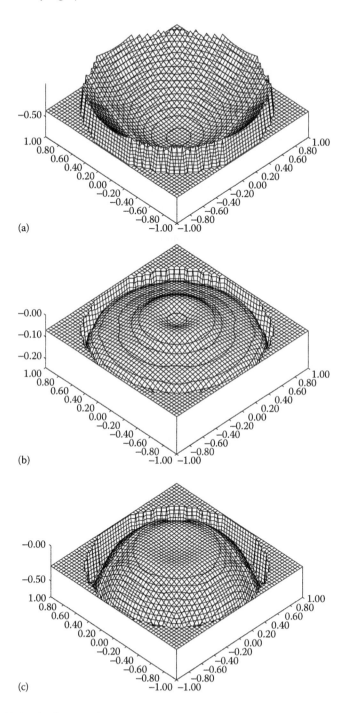

(a)

(b)

(c)

FIGURE 10.8 Surface profiles of an adaptive mirror: (a) for common focus–defocus electrode e1—applied voltage is 20 V, (b) for second ring e3—applied voltage is 40 V, and (c) for first ring of electrodes e2—applied voltage is 40 V.

10.3   FORMATION OF A DESIRED LASER BEAM AT THE LASER RESONATOR OUTPUT BY MEANS OF AN INTRACAVITY FLEXIBLE CORRECTOR

For many industrial applications, it is desirable to have the laser generating a single, mostly fundamental mode and to have good power extraction at the same time. Gaussian beams are relatively narrow and therefore result in poor energy extraction. We would thus expect that the formation of a wider super-Gaussian irradiance distribution inside the laser cavity would lead to an increase in the active medium gain extraction.

The algorithm for the desired irradiance formation as well as the numerical results will be discussed using a stable laser cavity (Figure 10.9). Such a cavity corresponds to the industrial continuous discharge CO_2 laser ILGN-704 produced by Istok (Fryazino, Russia). The geometry of the resonator consists of a plane output coupler and an active mirror separated from the coupler by the distance $L = 2$ m.

Azimuthal symmetry can be assumed, which allows us to use the one-dimensional Huygens–Fresnel integral equations[16,17] to calculate the amplitude of a mode in the empty laser resonator:

$$\gamma_2 \Psi_2(r_2) = \int_0^b K_1(r_1, r_2) \Psi_1(r_1) r_1 \, dr_1 \tag{10.1}$$

$$\gamma_1 \Psi_1(r_1) = \int_0^a K_2(r_2, r_1) \Psi_2(r_2) r_2 \, dr_2 \tag{10.2}$$

where:
  γ_i is the eigenvalue
  $\psi_i(r_i)$ is the eigenmode of the resonator
  r_i is a radial coordinate
  $i = 1$ indicates a plane output mirror of diameter $2b$
  $i = 2$ indicates an active mirror of diameter $2a$

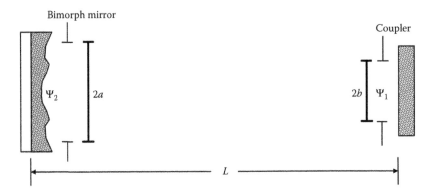

FIGURE 10.9   Schematic setup of the CW CO_2 laser with adaptive mirror.

The kernels of Equations 10.1 and 10.2 have the following expressions:

$$K_1(r_1, r_2) = \frac{j}{B} J_l \left(k \frac{r_1 r_2}{B} \right) \exp \left(-\frac{jk}{2B} \left(A r_1^2 + D r_2^2 \right) \right) \tag{10.3}$$

$$K_2(r_2, r_1) = \frac{j}{B} J_l \left(k \frac{r_1 r_2}{B} \right) \exp \left(-\frac{jk}{2B} \left(A r_1^2 + D r_2^2 \right) \right) \exp \left(jk \varphi_{\text{mirror}}(r_2) \right) \tag{10.4}$$

Here,

J_l is the Bessel function of order l (we take into account only the lowest transverse mode with $l = 0$)

A, B, C, and D are the constants determined by the ABCD ray matrix of the laser resonator. We consider an empty resonator, so that $A = 1$, $B = L$, $C = 0$, and $D = 1$.

The algorithm to form the desired irradiance distribution is the so-called inverse propagation method described in several references.[11,18–21] The desired output field distribution $\psi_1(r, \varphi)$ is specified on the output mirror. The back-propagation of the laser beam through all the resonator's elements to the active corrector is calculated using the Huygens–Fresnel integral equations (Equations 10.1 and 10.2). In the plane of the active corrector, the wavefront is calculated to determine the appropriate mirror phase profile $\varphi_{\text{mirror}}(r)$:

$$\varphi_{\text{mirror}}(r) = -\varphi_{\text{beam}}(r) \tag{10.5}$$

An ideal corrector (graded-phase mirror) could completely reconstruct such a phase profile.[11,18–21] In our case, bimorph mirrors can approximate the necessary phase profile with some small degree of error. This RMS error can be calculated using the experimentally measured response functions of the mirror. In other words, it can be stated as follows:

$$\left(Z(r) - \sum_{i=1}^{i=3} U_i F_i(r) \right)^2 \rightarrow \min \tag{10.6}$$

where:

$Z(r)$ is the profile to be reconstructed

$F_i(r)$ are response functions of the flexible mirror electrodes shown in Figures 10.7 and 10.8

U_i are weight coefficients corresponding to the voltages applied to each electrode

The left side of Equation 10.6 has a minimum when its first derivative equals zero as follows:

$$\frac{\partial}{\partial K_i} \left(Z(r) - \sum_{i=1}^{i=3} U_i F_i(r) \right)^2 = 0 \tag{10.7}$$

Equation 10.7 then determines the applied voltages to approximate the necessary shape of the laser beam.

The procedure described above gives us the reconstructed mirror surface $\varphi_{mirror}(r)$ which is substituted into Equation 10.4. To solve the integral equations (Equations 10.1 and 10.2), we used the Fox and Li iterative method of successive approximations[16,17] to take into account edge diffraction as well as nonideal reproduction of the necessary phase profile by the bimorph flexible mirror.

10.4 MAIN RESULTS

The main parameters of the laser resonator (Figure 10.9) are the Fresnel numbers $N_1 = b^2/(B\lambda)$ and $N_2 = a^2/(B\lambda)$, and the geometrical factor $G = (1 - L/R_2)$; where $\lambda = 10.6\ \mu m$ is the wavelength, $R_2 = 4$ m is the radius of curvature of the active mirror, and $L = 2$ m is the length of the resonator cavity. The initial field distribution on the planar output coupler is chosen as $\Psi(r) = \exp(-(r/W)^n)$, where n determines the order of the super-Gaussian function and W is the beam waist. The particular beam waist is chosen according to the methods of moments for laser beams as follows[21]:

$$w^2 = \left(M^2 \frac{\lambda L}{2\pi} \right) \left(\frac{1+G}{1-G} \right)^{1/2} \tag{10.8}$$

where:

M^2 is the beam quality factor, which is given in the following equation:

$$M^2 = \frac{(w\theta)_{mode}}{(w\theta)_{TEM_{00}}} \tag{10.9}$$

10.4.1 FORMATION OF A SUPER-GAUSSIAN BEAM OF THE FOURTH ORDER ($N = 4$)

From Equations 10.8 and 10.9, the super-Gaussian beam waist is calculated: $W = 3.1$ mm. Figure 10.10a represents the phase distribution of a super-Gaussian beam (curve 1) back-propagated through the resonator starting at the output coupler and going a distance of $L = 2$ m to the adaptive mirror. Curve 2 (Figure 10.10a) illustrates the phase profile of the active mirror reproducing the phase shape of laser beam with an RMS error of 0.3%. Figure 10.10b shows the beam irradiance distribution at the output coupler for various resonator conditions. Curve 1 corresponds to the fundamental Gaussian mode of the same resonator, but with a pure spherical mirror. Curve 2 (solid) shows the desired super-Gaussian relative irradiance profile, while curve 3 shows the profile produced with an ideal corrector (graded-phase mirror with no deviation from the necessary phase profile). Finally, curve 4 corresponds to the irradiance distribution in the resonator with an adaptive mirror.

One may see from Figure 10.10b that applying the active corrector (curve 4) increases the output mode volume by 1.5 times in comparison with a pure Gaussian beam (curve 1). At the same time, diffraction losses per transit decrease by 1.7 times.

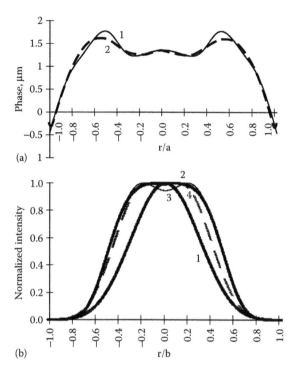

FIGURE 10.10 Formation of a super-Gaussian beam of the fourth order, $N_1 = 1$, $N_2 = 4.7$, $G = 0.5$: (a) (1) (solid curve)—the phase profile of the laser beam to be reconstructed and (2) (dashed)—the phase profile of the flexible mirror and (b) normalized irradiance distributions on plane coupler: (1) Gaussian irradiance distribution, corresponding to the same geometry of resonator but with a spherical mirror, (2) the desired super-Gaussian irradiance profile, (3) irradiance formed by graded-phase mirror, and (4) (dashed)—by bimorph flexible one.

TABLE 10.3

Voltages (V) Applied to the Electrodes of a Bimorph Flexible Mirror

n, Order of the Given Super-Gaussian Beam Formation	e1	e2	e3
4	−64	−116	71
6	−107	−300	−240
8	−43	−214	−171

The voltages applied to each electrode were calculated from Equation 10.7 and are presented in Table 10.3.

Users of lasers often dislike super-Gaussian irradiance distributions for their side lobes in far-field patterns. However, the shaped super-Gaussian distribution is not exactly the super-Gaussian function: its form has been changed by edge diffraction and by the nonideal behavior of the active mirror in forming the necessary phase

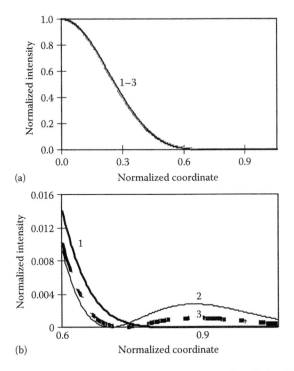

(a)

(b)

FIGURE 10.11 Formation of a super-Gaussian beam profile of the fourth order. (a) Normalized irradiance distributions in the far-field zone: (1) Gaussian beam, (2) beam formed by ideal corrector (graded-phase mirror), and (3) (dashed curve)—beam formed by active mirror; and (b) fragment near the edge of the same irradiance distributions.

profile. That is why the side lobes of the beam formed by an active mirror contain only 2% of the total energy (dashed curve 3 in Figure 10.11). This irradiance profile is thus very attractive for industrial applications.

For comparison, curve 2 in Figure 10.11a and b represents the far-field pattern for a super-Gaussian beam formed by an ideal (graded-phase) corrector.

10.4.2 FORMATION OF A SUPER-GAUSSIAN BEAM OF THE SIXTH ORDER (N = 6)

In this case, the beam waist of the desired super-Gaussian distribution is chosen as 3.3 mm. Figure 10.12a represents the exact phase distribution of the desired super-Gaussian beam and the phase profile of a flexible mirror (RMS error of the approximation is about 0.1%). Figure 10.12b shows irradiance distributions formed by the flexible mirror at the output coupler. The far-field results are very close to those for the fourth-order super-Gaussian beam represented in Figure 10.11a and b. Mirror electrode voltages calculated from Equation 10.7 are presented in Table 10.3.

For this case, the output mode volume increases by a factor of 1.3 in comparison with a pure Gaussian beam while diffraction losses per transit decrease by 1.5 times.

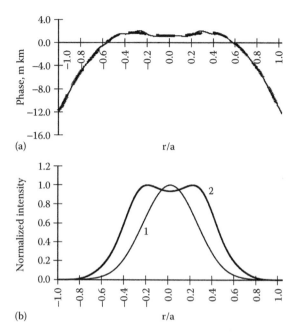

FIGURE 10.12 Formation of a super-Gaussian beam profile of the sixth order, $N_1 = 1$, $N_2 = 14.1$, and $G = 0.5$. (a) (solid)—phase profile of laser beam to be reconstructed and (dashed)—phase profile of bimorph mirror and (b) normalized irradiance distributions: (1) Gaussian TEM_{00} mode for the same resonator but with a pure spherical bimorph mirror and (2) sixth-order super-Gaussian beam formed by the bimorph flexible mirror.

10.4.3 FORMATION OF A SUPER-GAUSSIAN BEAM OF THE EIGHTH-ORDER ($N = 8$)

For this case, the beam waist of the desired eighth-order super-Gaussian fundamental mode is equal to 3.5 mm. The main results are shown in Figure 10.13a and b. The output mode volume increased 1.6 times in comparison with a pure Gaussian beam, while diffraction losses per transit decrease by a factor of 1.7. Although the flexible mirror error in producing the desired phase shape is relatively low (RMS error < 0.1%), it could not exactly reproduce the two main humps (Figure 10.13a). However, if the controlling electrode's position could be matched to the coordinates of the two local maxima (Figure 10.13a, solid curve), such an active corrector would be able to reproduce the shape more accurately. Applied mirror electrode voltages are again presented in Table 10.2.

It should be mentioned that active mirrors sometimes produce a smoother phase profile than is desirable for higher order modes. This situation can lead to higher diffraction losses for the TEM_{00} mode than for the TEM_{01} mode, which causes the laser to resonate in the TEM_{01} or some combination of higher order modes. In contrast, an ideal corrector, such as a graded-phase mirror, tends to disturb the higher order modes to a greater extent, which causes higher diffraction losses and thus suppresses their amplitude in the cavity.[18] This is confirmed by Figure 10.14, which

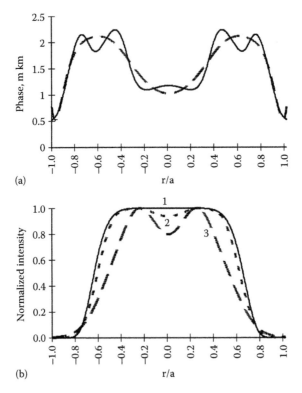

(a)

(b)

FIGURE 10.13 Formation of a super-Gaussian beam profile of order eight, $N_1 = 1$, $N_2 = 4.7$, and $G = 0.5$. (a) (Solid)—the phase profile of the given laser beam at the position of bimorph mirror and (dashed)—phase profile of the mirror and (b) normalized irradiance distributions: (1) (solid)—the desired super-Gaussian beam, (2) (dotted)—irradiance produced with a graded-phase mirror, and (3) (dashed)—irradiance produced with bimorph flexible mirror.

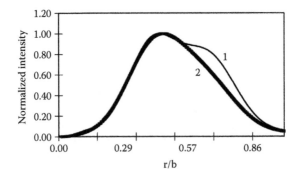

FIGURE 10.14 Normalized irradiance distribution of TEM_{01} mode while forming an eighth-order super-Gaussian fundamental mode: (1) ideal corrector (such as a graded-phase mirror) and (2) bimorph flexible mirror.

shows the irradiance distribution of a TEM_{01} mode on the output coupler, where an eighth-order super-Gaussian beam is desired. The TEM_{01} mode, shaped by an ideal corrector (curve 1 in Figure 10.14) and by an active corrector (curve 2), is shown. One may see that the irradiance of the TEM_{01} mode in the first case is more distorted; hence, its diffraction losses are higher. As an example, for the previously given resonator parameters and a bimorph active corrector, diffraction losses per transit of the super-Gaussian mode formed by a bimorph mirror are $(1-|\gamma_1\gamma_2|) = 4.4 \times 10^{-5}\%$, while for the TEM_{01}, the losses are $(1-|\gamma_1\gamma_2|) = 2 \times 10^{-5}\%$. To increase the discrimination between transverse modes, one needs to perform an optimization procedure.

10.4.4 Optimization of Laser Parameters

The optimization of the laser resonator parameters N_2 (Fresnel number) and G (geometric factor) was carried out in order to produce the closest match to the desired irradiance distribution. Figure 10.15 shows the normalized output irradiance profiles produced by varying N_2 from 4.7 to 14.1 ($N_1 = 1$, $G = 0.5$). Note that for the graded-phase mirror, the best result is reached when the Fresnel number N_2 is a maximum. This is understandable: Increasing N_2 corresponds to enlarging the size of an aperture in front of the graded-phase mirror (see Figure 10.9), which causes a more complete approximation of the incident resonator mode shape by this mirror. As a result, the output profile is closer to what is desired. In contrast, for an active corrector with given electrode axial positions, an increase in N_2 does not necessarily lead to better performance. For example, one may see from Figure 10.15 that for $N_2 = 10$ (Series 4), the obtained irradiance distribution is very close to the desired one. Also, a rather good approximation to the desired super-Gaussian curve was obtained for $N_2 = 4.7$ (Series 2, Figure 10.15). However, for the highest Fresnel number $N_2 = 14.1$ (Series 5, Figure 10.15), the shape of the obtained output profile is not as good as for lower Fresnel numbers. This is because the fixed electrodes may have

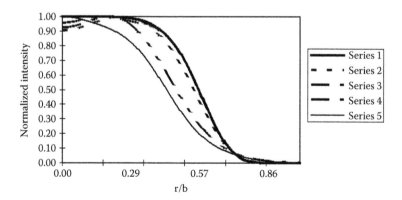

FIGURE 10.15 Normalized irradiance profiles on the plane output coupler for the sixth-order super-Gaussian fundamental mode. Series 1: the desired beam; Series 2: $N_2 = 4.7$; Series 3: $N_2 = 6.8$; Series 4: $N_2 = 10$ (very close to the desired beam except at the center); and Series 5: $N_2 = 14.1$, $N_1 = 1$, and $G = 0.5$.

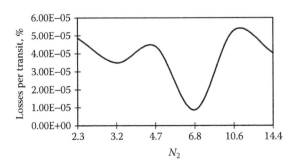

FIGURE 10.16 Diffraction losses per transit for an eighth-order super-Gaussian fundamental mode in a resonator with an active corrector. $N_1 = 1$, $G = 0.5$.

difficulty replicating a larger incident mode profile, and in fact may have to make compromises in the center of the mirror where the majority of the incident mode amplitude is concentrated.

Another parameter that must be taken into consideration is diffraction losses per transit. We calculated such losses for the eighth-order super-Gaussian fundamental mode and the results are presented in Figure 10.16. The curve has a minimum for $N_2 = 6.8$. In that case, the bimorph flexible mirror does the best job of reproducing the resonator mode shape. At this minimum point, the diffraction losses of TEM_{01} mode are approximately 10 times higher than the TEM_{00} mode. The obtained irradiance curves do not depend upon the variation of G, because the active mirror can reproduce focus (or defocus) very well (mainly by applying a voltage to the common focus—defocus electrode; Figures 10.7 and 10.8).

10.5   FORMATION OF AN ANNULAR BEAM

To solve various tasks in modern laser physics and nonlinear optics, it is sometimes necessary to have beam profiles that differ from a uniform profile. For example, to deliver high-power laser radiation through the atmosphere, it is desirable to form an annular laser beam with a planar phase distribution because it has less nonlinear distortion in nonlinear and turbulent media than other forms of laser beams.[22,23] The traditional way to form such a beam is to obscure the central part of a Gaussian fundamental mode. However in this case, we have additional power losses. As is shown below, an intracavity active mirror allows us to form an annular beam without any additional power losses and with less diffraction losses than for a traditional Gaussian fundamental mode.

The desired initial field distribution for the annular beam at the plane output coupler of a laser resonator (Figure 10.9) is chosen as follows: $\psi_1(r_i) = \{(r + 0.1)/3.1\}2 \exp(-((r + 0.1)/3.1)^4)$. A small decentration value, in this case 0.1 mm, had to be added so that diffraction did not drastically change the desired annular irradiance shape of the fundamental mode during Fox–Li numerical calculations.[16,17] The main parameters of the laser resonator were Fresnel numbers $N_1 = b^2/(B\lambda) = 1$, $N_2 = a^2/(B\lambda) = 4.7$ and geometrical factor $G = (1 - L/R_2) = 0.5$,

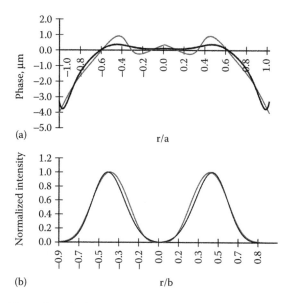

FIGURE 10.17 Formation of a doughnut-like beam: $N_1 = 1$, $N_2 = 4.7$, and $G = 0.5$: (a) (gray curve) the phase profile of the laser mode to be reconstructed and (black) the phase profile of the bimorph flexible mirror and (b) normalized irradiance distributions on the plane output coupler: (black) the desired irradiance profile and (gray) the irradiance profile formed by the flexible mirror.

where $R_2 = 4$ m is the radius of curvature of the flexible mirror and $L = 2$ m is the length of the resonator cavity.

Figure 10.17a represents the phase distribution of the annular beam propagated back to the flexible mirror at the distance $L = 2$ m from the output plane mirror (gray curve). The smoother, black curve of Figure 10.17a illustrates the phase profile of the bimorph flexible mirror reproducing the phase shape of the laser beam with RMS error 0.7%. Figure 10.17b shows irradiance distributions on the plane output coupler. The black curve corresponds to the desired irradiance profile and the gray curve to the profile formed by the intracavity flexible mirror. The diffraction losses of the annular beam were estimated as $\delta = |1 - \gamma_1\gamma_2|$, where γ_1, γ_2 are the eigenvalues defined from Equations 10.1 and 10.2. Such losses are decreased by a factor of 1.4 in comparison with a Gaussian TEM$_{00}$ mode.

The far-field pattern of this annular beam contains about 96% of its total energy in the main lobe (as shown by the gray curve in Figure 10.18). With a beam quality factor $M^2 = 1.2$, this irradiance profile is very attractive for industrial applications. For comparison, we plotted the far-field pattern for a Gaussian fundamental mode from the same resonator. The edges of the far-field intensities are shown in Figure 10.19 in expanded scale, showing a small amount of energy in the second diffraction ring for the annular beam. The voltages applied to each electrode to form the annular fundamental mode are presented in Table 10.4.

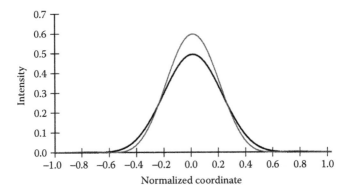

FIGURE 10.18 Irradiance distributions in the far field: (black) far-field pattern of a Gaussian fundamental mode and (gray) far-field pattern of the doughnut-like beam.

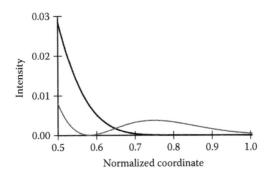

FIGURE 10.19 The fragment near the beam edge for the irradiance distributions shown in Figure 10.18.

TABLE 10.4
Voltages (V) Applied to the Electrodes of a Flexible Mirror to Form an Annular Beam

e1	e2	e3	e4
−47.5	−254	0	0.7

10.6 EXPERIMENTAL FORMATION OF A SUPER-GAUSSIAN BEAM BY MEANS OF BIMORPH FLEXIBLE MIRROR

The experimental formation of a specified beam, namely a super-Gaussian of fourth and sixth orders, was performed using an industrial fast axial flow continuous-discharge CO_2 laser with a stable resonator, produced by IPLIT, Russian Academy

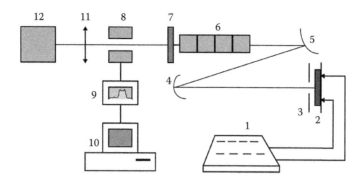

FIGURE 10.20 Schematic of the experimental set up to form a super-Gaussian TEM_{00} mode: (1) variable-voltage power supply for controlling mirror electrodes, (2) semipassive bimorph mirror, (3) diaphragm, (4) concave mirror $R = 2200$ mm, (5) convex mirror $R = -800$ mm, (6) active CO_2 gain medium, (7) ZnSe output mirror with coefficient of reflectivity 69%, (8) LBA-2A (laser beam analyzer), (9) oscilloscope, (10) computer, (11) lens $f = 275$ mm, and (12) MAC-2 (mode analyze computer).

of Sciences, model TLA-600.[24] The laser resonator (Figure 10.20) consists of a plane output coupler (7), a CO_2 gain tube (6), convex (5) and concave (4) mirrors, and the bimorph flexible mirror (2).

First of all, in order to model the behavior of the cavity with our flexible mirror and to determine the mirror electrode voltages to form super-Gaussian beams of fourth and sixth orders, we perform all of the numerical calculations described in Equations 10.1 through 10.7. However, in the case of a real laser having an active medium (not just an empty resonator as it was before), we must take into account active medium saturation caused by the intense beam. That is why the kernels of integral Equations 10.1 and 10.2, given by Equations 10.3 and 10.4, have the additional multiplier $H(I)$ taking into account this effect as follows:

$$K_1(r_1, r_2) = 2\pi \frac{j}{\lambda B} J_0\left(k \frac{r_1 r_2}{B}\right) \exp\left(-\frac{jk}{2B}(Ar_1^2 + Dr_2^2)\right) H(I) \qquad (10.10)$$

and

$$K_2(r_2, r_1) = 2\pi \frac{j}{\lambda B} J_0\left(k \frac{r_1 r_2}{B}\right) \exp\left(-\frac{jk}{2B}(Ar_1^2 + Dr_2^2)\right)$$

$$\times \exp(jk\varphi_{mirror}(r_2)) H(I) \qquad (10.11)$$

where:

$$H(I) = \left(1 + \frac{g_0 L_{am}}{2\left(1 + \frac{I(r)}{I_s}\right)}\right) \qquad (10.12)$$

Here g_0 is the small-signal gain coefficient; $g_0 = 110$ cm^{-1} for this particular type of laser.[24] L_{am} is the length of the active medium ($L_{am} = 80$ cm), $I(r)$ is the transverse irradiance distribution of the laser beam, and I_s is the saturation irradiance ($I_s = 110$ W/mm^2). If the irradiance of the laser beam $I(r)$ becomes comparable with I_s (saturation irradiance of the active medium), then the overall irradiance of the laser beam becomes lower.

The main resonator parameters for the CO_2 laser shown in Figure 10.20 are Fresnel numbers $N_1 = b^2/\lambda B = 0.66$ and $N_2 = a^2/\lambda B = 6.47$, and stability factor $G = 0.51$ (here $b = 8$ mm, the radius of the plane output mirror; $a = 25$ mm, the radius of bimorph mirror; $\lambda = 10.6$ μm, the wavelength; and $B = 9104$ mm, the effective length of the resonator). Parameters of the super-Gaussian function $\Psi(r) = \exp(-(r/W)^n)$ are chosen as $W = 4.8$ mm and $W = 5.1$ mm for $n = 4$ and $n = 6$, respectively. The particular beam waists were chosen according to the theory of moments and are well described in Bélanger and Paré.[21] For this laser, we can assume $H(I) = $ constant, since the irradiance of the beam is much smaller than the saturation irradiance I_s of the active medium. The voltages at the flexible mirror electrodes needed to form the super-Gaussian beams were calculated according to the algorithm described earlier in Chapter 3.

Figure 10.21 shows the evolution of the fourth-order super-Gaussian beam at the surface of output mirror as a function of the round trip number, based on numerical calculations. Curve number 1 represents the original beam, while the other curves show beam profiles after a corresponding number of round trip passes calculated from Equations 10.2 to 10.5. Given the resonator parameters mentioned previously in this section, we need to make about 130 iterations to reach a convergence point where successive calculations differ by less than 10^{-7}.

Figure 10.22a and b shows the main results of the calculations: the Gaussian (curve 1) and super-Gaussian of the fourth- and sixth-order fundamental mode

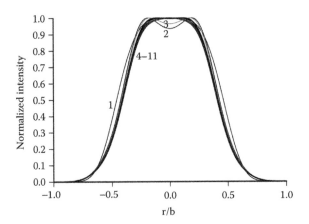

FIGURE 10.21 Evolution of the fourth-order super-Gaussian beam at the surface of the output mirror: (1) initial beam, (2) the beam after 1 round trip, (3) after 2 round trips, (4) after 3 round trips, and so on, and (11) after 10 round trips.

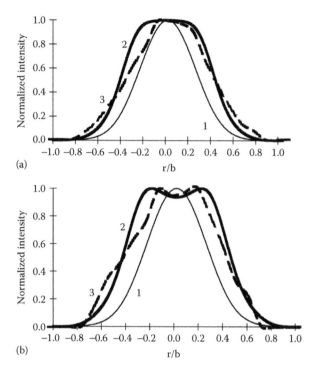

FIGURE 10.22 Formation of the super-Gaussian fundamental mode: (a) super-Gaussian fourth-order mode. (1) Gaussian mode, (2) theoretically obtained super-Gaussian fourth-order mode, and (3) experimentally obtained super-Gaussian fourth-order mode; (b) super-Gaussian sixth-order mode. (1) Gaussian mode, (2) theoretically obtained super-Gaussian sixth-order mode, and (3) experimentally obtained super-Gaussian sixth-order mode.

irradiance distributions (curve 2) on the plane output mirror. The far-field pattern for the calculated fourth-order super-Gaussian beam is shown in Figure 10.23a. As one may see, the fourth-order super-Gaussian increases the peak value of far-field irradiance by 1.6 times compared to the Gaussian TEM_{00} mode of the resonator. At the same time, however, the super-Gaussian causes side lobes in the far-field irradiance distribution. However, they are not very significant, so that the experimental M^2 factor for the fourth-order super-Gaussian mode is $M^2 = 1.36$. We would like to mention that for an *ideal* fourth-order super-Gaussian beam having the same near-field beam waist $\omega = 4.8$ mm, the M^2-factor is generally higher ($M^2 = 1.46$), as illustrated by curve 2 in Figure 10.11. In this case, it is seen from Figure 10.11 that the width of the far-field beam increases because of the side lobes. For sixth-order super-Gaussian beams, the difference is higher: the beam formed by the flexible mirror has $M^2 = 1.38$, while the ideal beam has $M^2 = 1.8$. So, the fact that the resonator with a corrector does not ideally reproduce the super-Gaussian function in the near-field (Figure 10.22) results in a positive effect in the far field: the side lobes are reduced which improves the M^2-factor. We define the M^2-factor according to the international standard ISO 11146[25] as $M^2 = \pi d_0 \theta/(4\lambda)$, where λ is the wavelength, d_0 is the near-field waist diameter calculated as the second moment of the irradiance distribution at the waist location (in our case at the plane of the output resonator mirror) which is shown as follows:

$$d_0 = 2\sqrt{2} \, \frac{\iint r^2 I(r,z) r \, dr \, d\varphi}{\iint I(r,z) r \, dr \, d\varphi}$$

θ is the divergence angle defined as $\theta - d_f/f$.

where:
  f is the focal length of lens
  d_f is the beam width defined as the second moment of the focal plane irradiance
     distribution

$$I_f(r,z): d_f = 2\sqrt{2} \, \frac{\iint r^2 I_f(r,z) r \, dr \, d\varphi}{\iint I_f(r,z) r \, dr \, d\varphi}$$

Now, we can come back to the experiment itself. To apply voltages to the electrodes of the flexible mirror, we used a variable-voltage power supply (Figure 10.20, 1). The near-field irradiance distribution was observed with the help of a laser beam analyzer (LBA-2A; Figure 10.20, 8), and the far-field pattern (in the focal plane of lens; Figure 10.20, 11; $f = 275$ mm) was analyzed by a mode analyze computer (MAC-2; Figure 10.20, 12). The result of experimentally forming a fourth-order super-Gaussian beam is presented in Figure 10.22a, curve 3. The total power of the formed super-Gaussian beam was about 10% higher than the power contained in the Gaussian beam, and the waist was widened by 1.26 ± 0.05 times (the calculations showed that it should have increased by 1.29 times). In the focal plane of lens (Figure 10.20, 11; far field), the experimental irradiance profile for the Gaussian mode is shwon in Figure 10.23b. The peak value of the far-field irradiance distribution for the fourth-order super-Gaussian beam (Figure 10.23c) is 1.6 times higher than that for the Gaussian beam. This fact is in good agreement with the theory as shown in Figure 10.23a. The shape of the far-field pattern becomes narrower, but the side lobes that should exist are not distinguishable from the noise level. The near-field irradiance distribution for the experimentally formed sixth-order super-Gaussian beam is shown in Figure 10.22b, curve 3. In this case, we observed a 12% total power increase in comparison with the Gaussian fundamental mode. The observed far-field pattern for the formed sixth-order super-Gaussian mode is very similar to the one for the fourth-order mode.

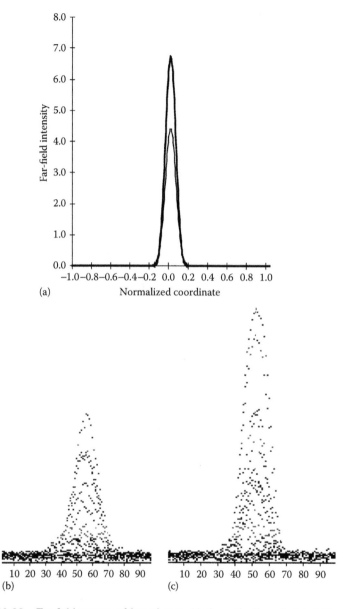

(a)

(b)

(c)

FIGURE 10.23 Far-field pattern of laser beam: (a) theoretically calculated Gaussian and fourth-order super-Gaussian modes, (b) experimentally obtained Gaussian TEM$_{00}$ mode, and (c) experimentally obtained super-Gaussian fourth-order TEM$_{00}$ mode.

10.7   YAG:Nd³⁺ LASER. FORMATION OF A SUPER-GAUSSIAN OUTPUT BEAM—NUMERICAL RESULTS

This section presents calculations aimed at investigating the formation of specified irradiance profiles for a YAG:Nd³⁺ laser. Active mirrors tend to have rather large apertures—their diameter is 20 mm or larger. It is difficult to use such deformable mirrors in the cavities of industrial CW solid-state lasers because of the relatively small beam apertures in stable resonators. To solve this problem, we suggest expanding the beam inside the laser cavity up to the diameter of the adaptive mirror using a meniscus on the one end of the gain element[26] (Figure 10.24). At the same time, the bimorph flexible mirror has a concave spherical profile. Such a laser resonator permits the use of wide aperture mirrors without any supplementary optical elements and therefore without undesirable loss. We considered the case of forming a super-Gaussian output with a resonator having the main parameters of Fresnel numbers $N_1 = b^2/B\lambda = 0.3$ and $N_2 = a^2/B\lambda = 12.3$, and geometric factor $G = 0.58$. Here, $2a = 20$ mm is the diameter of the bimorph deformable mirror, $2b = 6$ mm is the diameter of the plane output mirror, $\lambda = 1.06$ μm is the wavelength, and $B = 6200$ mm is the effective length of the telescopic type of resonator (Figure 10.24). A, B, and D are the elements of the ray $ABCD$ matrix for the laser resonator.

Figure 10.25 shows the results of the diffraction calculations. Curve 1 represents the Gaussian fundamental TEM$_{00}$ mode at the resonator output while curve 2 shows

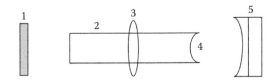

FIGURE 10.24   Schematic of a telescopic-type stable resonator for a YAG:Nd³⁺ laser with a wide-aperture mirror: (1) output coupler, (2) active medium, (3) thermal lens, (4) meniscus, and (5) bimorph flexible mirror.

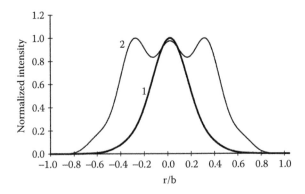

FIGURE 10.25   Formation of super-Gaussian fundamental modes at the output of a stable resonator in a YAG:Nd³⁺ laser: (1) Gaussian mode and (2) fourth-order super-Gaussian modes.

the fourth-order super-Gaussian beam profile. Calculations show that the mode volume for the super-Gaussian beam increases by a factor of 2.1–2.2, diffraction losses decrease by 1.1–1.2 times, and the far-field peak irradiance increases by a factor of 2 in comparison with the Gaussian TEM_{00} mode. The mode volume was estimated as the average of transverse irradiance distributions calculated at five points inside the cavity and at the mirrors as well. Power losses δ were calculated as follows:

$$\delta = |1 - \gamma_1\gamma_2|.$$

10.8 CONCLUSION

This chapter has shown the ability to form a specified irradiance output in both $YAG:Nd^{3+}$ and CO_2 laser resonators by means of an intracavity flexible mirror. The experiment with a CW CO_2 laser has shown that while remaining in the TEM_{00} regime, we were able to increase the total output power by 10%–12% and to increase the peak value of far-field irradiance by 1.6 times in comparison with a Gaussian fundamental mode. This work opens the possibility of *intelligent* flexible lasers that generate irradiance distributions specified by the user.

ACKNOWLEDGMENTS

The authors would like to thank personnel from the Group of Adaptive Optics for Industry and Medicine of the Institute of Laser and Information Technologies, Russian Academy of Sciences, for their help in fabricating and testing the mirrors and in carrying out experiments. Also, we are very grateful to Prof. L. N. Kaptsov and Prof. S. S. Chesnokov from Moscow State University for their helpful discussions. Finally, we wish to thank Scott C. Holswade for his assistance in editing the manuscript.

REFERENCES

1. Abil'sitov, G., ed., *Technological Lasers, Moscow*, Nauka, 1991, (in Russian).
2. Koebner, H., ed., *Industrial Applications of Lasers*, Wiley-Interscience, New York, 1988.
3. Borghi, R., and Santarsiero, M., Modal structure analysis for a class of axially symmetric flat-topped laser beams, *IEEE J. Quantum Elect.*, 35, 745–750, 1999.
4. Santarsiero, M., and Borghi, R., Correspondence between super-Gaussian and flattened Gaussian beams, *J. Opt. Soc. Am A*, 16, 188–190, 1999.
5. Gory, F., Flattened Gaussian beams, *Opt. Comm.*, 107, 335–341, 1994.
6. Bollanti, S., Lazzaro, P.Di., Murra, D., and Torre, A., Analytical propagation of super-gaussian-like beams in the far-field, *Opt. Comm.*, 138, 35–39, 1997.
7. De Silvestri, S., Magni, V., Svelto, O., and Valentini, G., Lasers with super-Gaussian mirrors, *IEEE J. Quantum Elect.*, 26, 1500–1509, 1990.
8. Dainty, J.C., Koryabin, A.V., and Kudryashov, A.V., Low-order adaptive deformable mirror, *Appl. Opt*, 37(21), 4663–4668, 1998.
9. Anan'ev, Yu. A., in *Laser Resonators and the Beam Divergence Problem*, Higler, A., ed., CRC Press, Bristol, 1992.
10. McClure, E.R., Manufactures turn precision optics with diamond, *Laser Focus World*, 27, 95–105, 1991.

11. van Neste, R., Paré, C., Lachance, R.L., and Bélanger, P.A., Graded-phase mirror resonator with a super-Gaussian output in a CW-CO$_2$ laser, *IEEE J. Quantum Elect.*, 30, 2663–2669, 1994.

12. Kudryashov, A., and Shmalhausen, V., Semipassive bimorph flexible mirrors for atmospheric adaptive optics application, *Opt. Engng.*, 35, 3064–3073, 1996.

13. Kudryashov, A.V., and Samarkin, V.V., Control of high power CO$_2$ laser beam by adaptive optical elements, *Opt. Comm.*, 118, 317–322, 1995.

14. Baumhacker, H., Witte, K.-J., Stehbeck, H., Kudryashov, A., and Samarkin, V., in *Use of deformable mirrors in the 8-TW TiS-laser ATLAS*, Love, Gordon, ed., World Scientific, Singapore, 28–31, 2000.

15. Kudryashov, A.V., and Seliverstov, A.V., Adaptive stabilized interferometer with laser diode, *Opt. Comm.*, 120, 239–244, 1995.

16. Fox, A.G., and Li, T., Resonant modes in a maser interferometer, *The Bell Syst. Tech. J.*, 40, 453–488, 1961.

17. Li, T., Diffraction loss and selection of modes in maser resonators with circular mirrors, *The Bell Syst. Tech. J.*, 917–932, 1965.

18. Paré, C., and Bélanger, P.A., Custom laser resonators using graded-phase mirrors: Circular geometry, *IEEE J. Quant. Electron.*, 30, 1141–1148, 1994.

19. van Neste, R., Résonateurs laser à miroir de phase: Verification de principe, MS thesis, Université Laval, 1994.

20. Bélanger, P.A., Paré, C., Lachance, R.L., and van Neste, R., US patent 5,255,283, 1993.

21. Bélanger, P.A., and Paré, C., Optical resonators using graded phase mirrors, *Opt. Lett.*, 16, 1057–1059, 1991.

22. Zakharova, G., Karamzin, Yu. N., and Trofimov, V.A., Some problems of optical radiation nonlinear distortions compensation. Blooming and random phase distortions of profiled beams, *Atmos. Opt. Clim.*, 8, 706–712, 1995.

23. Ahmanov, S.A., Vorontsov, M.A., Kandidov, V.P., Syhorykov, A.P., and Chesnokov, S.S., Thermal self-action of light beams and methods of its compensation, *Izv. Visshih Uchebnih Vavedenii*, XXIII, 1–37, 1980, (in Russian).

24. Galushkin, M.G., Golubev, V.S., Zavalov, Yu. N., Zavalova, V.Ye., and Panchenko, V.Ya., Enhancement of small-scale optical nonuniformities in active medium of high-power CW FAF CO$_2$ laser, in *Optical Resonators—Science and Engineering*, Kossowsky, R., Jelinek, M., and Novak, J., eds., Kluwer Academic Publishers, New York, 289–300, 1998.

25. International Standard ISO 11146, Optics and optical instruments, Lasers and laser related equipment, Test methods for laser beam parameters: Beam widths, divergence angle and beam propagation factor. Document ISO/TC 172/SC 9/WC, 1995.

26. Cherezova, T.Yu., Kaptsov, L.N., and Kudryashov, A.V., CW industrial rod YAG:Nd^{3+} laser with an intracavity active bimorph mirror, *Appl. Opt.*, 35, 2554–2561, 1996.

11 Application of Laser Beam Shaping for Spectral Control of "Spatially Dispersive" Lasers

I. S. Moskalev, V. V. Fedorov, S. B. Mirov,
T. T. Basiev[†] and P. G. Zverev

CONTENTS

11.1 INTRODUCTION

Lasers, being oscillators at optical frequencies, usually produce laser radiation, which is characterized by an extremely high degree of monochromaticity, coherence, directionality, and brightness. These unique properties of laser light have inspired the widespread use of laser sources in numerous applications. Monochromaticity, or narrow spectrum output, which is usually much smaller than the gain profile of the amplification material, is considered to be one of the inherent laser features. Even media with a very broad, homogeneously broadened gain profile (e.g., laser dyes, Ti^{3+}:sapphire, alexandrite, forsterite, color center), being placed in a nondispersive cavity, provide a significant line narrowing of the output laser radiation. The physical mechanism of the line narrowing is a consequence of the spectral dependence of the gain profile of the amplification media: frequency modes with highest gain build up in the laser cavity faster than others, lower the total inversion, and eventually suppress other frequency modes with lower gain.[1–4]

However, there are many practical applications where a source of radiation combining spatial coherence and high intensity with a continuous or discreet ultra-broadband, multiwavelength spectrum is required. Among them are coherent-transient spectroscopy,[5] Fourier optical processing,[6,7] all-optical image transfer,[8,9] multiwavelength spectroscopy,[10–12] optical memory,[13–15] information processing,[16] and optical computing and optical communications.[17–21] Since the 1970s, there have been many attempts to build multiwavelength and ultrabroadband laser sources suit-able for the applications mentioned above. All these attempts are based on the idea of suppressing mode competition in the cavity, which is responsible for output spectrum narrowing. There are two possible ways to implement this idea: mode separation in the temporal or in the spatial domain. Laser cavities where mode separation was realized in temporal and spatial domains are denoted as *temporally dispersive* and *spatially dispersive* cavities, respectively.

The key idea of temporally dispersive lasers, also known as ultrafast lasers, is a mode-locked operation in which the output contains a multiplicity of temporally shifted, ultrashort optical pulse trains, each pulse train operating at a different wave-length.[22–28] All wavelengths are consequently generated in the same region of the laser gain element. The ultrafast, mode-locked lasers provide broadband output radia-tion, a large number of output spectral channels, high average power, and a spatially coherent output beam. However, the laser usually has a quite complex design and requires a complicated alignment procedure, which makes it a relatively expensive instrument.

While temporally dispersive lasers are limited to mode-locked regimes of opera-tion, the advantage of a spatially dispersive approach is its flexibility in terms of the regime of lasing: pure continuous wave[9,29,30] or pulsed nano-,[31–36] pico-, and femto-second[24,37] regimes of operation are feasible. Spatially dispersive cavities provide spatial separation of different wavelength channels in the laser gain medium accom-panied by effective suppression of gain competition between different frequency components. The construction of the cavities is flexible and allows the utilization of a variety of active media, including crystals,[9,31–40] glasses,[41] and dyes,[34,42–44] as well as single broad stripe,[29,30] multistripe diode chips,[17–19,45–48] and laser diode arrays.[49–51] Spatially dispersive lasers are capable of generating an arbitrary spectral structure of the output radiation, such as ultrabroadband continuous spectrum, or discreet, multiline spectrum (see, for example, Basiev et al.[31]; Danailov and Christov[43]). These lasers are usually much simpler when compared with the ultrafast lasers, and, depending on a particular cavity design, are quite simple in alignment.

This chapter focuses on flexible, spatially dispersive laser systems providing ultra-broadband or multiline spectral output. In appropriately designed systems, the mode competition is effectively suppressed by means of spatial separation of different fre-quency modes in the gain medium. Further, pump laser beam shaping, in accordance with the spatially separated modes in the gain medium, provides spectral control of the output radiation by means of the Fourier transformation of the spatial distribution of the pump radiation into the spectral domain of the output laser radiation.

The technology of spatially dispersive cavities was introduced in 1971 by Ashkin and Ippen,[52] where a dispersion apparatus inside the laser cavity was used to disperse different emitting wavelengths along distinct feedback paths in the resonator. It was proposed primarily for narrowband tunable oscillation, where tuning was achieved simply by scannable pumping beams without the tuning of the intracavity dispersive elements. In 1989, Danailov and Christov[42] proposed to utilize Ashkin's principle of spatially dispersive cavity for the generation of laser radiation with an ultrabroadband spectrum, and further developed the principle of ultrabroadband lasing for dyes in a series of their research papers.[41,43,44,53,54] Two new schemes of spatially dispersive resonators were introduced in the early 1990s. Danailov and Christov proposed a dispersive resonator with a pair of intracavity prisms,[43] and Basiev et al. introduced a dispersive scheme on the basis of a combination of an intracavity lens and a grating.[31,55] These cavities were utilized for realization of the first solid-state multiline[23,41] and ultrabroadband[31–33,55] lasers. In 1994, Basiev et al. had published the first experimental study of a room temperature $LiF:F_2^+$ ultrabroadband laser.[32] In 1995, Basiev et al. published a novel *white-color* laser, where the ultrabroadband multifrequency IR output spectrum from a $LiF:F_2^-$ color-center laser (CCL) in the range of 1.1–1.24 µm was frequency doubled with a simple, single lens phase-matching optical system to obtain a *white* light in the visible, red–green spectral interval of 0.55–0.62 µm.[33] In 1997, this group performed a thorough experimental and theoretical investigation of the temporal, spectral, and spatial features of the spatially dispersive cavity based on a $LiF:F_2^-$ CCL laser, and studied simultaneous frequency doubling of the ultrabroadband, multifrequency IR output spectrum with a single nonlinear crystal.[35] Later, in 2001, Fedorov et al. reported on simultaneous frequency doubling and frequency quadrupling of the ultrabroadband laser radiation from the $LiF:F_2^+$ and $LiF:F_2^-$ CCL lasers, and, in effect, obtained laser radiation with an ultrabroadband output spectrum that lasts from the infrared ($\lambda = 1.2$ µm) to the ultraviolet ($\lambda = 0.2$ µm) spectral regions.[36]

Since the 1990s, spatially dispersive laser cavities and a number of their modifications were used by various research groups to create multiwavelength lasers for different purposes, such as a three-wavelength $Nd:Y_3Al_5O_{12}$ laser,[23] an ultrabroadband DCM dye laser,[34] an ultrabroadband $LiF:F_2^+$ CCL laser,[32,36,38] an ultrabroadband Cr^{4+}:forsterite laser,[39] ultrabroadband Ti^{3+}:sapphire lasers,[40,56] and diode-pumped broadband Cr:LiSGAF and Cr:LiSAF lasers.[9]

In this chapter, in addition to the brief review of major developments of spatially dispersive cavities, we analyze operation of the systems in terms of the transformation of the spatial distribution of the pump radiation into the spectral domain of the output laser radiation. A theoretical model of this process is compared with experimental results on spatial-to-spectral transformation in CCLs.

11.2 PRINCIPLES OF OPERATION

To provide spectral mode suppression, the laser cavity is divided into a set of independent microcavities, which are laterally shifted in the gain media and operate at different wavelengths. Several possible configurations of these cavities are shown in Figure 11.1. The scheme shown in Figure 11.1a, proposed by Danailov,[42] illustrates

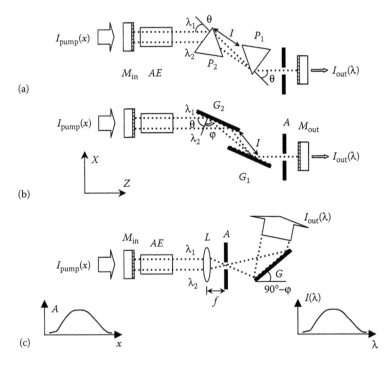

FIGURE 11.1 Schematic diagrams of *spatially dispersive* cavities: (a) The dispersive element is based on a pair of conjugate prisms P_1 and P_2, as proposed by Danailov and Christov[42]; (b) The dispersive element consists of a pair of conjugate diffraction gratings G_1 and G_2, as proposed by Ashkin and Ippen[52]; (c) The dispersive element is based on the combination of an intracavity lens and a Littrow mount grating, as proposed by Basiev et al.[31,55]

the tracing of two beams corresponding to different oscillation wavelengths, λ_1 and λ_2, in the cavity with a pair of conjugated identical prisms. The aperture A is placed in the common area of the beam. The intracavity radiations reflected by the output coupler M_{out} are spatially displaced by Δx after passing the prism pair, and therefore, each wavelength is amplified in its own unique region of the gain medium, and the frequency mode competition is eliminated. This results in each part of the crystal parallel to the laser axis working as an independent laser, with its own wavelength lying within the broad luminescence spectrum of the gain medium (GM). The dispersion of wavelengths $dx/d\lambda$ in the GM depends on prism dispersion $dn/d\lambda$ and the distance l between the prisms. For the Brewster prism pair, the value of dispersion may be estimated from a simple geometrical calculation[43,44] as

$$\frac{dx}{d\lambda} = l \times \frac{d\theta}{d\lambda} = 2l \frac{dn}{d\lambda} \tag{11.1}$$

One of the important issues in achieving ultrabroadband oscillation in the spatially dispersive cavity relates to maximization of the spatial separation between the beams in the GM. Thus, in the scheme shown in Figure 11.1b, proposed by Ashkin and Ippen,[52] the prism pair is replaced with a grating pair, which has significantly higher angular dispersion. The dispersion of wavelengths in the gain media for this cavity can be written as

$$\frac{dx}{d\lambda} = l \times \frac{d\varphi}{d\lambda}\frac{\cos(\theta)}{\cos(\varphi)} = \frac{lk}{t}\frac{\cos(\theta)}{\cos^2(\varphi)} \tag{11.2}$$

where:
  k is the order of diffraction
  t is the grating spacing period
  θ and φ are the incident angle and diffraction angle inside the grating pair at the central wavelength

It should be noted that lasers with intracavity prisms or grating pairs are also used for subpicosecond pulse laser oscillation. However, in this laser configuration, the GM is placed in the part of the cavity where radiations with different wavelengths propagate in one channel near the aperture.[24] The distance between dispersion elements is selected in such a way that the group velocity dispersion in the GM is compensated and mode locking at different oscillation wavelengths is provided. In the ultrabroadband schemes, the GM is placed in the part of the cavity where beams with different wavelengths propagate in different channels.

A cavity with a grating pair has a larger dispersion; however, due to the zero-order diffraction, it also has higher passive losses than a cavity with a pair of prisms. Furthermore, output radiation is combined into one beam at the diffraction grating. It raises the requirement for the optical damage threshold of the diffraction gratings. These shortcomings of the grating cavity are eliminated in the cavity proposed by Basiev et al.[31–33,55], whose optical scheme is depicted in Figure 11.1c. In this scheme, one of the selective elements is replaced by an intracavity lens, and the output coupler is combined with a Littrow mount diffraction grating operating in the auto-collimation regime. This laser works as follows. Emission from the active medium passes through the focusing lens L into the off-axis mode suppression element, aperture A, which separates from the amplified emission only the part of it that is spread parallel to the resonator axis. This separated radiation is diffracted on the diffraction grating G. The lens provides that beams propagating at different distances from the optical axis are incident on the diffraction grating at different angles θ. The Littrow mount grating works as a retroreflector in the first order of diffraction and returns a part of the radiation back to the aperture according to the following autocollimating condition:

$$k\lambda = 2t\,\sin(\theta) \tag{11.3}$$

where $k = 1$. In other words, the radiation with wavelength λ, incident on the diffraction grating at an angle θ, is retroreflected. The zeroth order of diffraction, reflected from the grating as from a conventional plane mirror, serves for the laser output. The off-axis mode suppression element, an aperture A, in turn extracts from the diffracted radiation only the radiation of the main laser modes. Secondary laser modes, which diverge from the optical axes, are expelled from the process of generation.

Hence, the aperture should simultaneously select the fundamental transverse modes for all existing channels in the cavity. This is done by positioning the aperture in the place where all the channels intersect, which corresponds to the focal plane of the lens L. The width of the aperture is estimated as the mode size in the focal plane of the lens.

The dispersion of wavelengths $dx/d\lambda$ in the active element[36] can be estimated as follows:

$$\frac{dx}{d\lambda} = \frac{f}{2t\cos(\theta)} \tag{11.4}$$

This scheme, featuring the same order of magnitude $dx/d\lambda$ as in the scheme shown in Figure 11.1b, has several advantages over it. First, it reduces the optical damage threshold of the diffraction grating. Second, the zeroth order of diffraction is used, whereas in the scheme B, the zeroth order of diffraction are considered as losses. Third, as it will be shown later, the angular dispersion of the output radiation in the scheme, as shown in Figure 11.1c, may be matched with the angular dispersion of the corresponding phase in a nonlinear crystal to provide ultrabroadband frequency mixing.

The key idea behind transforming the spatial domain of the pumping beam into the spectral domain of the output oscillation in the laser scheme, as shown in Figure 11.1c, is based on two fundamental cavity elements: the intracavity lens and the diffraction grating. The intracavity lens transforms the spatial distribution of the radiation in the active medium into an angular distribution. The diffraction grating transforms the angular distribution into a spectral distribution according to the autocollimation condition. Therefore, overall, this cavity provides a transformation of the spatial distribution of the pump radiation into a spectral distribution of the stimulated emission. In a regime of quasi-stationary lasing for pump intensities much higher than the threshold level, the intensity of the output radiation can be defined as follows[2–4]:

$$I_{osc}(\lambda) = \left(\frac{T}{T+L}\right)\eta_p\eta_a\eta_q I_{pump} \tag{11.5}$$

where:

η_p is the pump efficiency

η_a is the pump utilization efficiency

$\eta_q = \lambda_{pump}/\lambda_{osc}$ is the Stokes factor or quantum efficiency, which represents the ratio of photon energy at oscillation and pump wavelengths

T is transmission of the output coupler

L represents the cavity losses

I_{pump} is the intensity of the pump radiation

Equations 11.4 and 11.5 for the case of a constant quantum efficiency and the absence of a spectral dependence for the cavity losses and the output coupler transmission within the luminescence band of the active medium demonstrate a direct relationship between the spatial distribution of the pump radiation and the spectral distribution of the output oscillation in this spatially dispersive cavity. In other words, in the studied cavities, the spatial distribution of the pump radiation is transformed into the spectral distribution of the output laser oscillation:

$$I_{pump}(x) \rightarrow I_{osc}(\lambda) \qquad (11.6)$$

where, as will be shown in the next section, the output wavelength λ as a function of the transverse coordinate of the pump radiation x is determined by the following equation:

$$\lambda \approx \lambda_0 + x \frac{d\lambda}{dx} \qquad (11.7)$$

where λ_0 is the oscillation wavelength of the axial beam.

Active media with broad gain spectra and large gain coefficients are obviously the most appropriate for experimental realization of this spatial-to-spectral transformation. Among other gain media mentioned earlier, LiF color center crystals are of a special interest. Their advantages over the dye media include a wide spectral oscillation range, long operation lifetime, rigidity, and ease of handling. Due to their high gain, the color centers demonstrate some advantages over impurity-doped materials, such as low sensitivities to the quality of the cavity optical elements and to the angular and spectral characteristics of the pump radiation, and, what is most important, color center lasers feature practically no temporal delay between the pump and the output pulses. Recent progress in LiF CCL crystals stimulates the design of effective room-temperature-stable color center lasers tunable in the near-IR spectral region (Basiev and Mirov[57]; Mirov and Basiev[58]; Dergachev and Mirov[59]; Basiev et al.[60] and references herein). The F_2^{+**} CCL (LiF:F_2^{+**}) exhibits excellent photo and thermostable operation at room temperature when pumped by an alexandrite laser, and can provide efficient high power lasing tunable in the 800–1200 nm spectral range.[59,61] Another widely used color center in LiF crystal is F_2^- (LiF:F_2^-). These are used as passive Q-switchers of resonators of neodymium lasers and as active elements of near-IR tunable lasers. LiF:F_2^- crystals feature wide near-IR absorption (0.85–1.1 μm) and emission bands (1.0–1.3 μm), and a high (~50%) quantum efficiency of fluorescence at room temperature.[57,58,60] As a result, further analysis in this study will be based on using LiF:F_2^- and LiF:F_2^{+**} gain media in the spatially dispersive cavity, as shown in Figure 11.1c.

11.3 THEORETICAL ANALYSIS OF THE SPATIAL– SPECTRAL TRANSFORMATION IN THE SPATIALLY DISPERSIVE LASER CAVITY

The detailed optical scheme of the cavity is depicted in Figure 11.2. The cavity analysis is performed in the limits of the paraxial approximation. In that approximation, the angles of propagation of optical beams, with respect to the optical axis of a considered optical system, are small enough to assume that $\alpha \approx \sin(\alpha) \approx \tan(\alpha)$ and $\cos(\alpha) \approx 1$ (where the angle is measured in radians). The paraxial approximation allows the

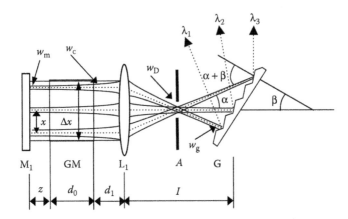

FIGURE 11.2 Schematic diagram and beam parameters of the spatially dispersive cavity shown in Figure 11.1c. The optical elements of the cavity are: M_1—input mirror, GM—gain medium, A—intracavity aperture, G—Littrow mount diffraction grating. The major radii of the modes are: w_m—on the input mirror; w_c—on the output facet of the gain crystal; w_D—in the plane of the intracavity aperture; w_g—on the diffraction grating.

application of the ray vector and matrix formalism, and the Gaussian beam approximation, as described in detail by Prokhorov[2] and Hodgson and Weber.[3]

In terms of the ray vector formalism, every beam passing through the active element is described by its initial transverse co-ordinate x with respect to the cavity optical axis. Only those beams that are initially parallel to the optical axis are considered, since all the others are blocked by the intracavity aperture A. After refraction in the intracavity lens L_1, the beams intersect the optical axis at the focal point of the lens at a paraxial angle a defined by the following equation:

$$\alpha = \frac{x}{f} \tag{11.8}$$

where f is the focal length of the lens. The normal to the diffraction grating forms an angle β to the optical axis of the cavity. Thus, each beam with initial coordinate x arrives at the diffraction grating at an angle $\theta = \beta + \alpha$. The angle β is found from the autocollimation condition 11.3 for the central wavelength λ_0, which corresponds to the axial beam with $x = 0$:

$$\beta = \arcsin\left(\frac{\lambda_0}{2t}\right) \tag{11.9}$$

Substituting the expression for the angle of incidence θ into 11.3 and using Equations 11.8 and 11.9, we find that each fraction of the laser crystal located at a distance x from the optical axis generates its own wavelength λ, determined as

$$\lambda = 2t \sin\left(\arcsin\left[\frac{\lambda_0}{2t}\right] + \frac{x}{f}\right) \tag{11.10}$$

This equation can be reduced to the following expression using simple trigonometric identities:

$$\lambda = \lambda_0 + x \frac{\sqrt{4t^2 - \lambda_0^2}}{f} \tag{11.11}$$

For a particular cavity configuration, λ_0, t and f in Equations 11.10 and 11.11 are parameters and x is a free variable. Differentiating Equation 11.11, the dispersion $d\lambda/dx$ of the wavelengths in the active element can be written as

$$\frac{d\lambda}{dx} = \frac{\sqrt{4t^2 - \lambda_0^2}}{f} \tag{11.12}$$

The coordinates x of the lasing wavelengths are fully determined by the spatial distribution of the pump beam radiation in the GM. Therefore, Equation 11.11 becomes identical to the functional dependence of the output wavelength λ on the spatial distribution of the pump in the laser crystal, represented by Equation 11.7. This illustrates the idea of control of the output spectrum of the ultrabroadband multiwavelength laser by the laser beam shaping of the pump radiation. Every ray with the initial coordinate x in the laser crystal and lasing wavelength λ represents a wavelength channel. The output spectrum of the laser consists of a large number of such wavelength channels. The beam shaping of the pump radiation can be used to select the desired wavelengths by placing a spatial mask in front of the input mirror. In this case, a discrete multiwavelength output spectrum is observed. On the other hand, a wide continuous pump beam can be used to obtain a multiwavelength continuous spectrum by exiting a certain portion of the GM.

So far, the propagation of light has been considered in terms of the geometrical optics, where every wavelength channel is represented by an infinitesimally thin ray. This approximation is applicable as long as the channels' diameters are much smaller than their optical path lengths. However, in order to study the behavior of the individual channels, and find the appropriate cavity parameters for a desired spectral distribution of the output radiation, it is necessary to apply the general laser theory based on wave optics. In the paraxial approximation, the matrix method for Gaussian beams can be used.[2,3]

In the framework of this method, the cavity of each channel is approximated by a cavity consisting of two plane mirrors, an intracavity lens and a dielectric medium. According to this model, the diffraction grating, operating in an autocollimation regime, is equivalent to a plane mirror. The matrix of the equivalent resonator is calculated as follows:

$$M = \begin{pmatrix} 1 & l \\ 0 & 1 \end{pmatrix} \begin{pmatrix} 1 & 0 \\ -\dfrac{1}{f} & 1 \end{pmatrix} \begin{pmatrix} 1 & d_{\text{eff}} \\ 0 & 1 \end{pmatrix} = \begin{pmatrix} g_1 & L \\ \dfrac{g_1 g_2 - 1}{L} & g_2 \end{pmatrix} \tag{11.13}$$

where $d_{\text{eff}} = z + d_0/n + d_1$ is the effective distance between mirror M_1 and the lens L_1, z is the separation between the mirror and the rear facet of the crystal, d_0 is the length of the crystal, d_1 is the distance from the output facet of the crystal to the lens, and n is the refraction index of the active medium. The resonator effective length L is given by

$$L = d_{\text{eff}} + l - \frac{l d_{\text{eff}}}{f} \qquad (11.14)$$

where l is the separation between lens L_1 and grating G, and f is the focal length of the lens. The parameters $g_1 = 1 - l/f$ and $g_2 = 1 - d_{\text{eff}}/f$ are the stability parameters of the resonator.

An unstable cavity configuration leads to a strong overlapping and coupling among different channels. Therefore, we consider only stable cavity configurations for which the following stability condition is satisfied:

$$0 < g_1 g_2 < 1 \qquad (11.15)$$

The beam radii of TEM_{00} modes for individual channels at the mirror w_{m}, the diffraction grating w_{G}, output crystal facet w_{c}, and the aperture w_{D} are calculated via the ABCD law for the appropriate cavity unit cells[2,3]:

$$w_{\text{m}}^2 = \left(\frac{L\lambda}{\pi}\right)\sqrt{\frac{g_2}{g_1(1 - g_1 g_2)}} \qquad (11.16)$$

$$w_{\text{G}}^2 = w_{\text{m}}^2 \frac{g_1}{g_2} \qquad (11.17)$$

$$w_{\text{c}}^2 = w_{\text{m}}^2 + \frac{\lambda^2 (z + d_0/n)^2}{\pi w_{\text{m}}^2} \qquad (11.18)$$

$$w_{\text{D}} = \frac{\lambda f}{\pi w_{\text{m}}} \qquad (11.19)$$

where a beam radius is defined as the radius where intensity of the mode is decreased to $1/e^2$ of its maximum.

In order to reduce crosstalk between adjacent channels, their overlap in the active medium must be minimized. It is assumed that two Gaussian beams do not interfere in the laser GM if they are separated by a distance larger than their diameter. The minimum channel's overlapping and maximum amount of oscillation lines can be realized by minimizing beam radius on the output facet of the laser crystal w_{c}. Minimizing w_{c}^2 in Equation 11.18 with respect to w_{m}^2 and z, all the other variables being parameters, we find that the minimum beam radius is achieved with $z = 0$ and:

$$w_{\text{c}}^2 = 2w_{\text{m}}^2 = \frac{2\lambda d_0}{n\pi} \qquad (11.20)$$

The maximum number of oscillation channels can be estimated as the ratio of the crystal width to the beam diameter on the output facet of the crystal $N = \Delta x/2w_{\text{c}}$. The spectral separation of the channels is found using Equation 11.3, where Δx is replaced by the spatial separation of the channels, which, in turn, equals the channels' diameters on the output facet of the crystal:

$$\delta\lambda_{ch} = \frac{4tw_c}{f}\cos(\beta) \qquad (11.21)$$

In order to avoid channel crosstalk, the diffraction grating must provide spectral resolution better than or equal to that given by Equation 11.21. The diffraction grating resolution is determined by the number of illuminated grooves (M), and can be calculated in terms of the beam diameter on the grating ($2w_m/\cos\beta$):

$$\delta\lambda_G = \frac{\lambda}{M} = \frac{\lambda t}{2w_G}\cos(\beta) \qquad (11.22)$$

The requirement on the diffraction grating resolution $\delta\lambda_G \leq 8\lambda_{ch}$ leads to the following condition on the minimum mode diameter on the diffraction grating:

$$w_G \geq \frac{\lambda f}{8w_c} \qquad (11.23)$$

The requirements on the laser cavity parameters discussed above can be conveniently summarized in three expressions in terms of the cavity stability parameters, g_1 and g_2, as follows. The requirement of the cavity stability (Equation 11.15):

$$g_1 < \frac{1}{g_2} \qquad (11.24)$$

The requirement of the minimum resolution of the diffraction grating (Equation 11.23):

$$g_1 \leq g_2 \frac{\pi^2}{128A^2} \qquad (11.25)$$

The requirement of the maximum number of channels (Equations 11.16, 11.17, and 11.20):

$$g_1 = \frac{g_2}{g_2^2 + A^2} \qquad (11.26)$$

where $A = d_0/(nf)$. An example of application of Equations 11.24 through 11.26 for determining the desired laser cavity parameters is shown in Figure 11.3. This figure shows a stability diagram calculated for the LiF:F$_2^-$ and LiF:F$_2^{+**}$ CC lasers used in our experiments. These lasers are based on 4 cm long LiF crystals, 1200 groves/mm holographic gratings, and 7 cm focal length intracavity lenses. The acceptable range of g_1, g_2 values is bounded by the stability region of the cavity (curves "a" and "b" in Figure 11.3 corresponding to conditions [11.24] and [11.25], respectively). Curve c shows the $g_1 = f(g_2)$ dependence (Equation 11.26) for the optimal beam waist w_m on the mirror M_1. As illustrated in Figure 11.3, both g_1 and g_2 parameters should be within the intervals (−2, 0) and (−0.87, 0), respectively. This places limits on the d_{eff} and l values: $3f < d_{eff} < f$ and $1.87f < d_{eff} < f$. Several cavity configurations calculated for the LiF:F$_2^-$ and LiF:F$_2^{+**}$ CC lasers are shown in Table 11.1. It follows from this table that for the laser crystals, diffraction gratings, and the intracavity lenses used

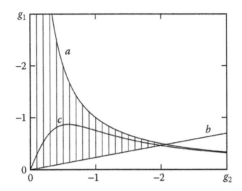

FIGURE 11.3 Stability diagram of the spatially dispersive cavity. (a) $g_1 = 1/g_2$, (b) $g_1 = g_2\pi^2/128A^2$, (c) $g_1 = g_2/(g_2^2 + A^2)$.

Table 11.1   Cavity Parameters of the LiF:F_2^- and LiF:F_2^{+**} CC Lasers Calculated According to Equations 11.20 through 11.22 for 1200 groves/mm Diffraction Grating

Gain media	λ_0 (gm)	$\Delta\lambda$ (gm)	w_m (gm)	f (mm)	$d\lambda/dx$ (nm/mm)	Δx (mm)	$\delta\lambda_{ch}$ (nm)	N
LiF:F_2^-	0.96	0.3	102	30	39	7.7	11.2	27
				50	23	12.8	6.7	44
				70	16.7	19.6	4.8	62
LiF:F_2^{+**}	1.14	0.4	94	30	43	9.4	11.3	35
				50	26	15.6	6.8	59
				70	19	21.9	4.9	82

in the experiments, the spatial-to-spectral transformation $I_{pump}(x) \rightarrow I_{osc}(\lambda)$ will be defined by the following relationship:

$$\lambda \approx \lambda_0 + 19\frac{nm}{mm} \times x[mm] \qquad (11.27)$$

It follows from Equation 11.20 that for a given lasing wavelength, the optimal size of the beam waist on the mirror is defined by the length and the index of refraction of the laser crystal. For the laser crystals used in the experiments and the central wavelength of their amplification bands, the beam waist radius w_m calculated from this equation equals approximately 94 μm. The cavity spectral resolution defined by Equation 11.21 is $\delta\lambda_{ch} = 4.9$ nm. It is noteworthy that the spectral resolution determines the minimum spectral separation between two adjacent channels, while the width of the oscillation spectrum of an individual channel can be much smaller than the cavity spectral resolution. In this case, when many individual channels are simultaneously excited, the output oscillation spectrum will be composed of the separate lines, with the spectral separation between adjacent lines equal to the spectral resolution. The full width of the LiF:F_2^{+**} gain spectrum $\Delta\lambda$ is 400 nm (from 800 to 1200 nm). Hence, the cavity configuration allows the $I_{pump}(x) \rightarrow I_{osc}(\lambda)$ transformation to be obtained with the maximum achievable number of the output wavelength channels $N = \Delta\lambda/\delta\lambda \approx 82$. The required overall width of the pump beam Δx, found from Equation (11.27), is approximately 22 mm.

11.4 UTILIZATION OF SINGLE NONLINEAR CRYSTAL FOR THE TRANSFORMATION OF THE SPATIAL DISTRIBUTION OF THE ULTRABROADBAND RADIATION INTO THE SPECTRAL DOMAIN OF THE FREQUENCY UP-CONVERTED RADIATION

The technology of the spatially dispersive cavities provides another important spatial-to-spectral transformation method realized with nonlinear crystals. This is a novel method of obtaining second (and similarly fourth) harmonic generation of the ultra-broadband radiation in a single nonlinear crystal. This method is based on a simultaneous realization of phase-matching conditions for all oscillating wavelengths of the ultrabroadband spatially dispersive laser. This is achieved by means of an appropriate beam shaping of the output radiation of the ultrabroadband laser. In other words, a proper shaping of the ultrabroadband laser output beam could provide simultaneous phase-matching conditions for frequency up- and even down-conversion processes for all output wavelengths of the laser in a single nonlinear crystal. For example, due to a nonlinear dependence of the second harmonic generation (SHG) intensity versus incident radiation intensity (I_{osc}), the total transformation of the pump beam after frequency doubling can be written as follows:

$$I_{pump}(x)^2 \rightarrow I_{osc}(v)^2 \rightarrow I_{shg}(2v) \tag{11.28}$$

The sum frequency generation (SFG) process provides the possibility for an even wider class of transformations. In this case, the transformation of the pump beam is proportional to the spatial autocorrelation function of the pumping beam and can be presented as follows:

$$\int I(x)I(x - x_0)dx \rightarrow \int I_{OSC}(v)I_{OSC}(v_\Sigma - v)dv \rightarrow I_{sfg}(v_\Sigma) \tag{11.29}$$

To realize these transformations (Equations 11.28 and 11.29), it is necessary to provide phase-matching conditions for all the oscillating wavelengths. Due to a small spectral and angular acceptance bandwidth in the crystal, the nonlinear frequency conversion is a difficult problem. The feasibility of SHG of broadband laser radiation with a spectral width of 10 nm was demonstrated by Volosov and Goryachkina (1976) and Szabo and Bor (1990). The spectral width of the output radiation from the LiF:F_2^- and LiF:F_2^{+**} color center lasers can exceed 300 nm. One of the possible ways for SHG of such an ultrabroadband radiation is matching the angular dispersion of the output radiation with the angular dispersion of the phase-matching angle by means of an additional lens (Figure 11.4).

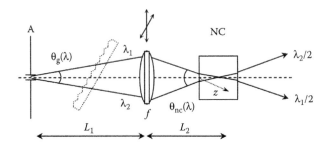

FIGURE 11.4 Schematic diagram demonstrating optimization of the angular dispersion of the output radiation ($\theta_g(\lambda)$) and phase-matching angular dependence ($\theta_{nc}(\lambda)$) in the nonlinear crystal ($\Gamma = (d\theta_g/d\lambda)/(d\theta_{nc}/d\lambda) = L_1/L_2$.

The angular distribution of oscillating wavelengths of the CC laser output radiation after the grating is described by Equation 11.10:

$$\theta_g(\lambda) = \arcsin\left(\frac{\lambda_0}{2t}\right) - \arcsin\left(\frac{\lambda}{2t}\right) \tag{11.30}$$

On the other hand, the required angular dependence for nonlinear conversion could be written as

$$\theta_{nc}(\lambda) = \arcsin\left(\frac{\sin(\theta_c - \theta_{ph}(\lambda))}{n(\lambda)}\right) \tag{11.31}$$

where θ_c is the cut angle between the crystal axis and the input facet of the nonlinear crystal, n is the refractive index at an incident wavelength, and θ_{ph} is the phase-matching angle of the nonlinear frequency conversion. The dependence (Equation 11.31) can be found from the theory of nonlinear optics elsewhere.[62] As an example, for $1/\lambda_1 + 1/\lambda_2 \rightarrow 1/\lambda_3$ type I interaction in the negative uniaxial crystal, the phase-matching can be calculated as follows:

$$\theta_{ph}(\lambda) = \arctan\left(\frac{1-U}{W-1}\right) \tag{11.32}$$

where $U = (A + B)^2/C^2$; $W = (A + B)^2/F^2$; and $A = n_{o1}/\lambda_1$; $B = n_{o1}/\lambda_2$; $C = n_{o1}/\lambda_3$; $F = n_{e1}/\lambda_3$.

In the linear approximation, the required angular magnification of the optical system is given by the expression:

$$\Gamma = \frac{d\theta_g/d\lambda}{d\theta_{nc}/d\lambda} \tag{11.33}$$

This scheme operates efficiently only in the crystals with phase matching of the "ooe" or "eeo" type (see, for example, Dmitriev et al.[62] for a detailed description of these processes). The optimal nonlinear crystal can be selected by calculating the spectral dependence of the phase-matching angles for different crystals. It is assumed that the crystal is cut for the normal incidence of radiation at a wavelength corresponding to

Table 11.2   Parameters of Nonlinear Crystals (1 cm long) Used for Frequency Doubling of Broadband LiF:F$_2^-$ and LiF:F$_2^{+**}$ Lasers, and the Calculated Widths of the Spectral ($\Delta\lambda$), Angular ($\Delta\theta_{pm}$), and Temperature (ΔT) Phase-Matching

Crystal	θ_c (degree)		$d\theta_{pm}/d\lambda$ (degree/μm)		$\Delta\lambda$ (nm)	$\Delta\theta_{pm}$ (mrad)	ΔT (K)	$D_{eff} \times 10^{-12}$ (mV^{-1})
	0.96 μm	1.14 μm	0.965 μm	1.14 μm				
KDP	41.4	41.7	11	1.5	28.3	1.7	25.1	0.29
LiIO$_3$	33.5	27.9	75	4.5	0.7	0.6		2.75
LiNbO$_3$	–	69.3	–	24	0.3	3.1	1.1	5.32
BBO	24.5	22.8	33	1.7	2.1	0.52	39.8	1.69–2.1

Source: Hodgson, N. and Weber, H., *Optical Resonators: Fundamentals, Advanced Concepts and Applications,* Springer-Verlag, Heidelberg, Germany, 1997.

the middle of the spectral range of lasing (at 0.965 nm for the LiF:F$_2^{+**}$ laser and at 1.14 μm for the LiF:F$_2^-$ laser). Table 11.2 shows the calculated phase-matching angles θ_c, the angular matching dispersion $d\theta/d\lambda$ outside the crystal, the nonlinearity coefficient D_{eff}, and the widths of the spectral $\Delta\lambda$, angular $\Delta\theta$, and temperature ΔT phase-matching for some crystals. Figure 11.5 presents the calculated dispersion curves for the angle of incidence of the radiation from the broadband LiF:F$_2^{+**}$ and LiF:F$_2^-$ lasers on the nonlinear crystal (solid curves), and the phase-matching angles in KDP, BBO, LiNbO$_3$, and LiIO$_3$ crystals recalculated for the optimal angular magnification Γ.

Among the nonlinear crystals considered, LiNbO$_3$ has the largest nonlinearity coefficient ($D_{eff} = 5.32 \times 10^{-12}$ mV^{-1}). This crystal exhibits a strong nonlinear dependence of the phase-matching angle on the wavelength, which makes it possible to obtain 90° phase matching by decreasing the wavelength to 1.05 μm. Hence, this crystal is suitable only for output frequency doubling of the LiF:F$_2^-$ laser. However, strong nonlinearity leads to a substantial discrepancy between the angular matching curves and the wavelength dependence of the angle of incidence of the radiation from a broadband laser on the nonlinear crystal (angular mismatch). The dispersion dependence of the phase-matching angle of the KDP crystal has a flat maximum around 1.05 μm. As a result, a strong angular mismatch was observed for the LiF:F$_2^{+**}$ laser. The KDP crystal has the smallest nonlinearity coefficient ($D_{eff} = 0.29 \times 10^{-12}$ mV^{-1}) among the suggested crystals. The spectral dependences of the phase-matching angles for the LiIO$_3$ and BBO crystals are similar (Figure 11.5) when the angular magnifications Γ are optimal ($\Gamma_{BBO} = 0.8$ and $\Gamma_{LiIO3} = 1.8$ for the 0.87–1.1 μm range, and $\Gamma_{BBO} = 0.4$ and $\Gamma_{LiIO3} = 1$ for the 1.1–1.25 μm range). The LiIO$_3$ crystal has a somewhat better angular matching and a larger nonlinearity coefficient ($D_{eff} = 2.75 \times 10^{-12}$ mV^{-1}) compared with the BBO crystal ($D_{eff} = 1.69 \times 10^{-12}$ mV^{-1}). Thus, among the crystals we examined, the LiIO$_3$ crystal proved to be the best for frequency doubling of the infrared radiation emitted by an ultrabroadband laser.

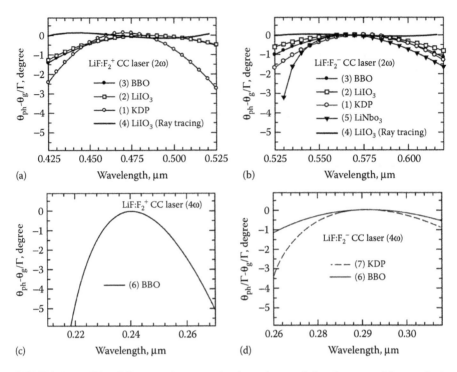

FIGURE 11.5 The differences between the dependence of the phase-matching angle in nonlinear crystals and angular dispersion of the output radiation of the ultrabroadband laser for SHG (a,b) and FHG (c,d): (1) KDP crystal ($\Gamma = 0.33$ for LiF:F_2^- and $\Gamma = 0.4$ for LiF:F_2^{+**}), (2) LiIO$_3$ crystal ($\Gamma = 1$ for LiF:F_2^- and $\Gamma = 1.8$ for LiF:F_2^{+**}), (3) BBO crystal ($\Gamma = 0.36$ for LiF:F_2^- and $\Gamma = 0.77$ for LiF:F_2^{+**}), (4) LiIO$_3$ crystal (ray tracing), (5) LiNbO$_3$ crystal ($\Gamma = 5.9$ for LiF:F_2^-), (6) BBO crystal ($\Gamma = 1.7$ for LiF:F_2^- and $\Gamma = 0.25$ for LiF:F_2^{+**}), (7) KDP crystal ($\Gamma = 2.5$ for LiF:F_2^-).

An additional opportunity to match angular dispersions is to use a nonparaxial approximation for beam propagation. In this case, the additional parameters such as lens thickness, lens tilt, off-axial lens shift, and radii of curvatures of the lens surfaces allow us to obtain better matching. For example, the curve (4) in Figure 11.5 shows the differences between the angular dependence of the nonlinear phase-matching in LiIO$_3$ crystal and the output radiation of the ultrabroadband laser calculated using ray tracing for nonparaxial approximation. The curve 4 in Figure 11.5b was calculated using the following parameters: focal length is 50 mm, lens thickness is 10 mm, off-axial lens shift is 10 mm, and lens tilt is 4°. As one can see from the figure, this approximation demonstrates essentially better results than geometric optics for the same crystal (curve 2).

The broadband radiation produced in the visible region can be converted into the UV range by doubling its frequency one more time. After the frequency doubling in a nonlinear crystal with the ooe phase-matching, the polarization of the second-harmonic radiation lies in the diffraction plane of the grating. So, for further conversion of the radiation to the fourth harmonic, one should use eeo interaction. However, the positive nonlinear crystals operating in the UV region are not available now. Other way to obtain an ultrabroadband UV radiation is to rotate the polarization of the ultrabroadband second harmonic by 90° (for example, an optically active quartz crystal could be utilized) and to use a negative nonlinear crystal. Among the available nonlinear crystals, the BBO is the most efficient crystal for generating the ultrabroadband fourth harmonic radiation.

11.5   EXPERIMENTAL RESULTS

Laser experiments of pump beam transformation were realized with gain-switched $LiF:F_2^{+**}$ and $LiF:F_2^-$ CCLs. Due to the high gain coefficients, the CCLs demonstrate effective performance in flat cavities.[32,33,35,36] Hence, to simplify the scheme of the spatially dispersive cavity, a cylindrical intracavity lens operating in the plane of dispersion of the diffraction grating can be utilized. The schematic of this laser is similar to that previously described, and is shown in Figure 11.6. Cylindrical lenses with focal lengths $f = 30$ or 50 mm were used as the intracavity lenses. The 1200 groves/mm diffraction grating with 20%–50% reflectivity was utilized as a dispersive element. The input dichroic mirror M_1 had a transmission of >80% at the pumping wavelength and a reflection of >95% in the oscillation wavelength range. The active elements used in the experiments were LiF crystals 4 cm long, cut at the Brewster angle. To produce a high concentration of stable laser-active F_2^- and F_2^{+**} color centers in LiF, the crystals were γ-irradiated using a ^{60}Co source. The $LiF:F_2^{+**}$ active elements had a 2.5–3 cm^{-1} coefficient of absorption at 610 nm. The absorption coefficient of the $LiF:F_2^-$ crystal was 0.7 cm^{-1} at 1047 nm. Due to a wide absorption band of CC's, a large number of different solid-state lasers can be used as a pump source. Some of those used in the experiments are displayed in Table 11.3. A pumping beam, expanded to a diameter of 10 mm, was further reshaped into an elliptical profile with a cylindrical lens. In order to demonstrate spatial–spectral transformation, the pump beam was further modulated by the spatial mask. Laser spectra of single output pulses were measured with a CCD camera coupled to an imaging grating spectrograph. The averaged lasing spectra were recorded with a spectrometer, a photodiode, and a boxcar integrator.

Table 11.3   The Pump Laser Source Used for Spatial–Spectral Beam Transformation in $LiF:F_2^{+**}$ and $LiF:F_2^-$ CCLs Experiments

Pump source		
Laser	**Wavelength (μm)**	**Gain media**
SHG of Nd^{3+}:YAG (1.32 μm)	0.66	$LiF:F_2^{+**}$
Alexandrite laser	0.74	
Nd^{3+}:YAG	1.064	$LiF:F_2^-$
Nd^{3+}:YLF	1.047	

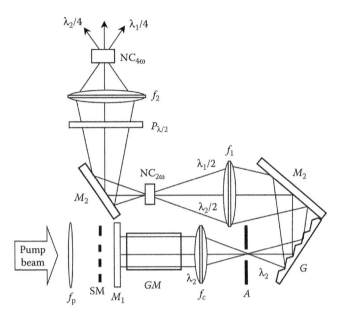

FIGURE 11.6 Experimental setup of the spatially dispersive laser, where the spatial profile of the pumping beam is transferred into spectral output of ultrabroadband fundamental, SHG, and FHG oscillation: f_p—cylindrical lens, SM—spatial mask, M_1—input mirror, GM—gain media (LiF:F_2^{+**} or LiF:F_2^- CC crystal), f_c—intracavity lens, A—aperture, G—diffraction grating, f_1, f_2—collimating lens for SHG and FHG, NC$_{2\omega}$, NC$_{4\omega}$—nonlinear crystal for SHG and FHG, $P_{\lambda/2}$—$\lambda/2$ plate, M_2—turning mirror.

The dependence of the overall energy efficiency of the ultrabroadband CC lasers on the pump energy is depicted in Figure 11.7. The maximum efficiency of the LiF:F_2^- laser, equal to 16%, was reached using Nd^{3+}:YLF pump laser at a pump energy equal to $E = 25$ mJ. Figure 11.8a shows the experimental spectrum of broadband lasing of the LiF:F_2^- laser with a width of more than 0.15 μm in the near-infrared range (1.08–1.23 μm). The infrared output radiation has a high divergence in the horizontal plane determined by the autocollimation condition (Equation 11.9). In the absence of any collimating optics, its divergence is approximately 10° for the whole oscillation spectra from 1.08 to 1.23 μm. However, the divergence of each channel is low and can be determined by the laser mode structure. By using additional cylindrical lenses, we were able to compensate for the angular divergence of these beams, and to collimate the output radiation to a single beam with a divergence of about 1 mrad.

An ultrabroadband oscillation of the LiF:F_2^{+**} laser with a spectral bandwidth in the 0.89–1.04 μm range was obtained using SHG of Nd^{3+}:YAG laser (0.66 μm) as a pump source (Figure 11.9f). The maximum efficiency of the ultrabroadband LiF:F_2^{+**} laser was equal to 20% and was obtained at $E = 9$ mJ of pump energy (Figure 11.7).

For experimental demonstrations of the spatial–spectral relationships, the spatial masks modulated the pump beam shape in front of the input mirror. Figure 11.9a–d demonstrates spatial–spectral pump-output transformations with a two-aperture

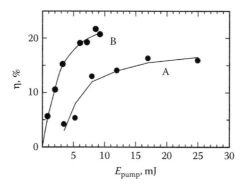

FIGURE 11.7 Output efficiencies of the ultrabroadband LiF:F$_2^-$ (a) and (b) LiF:F$_2^{+**}$ CC lasers.

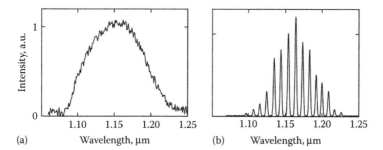

FIGURE 11.8 Output spectrum of the ultrabroadband (a) and multiline (b) LiF:F$_2^-$ CC laser.

spatial filter arrangement into the output radiation of the broadband LiF:F$_2^{+**}$ laser. When these spatial filters, with circular apertures of 0.5 mm diameter separated by distances 0.86, 2, and 2.8 mm (Figure 11.9b–d, respectively), were used to reshape the pump beam, it resulted in a two-line spectral output of the LiF:F$_2^{+**}$ laser with a variable spectral separation between two lines, increasing from 20 to 60 nm. As one can see from Figure 11.9b–d, we experimentally realized spatial–spectral transformation with a coefficient of transformation $\Delta\lambda/\Delta x = 19$ nm mm^{-1}, which is in good agreement with the theoretical value defined by Equation 11.5. The narrow-line oscillation was realized by the utilization of a spatial filter—a metal plate with 0.5 mm aperture placed in front of the input mirror (see Figure 11.9a). Tuning the narrow-line oscillation wavelength was realized by moving the aperture in parallel across the pump beam without any tuning of the laser cavity components. The experimentally measured range of tunability was from 900 to 1150 nm. Figure 11.9e presents another variant of spatial–spectral transformation. It demonstrates a fragment of an experimental LiF:F$_2^{+**}$ output spectrum obtained when a spatial periodic structure (seven rectangular shape holes) with a shadow width of 100 μm and a period of 400 μm was installed into the pump beam. In this case, a spectral output consisted of seven narrow lines separated by ~18 nm with a linewidth of each line

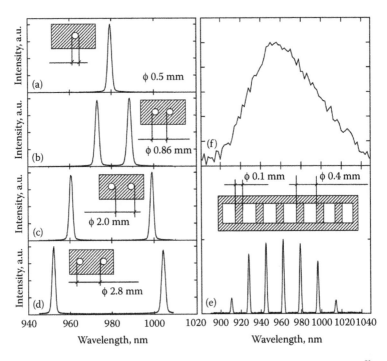

FIGURE 11.9 Output spectra of the ultrabroadband (f) and multiline (a–e) LiF:F$_2^{+**}$ CC laser formed due to different spatial filtering elements installed into the pumping beam: (a) single hole spatial filter, (b) two circular aperture (diameter is 0.5 mm) filter with separation between holes of 0.86 mm, (c) similar to (b), but with separation of 2.0 mm, (d) similar to (b), but with separation of 2.8 mm, (e) spatial mask with a periodic structure (shadow width is 100 μm and a period of 400 μm).

as narrow as 0.5–0.6 nm was realized. In this particular experiment, a 30 mm focal length intracavity lens was utilized.

For the LiF:F$_2^-$ CC laser, the maximum number of equidistant lines obtained with this mask was 15, covering the spectral range from 1.095 to 1.23 μm (see Figure 11.8b). It is noteworthy that the intensity envelope of the pumping beam defines the envelope of the output spectrum.

For frequency doubling of the broadband lasing, a 20 mm long LiIO$_3$ nonlinear crystal was used. The crystal was cut for frequency doubling at 1.064 μm. Matching the angular synchronism dispersion and the spectral dependence of the angle of incidence of radiation on the nonlinear crystal was achieved using the spherical lens f_{nc} with a focal length equal to 50 mm. The selection of the proper angular magnification of the lens provided a 12% overall conversion efficiency of broadband infrared radiation to the second harmonic for LiF:F$_2^{+**}$ and LiF:F$_2^-$ CC lasers. The emission spectrum of the second harmonic in the visible (green–yellow–red) spectral range (0.545–0.615 μm) and green–blue spectral range (0.45–0.51 μm) are shown in Figure 11.10. The linewidth in the visible range was 0.38 nm for the pump energy $E = 14$ mJ, and 1.4 nm for the pump energy $E = 25$ mJ. The increase in the linewidth

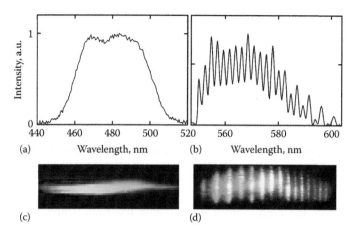

FIGURE 11.10 Spectra of SHG of the LiF:F_2^{+**} (a, ultrabroadband) and LiF:F_2^- (b, multiline) CCLs accompanied by photos of their rainbow like output radiation (c, LiF:F_2^{+**}; d, LiF:F_2^-).

can be explained by the increase in the inversion and gain, especially in the mask's half-shadow region, and by the increase in the SHG efficiency in the wings of the spectral lines. Due to the use of cylindrical intracavity and spherical collimating lenses, the waist of the beams and spatial channels intersection were located at different points of the optical axis. The shift of the nonlinear crystal into the second position provided not only efficient frequency doubling, but also the generation of the sum frequencies from various spectral regions of multifrequency lasing, which doubled the number of lines in the second-harmonic lasing spectrum (Figure 11.10b). The photos of the broadband SHG outputs of LiF:F_2^- and LiF:F_2^{+**} lasers are depicted in the bottom part of Figure 11.10c,d.

For fourth harmonic generation (FHG) of broadband visible radiation, the BBO ($5 \times 7 \times 10$ mm) nonlinear crystal was used. Polarization of the visible radiation from SHG of the LiF:F_2^- laser was rotated by a broadband $\lambda/2$ quartz plate. The visible radiation was focused into the nonlinear crystal using a spherical lens with a focal distance of $f = 70$ cm. The filter installed behind the nonlinear crystal blocked visible radiation and transmitted FHG in UV. Under the same pumping conditions, the maximum conversion efficiency of FHG was obtained using BBO crystal and was measured to be about 7% with respect to the visible SHG radiation.

11.6 CONCLUSIONS

This chapter is focused on the theoretical description and the experimental realization of a spatially dispersive laser system providing ultrabroadband or multiline spectral output. Spatially dispersive cavities can provide either ultrabroadband spectral output, which is a combination of hundreds of independent lasing channels of the laser, or a preassigned multiline spectral composition. The construction of the proposed cavity is flexible and allows the utilization of a variety of active media, including crystals, glasses, and dyes, as well as single broad stripe, multistripe diode chips, and diode arrays.

In an appropriately designed system, the mode competition is effectively suppressed by means of the spatial separation of different frequency modes in the GM. Further, pump laser beam shaping, in accordance with the spatially separated modes in the GM, provides spectral control of the output radiation by means of the transformation of the spatial distribution of the pump radiation into the spectral domain of the output laser radiation. There is a direct relationship between the output spectral composition of the spatially dispersive laser and the spatial filtration (shaping) of the pump or lasing oscillation. In this chapter, we have theoretically analyzed and demonstrated experimentally this relationship.

Several spatially dispersive laser cavities were designed and studied. Ultrabroadband and multifrequency oscillation has been achieved with LiF:F_2^- and LiF$_2^{+**}$ color center lasers with a spectral width practically coinciding with the luminescence spectrum of the GM. Experimental results were in good agreement with theoretical description.

A novel principle of obtaining second (and similarly fourth) harmonic generation of ultrabroadband radiation in a single nonlinear crystal was theoretically considered and experimentally realized. The principle is based on a simultaneous realization of phase matching conditions for all oscillating wavelengths by means of the compensation of phase-matching angular dispersion in a nonlinear crystal with the spatial angular dispersion of different wavelengths forming an output ultrabroadband continuum. In another words, proper shaping of the ultrabroadband laser output beam can provide simultaneous phase matching conditions for frequency up-conversion.

The important feature of the proposed cavity with a wideband SHG and FHG is that both nonlinear crystals are installed in the optical scheme to satisfy the phase matching conditions simultaneously for all oscillating wavelengths from the whole amplification spectral region. This allows the efficient frequency conversion of the oscillation to be obtained without any additional tuning of nonlinear crystals.

Today, the active exploration of applications of spatially dispersive laser sources are taking place in coherent-transient spectroscopy, Fourier optical processing, all-optical image transfer, multiwavelength spectroscopy, optical memory, information processing, optical computing, and optical communications.

REFERENCES

1. Casperson, L.W. and Yariv, A., Spectral narrowing in high-gain lasers, *IEEE J. Quantum Electron.*, QE-8(2), 80–85, 1972.
2. Prokhorov, A.M., *Laser Handbook*, Nauka, Moscow, 1976.
3. Hodgson, N. and Weber, H., *Optical Resonators: Fundamentals, Advanced Concepts and Applications*, Springer-Verlag, Heidelberg, Germany, 1997.
4. Svelto, O., *Principles of Lasers*, 4th ed., Plenum Publishing Corporation, New York, 1998.
5. Morita, N. and Yajima, T., *Phys. Rev. A*, 30, 2525, 1984.
6. Stark, H., *Applications of Optical Fourier Transforms*, Academic Press, New York, 1982.

7. Azana, J. and Chen, L.R., Multiwavelength optical signal processing using multistage ring resonators, *IEEE Photon. Tech. Lett.,* 14(5), 654–656, 2002.

8. Tagliaferri, A.A., Calatroni, J., and Froehly, C., *Opt. Commun.,* 67, 180, 1988.

9. Parsons-Karavassilis, D., Gu, Y., Ansary, Z., French, P.W.M., and Taylor, J.R., Diode-pumped spatially dispersed broadband Cr:LiSGAF and Cr:LiSAF c.w. laser sources applied to short-coherence photorefractive holography, *Opt. Commun.,* 181, 361–367, 2000.

10. Duncan, A., Whitlock, T.L., Cope, M., and Delpy, D.T., Multiwavelength, wideband, intensity-modulated optical spectrometer for near-infrared spectroscopy and imaging, *Proc. SPIE,* 1888, 248–257, 1993.

11. Bublitz, J. and Schade, W., Multiwavelength laser-induced fluorescence spectroscopy for quantitative classification of aromatic hydrocarbons, *Proc. SPIE,* 2504, 265–276, 1995.

12. Mattley, Y.D. and Garcia-Rubo, L.H., Multiwavelength spectroscopy for the detection, identification and quantification of cells, *Proc. SPIE,* 4206, 64–71, 2001.

13. Zhmud, A.A., Multilevel optical logic and optical memory based on interferential scanners with tunable diode lasers, *Optoelectron.—Devices Technol.,* 6(1), 97–109, 1991.

14. Zhmud, A.A., Optical memories based on tunable semiconductor lasers, *Telecommun. Radio Eng.,* 47(12), 105–109, 1992, English translation of Elektrosvyaz and Radiotekhnika.

15. Mataki, H., Laser diodes for optical memories and data communications, *J. Electron. Eng.,* 30(321), 88–92, 1993.

16. Mathason, B.K. and Delfyett, P.J., Simultaneous multiwavelength switching for parallel information processing, *Proc. SPIE,* 3714, 57–64, 1999.

17. White, I.H., Nyairo, K.O., Kirkby, P.A., and Armistead, C.J., Demonstration of a 1×2 multichannel grating cavity laser for wavelength division multiplexing (WDM) applications, *Electron. Lett.,* 26(13), 832–833, 1990.

18. White, I.H., A multichannel grating cavity laser for wavelength division multiplexing applications, *J. Lightwave Tech.,* 9(7), 893–899, 1991.

19. Soole, J.B.D., Poguntke, K., Scherer, A., LeBlank, H.P., Chang-Hasnain, C., Hayes, J.R., Caneau, C., Bhat, R., and Koza, M.A., Multistripe array grating integrated cavity MAGIC laser: A new semiconductor laser for (WDM) applications, *Electron. Lett.,* 28(19), 1805–1807, 1992.

20. Louri, A. and Sung, H., Scalable optical hypercube-based interconnection network for massively parallel computing, *Appl. Opt.,* 33(32), 7588–7598, 1994.

21. Tench, R.E., Nagel, J.A., and Delavaux, J.-M.P., Challenges in the design and packaging of optical amplifiers for multiwavelength lightwave communication systems, *Proc. Electron. Components Technol. Conf.,* 739–750, 1995.

22. Takara, H., Kawanishi, S., Saruwatari, M., and Schalager, J.B., Multiwavelength birefringent-cavity mode-locked fibre laser, *Electron. Lett.,* 28(25), 2274–2275, 1992.

23. Milev, I.Y., Ivanova, B.A., Danailov, M.B., and Saltiel, S.M., Multi-pulse operation of three-wavelength pulsed Q-switched Nd:$Y_3Al_5O_{12}$ laser, *Appl. Phys. Lett.,* 64(10), 1198–1200, 1994.

24. Kopf, D., Spühler, G.J., Weingarten, K.J., and Keller, U., Mode-locked laser cavities with a single prism for dispersion compensation, *Appl. Opt.,* 35(6), 912–915, 1996.

25. Zhu, B. and White, I.H., Multiwavelength picosecond optical pulse generation using an actively mode-locked multichannel grating cavity laser, *J. Lightwave Tech.,* 13(12), 2327–2335, 1995.

26. Sanjoh, H., Yasaka, H., Sakai, Y., Sato, K., Ishii, H., and Yoshikuni, Y., Multiwavelength light source with precise frequency spacing using a mode-locked semiconductor laser and an arrayed waveguide grating filter, *IEEE Photon. Tech. Lett.,* 9(6), 818–820, 1997.

27. Li, S. and Chan, K.T., Electrical wavelength tunable and multiwavelength actively mode-locked fiber ring laser, *Appl. Phys. Lett.,* 72(16), 1954–1956, 1998.

28. Park, E.D., Croeze, T.J., Delfyett, P.J., Braun, A., and Abeles, J., Multiwavelength mode-locked InGaAsP laser operating at 12ch $\times$ 2 GHz and 16ch $\times$ 10 GHz, *IEEE Photon. Tech. Lett.,* 14(6), 837–839, 2002.

29. Moskalev, I., Mirov, S., Fedorov, V., Basiev, T., Grimes, G., and Berman, E., External cavity multiwavelength or ultrabroadband diode laser for wavelength division multiplexing applications, *Bull. Am. Phys. Soc., Ser. II,* 45, 19, 2000.

30. Moskalev, I.S., Mirov, S.B., Basiev, T.T., Fedorov, V.V., Grimes, G.J., and Berman, I.E., External cavity multiline semiconductor laser for WDM applications, *Proc. SPIE,* 4287, 128–137, 2001.

31. Basiev, T.T., Mirov, S.B., Zverev, P.G., Fedorov, V.V., and Kuznetsov, I.V., Ultrabroadband and synchronized multiline oscillation of LiF:F_2^- color center laser, in *Proceedings of the 11th International Congress Laser 93,* Laser in Engineering, pp. 919–924. Springer-Verlag, Munich, Germany, 1993.

32. Basiev, T.T., Ermakov, I.V., Fedorov, V.V., Konushkin, V.A., and Zverev, P.G., Laser oscillation of LiF:F_2^+—stabilized color center crystals at room temperature. In *International Conference on Tunable Solid State Lasers,* pp. 27–29, Institute of Molecular and Atomic Physics of the Academy of Science of Belarus, Minsk, Belarus, 1994.

33. Basiev, T.T., Zverev, P.G., Fedorov, V.V., and Mirov, S.B., Solid state laser with ultrabroadband or control generation spectrum, in *Proceedings of the SPIE,* Vol. 2379 of Solid State Lasers and Nonlinear Crystals, San Jose, CA, pp. 54–61, 1995.

34. Astinov, V.H., Spatial modulation of the pump in the ultrabroadband dye laser with a "spatially-dispersive" resonator, *Opt. Commun.,* 118, 297–301, 1995.

35. Basiev, T.T., Zverev, P.G., Fedorov, V.V., and Mirov, S.B., Multiline, ultrabroadband and sun-color oscillation of a LiF:F_2^- color-center laser, *Appl. Opt.,* 36(12), 2515–2522, 1997.

36. Fedorov, V.V., Zverev, P.G., and Basiev, T.T., Broadband lasing and nonlinear conversion of radiation from LiF:F_2^+ and LiF:F_2^- colour centre lasers, *Kvantovaya Elektronika,* 31(4), 285–289, 2001.

37. Ramaswamy-Paye, M. and Fujimoto, J.G., Compact dispersion-compensating geometry for Kerr-lens mode-locked femtosecond laser, *Opt. Lett.,* 19(21), 1756–1758, 1994.

38. Ter-Mikirtychev, V.V. and Tsuboi, T., Ultrabroadband LiF:F_2^+ color center laser using two-prism spatially-dispersive resonator, *Opt. Commun.,* 137, 74–76, 1997.

39. Ter-Mikirtychev, V.V., McKinnie, I.T., and Warrington, D.M., Ultrabroadband operation of Cr^{4+}: Forsterite laser by use of spatially dispersed resonator, *Appl. Opt.,* 37(12), 2390–2393, 1998.

40. Ter-Mikirtychev, V.V. and Arestova, E.L., Temporal and energetic characteristics of ultrabroadband Ti^{3+}: Sapphire laser, *Opt. Eng.,* 38(4), 641–645, 1999.

41. Danailov, M.B. and Milev, I.Y., Simultaneous multiwavelength operation of Nd:YAG laser, *Appl. Opt. Lett.,* 61(7), 746–748, 1992.

42. Danailov, M.B. and Christov, I.P., A novel method of ultrabroadband laser generation, *Opt. Commun.,* 73, 235–238, 1989.

43. Danailov, M.B. and Christov, I.P., Ultrabroadband laser using prism-based "spatially-dispersive" resonator, *Appl. Phys.,* B51, 300–302, 1990.

44. Danailov, M.B. and Christov, I.P., Amplification of spatially-dispersed ultrabroadband laser pulses, *Opt. Commun.,* 77, 397–401, 1990.

45. Soole, J.B.D., Poguntke, K.R., Scherer, A., LeBlanc, H.P., Chang-Hasnain, C., Hayes, J.R., Caneau, C., Bhat, R., and Koza, M.A., Wavelength-selectable laser emission from a multistripe array grating integrated cavity laser, *Appl. Phys. Lett.,* 61(23), 2750–2752, 1992.

46. Poguntke, K.R., Soole, J.B.D., Scherer, A., LeBlanc, H.P., Caneau, C., Bhat, R., and Koza, M.A., Simultaneous multiple wavelength operation of a multistripe array grating integrated cavity laser, *Appl. Phys. Lett.,* 62(17), 2024–2026, 1993.

47. Ferrari, G., Mewes, M.O., Schreck, F., and Salomon, C., High-power multiple-frequency narrow-linewidth laser source based on a semiconductor tapered amplifier, *Opt. Lett.,* 24(3), 151–153, 1999.

48. Moskalev, I.S., Mirov, S.B., Fedorov, V.V., Grimes, G.J., Basiev, T.T., Berman, E., and Abeles, J., External cavity multiwavelength semiconductor laser, in *CLEO—Technical Digest,* Vol. 73, pp. 410–411, OSA, Washington, DC, 2002.

49. Papen, G.C., Murphy, G.M., Brady, D.J., Howe, A.T., Dallesasse, J.M., Dejule, R.Y., and Holmgren, D.J., Multiple-wavelength operation of a laser-diode array coupled to an external cavity, *Opt. Lett.,* 18(17), 1441–1443, 1993.

50. Wang, C.L. and Pan, C.L., Tunable dual-wavelength operation of a diode array with an external grating-loaded cavity, *Appl. Phys. Lett.,* 64(23), 3089–3091, 1994.

51. Wang, C.L. and Wang, C.L., A novel tunable dual-wavelength external-cavity laser diode array and its applications, *Opt. Quantum Electron.,* 28, 1239–1257, 1996.

52. Ashkin, A. and Ippen, E.P., Wavelength selective laser apparatus, U.S. Patent No 3,774,121, 1973.

53. Christov, I.P., Michailov, N.I., and Danailov, M.B., Mode locking with spatial dispersion in the gain medium, *Appl. Phys., B,* 53, 115–118, 1991.

54. Christov, I.P. and Danailov, M.B., Theory of ultrabroadband light amplification in a spatially-dispersive scheme, *Opt. Commun.,* 84, 61–66, 1991.

55. Basiev, T.T., Mirov, S.B., Zverev, P.G., Kuznetsov, I.V., and Tedeev, R.Sh, Solid state laser with Ultrabroadband or Control Generation Spectrum, U.S. Patent No 5,461,635, 1995.

56. Faure, J., Itatani, J., Biswal, S., Chériaux, G., Bruner, L.R., Templeton, G.C., and Mourou, G., A spatially dispersive regenerative amplifier for ultrabroadband pulses, *Opt. Commun.,* 159, 68–73, 1999.

57. Basiev, T.T. and Mirov, S.B., Room temperature tunable color center lasers, in *Laser Science & Technology Book Series,* Vol. 61, Letokhov, V.S., Shank, C.V., Shen, Y.R., and Walter, H., eds., pp. 1–160, Gordon and Breach Science Publication/Harwood Academic Publication, New York, 1994.

58. Mirov, S.B. and Basiev, T.T., Progress in Color Center Lasers, *IEEE Journal of Special Topics in Quantum Electronics,* Vol. 1, Esterowitz, L., ed., pp. 22–30, 1995, Invited paper.

59. Dergachev, A.Y. and Mirov, S.B., Efficient room temperature LiF:F_2^{+**} color center tunable laser tunable over 820–1210 nm range, *Opt. Commun.,* 145, 107–112, 1998.

60. Basiev, T.T., Zverev, P.G., Papashvili, A.G., and Fedorov, V.V., Temporal and spectral characteristics of tunable LiF:F_2^- color center laser, *Rus. Quantum Electron. (Kvantovaya Elektronika),* 27(7), 574–578, 1997.

61. Jenkins, N.W., Mirov, S.B., and Fedorov, V.V., Temperature-dependent spectroscopic analysis of F_2^{+**} and F_2^+ like color centers in LiF, *J. Lumin.,* 91, 147–153, 2000.

62. Dmitriev, V.G., Gurzadyan, G.G., and Nikogosyan, D.N., *Handbook of Nonlinear Optical Crystals, Springer Series in Optical Sciences,* p. 64, Springer-Verlag, Berlin, Germany, 1991.

12 Beam Shaping
A Review

Fred M. Dickey and Scott C. Holswade

CONTENTS

12.1   INTRODUCTION

Beam shaping is the process of redistributing the irradiance and phase of a beam of optical radiation. The irradiance distribution defines the profile of the beam, such as Gaussian, circular, rectangular, annular, or multimode. The phase of the shaped beam determines its propagation properties to a large extent. For example, a reasonably large beam with a uniform phase front will maintain its shape over a considerable propagation distance. On the other hand, a beam with large variations in its phase front will tend to expand and change shape as it propagates. Applications of beam shaping include laser/material processing, laser/material interaction studies, laser weapons, optical data/image processing, lithography, printing, and laser art patterns. Beam shaping technology can be applied to beams with an arbitrary degree of coherence.

Figure 12.1 illustrates the general beam shaping problem. A beam is incident upon an optical system that may consist of one or more elements. The optical system must operate upon the beam to produce the desired output. For coherent beams, the

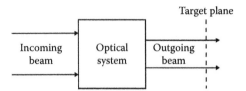

FIGURE 12.1 General beam shaping problem. (From Dickey, F. M. and Holswade, S.C., 2000. *Laser Beam Shaping: Theory and Techniques.* Marcel Dekker, New York. With permission.)

designer may or may not want to constrain the phase of the beam at the output plane. For example, if a collimated output beam is desired, the phase front of the beam exiting the optical system must be uniform. If the design only requires a certain irradiance distribution at the target plane, then the optical system is usually simpler if the phase is left unconstrained. For incoherent beams, phase considerations do not apply, and only the desired irradiance distribution is considered.

Laser beam shaping techniques can be divided into three broad classes. The first is the trivial, but useful, aperturing of the beam illustrated in Figure 12.2. In this case, the beam is expanded and an aperture is used to select a suitably flat portion of the beam. The resulting irradiance pattern can be imaged with magnification to control the size of the output beam. The major disadvantage of this technique is that much of the input beam power or energy is lost. In most cases, it is desirable, for obvious reasons, that the beam shaping operation conserves energy. Further, if the input beam irradiance is not suitably smooth, it might not be possible to find an aperture size and position that gives the desired result. In that case, some form of input beam homogenization might be required.

The second major technique for beam shaping is what could be called field mapping. Field mappers transform the input field into the desired output field in a controlled manner. The basic field mapper concept is illustrated in Figure 12.3 for the case of mapping a single-mode Gaussian beam into a beam with a uniform

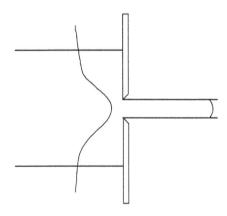

FIGURE 12.2 Uniform irradiance obtained by aperturing the input beam.

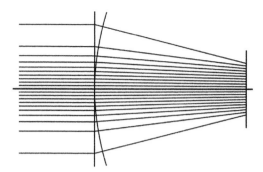

FIGURE 12.3 Schematic of the field mapping concept. (From Dickey, F.M. and Holswade, S.C., 2000. *Laser Beam Shaping: Theory and Techniques.* Marcel Dekker, New York. With permission.)

irradiance. In the figure, Gaussian-distributed rays are bent in a plane so that they are uniformly distributed in the output plane. The ray bending described in Figure 12.3 defines a wavefront that can be associated with an optical phase element. Field mappers can be made effectively lossless. The field mapping approach to beam shaping is applicable to well-defined single-mode laser beams.

The remaining class of beam shapers is beam integrators, also known as beam homogenizers. A representative example of a beam integrator is shown in Figure 12.4. In this configuration, the input beam is broken up into beamlets by a lenslet array and superimposed in the output plane by the primary lens. The term integrator comes from the fact that the output pattern is a sum of diffraction patterns determined by the lenslet apertures. Beam integrators are especially suited to multimode lasers with a relatively low degree of spatial coherence. They can also be designed to be effectively lossless.

In the next section, we discuss the basic system considerations for beam shaping. Field mapping techniques are presented in Section 12.3. Beam integrator (homogenizer) systems are treated in Section 12.4. Diffusive beam shapers, which are a

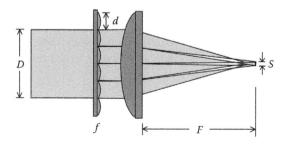

FIGURE 12.4 Multiaperture beam integrator. (From Dickey, F.M. and Holswade, S.C., 2000. *Laser Beam Shaping: Theory and Techniques.* Marcel Dekker, New York. With permission.)

type of field mapper, are discussed in Section 12.5. The problem of choosing a beam shaping technique for a given application is treated in Section 12.6.

This chapter is based on the authors' book, *Laser Beam Shaping: Theory and Techniques*.[1] In addition, several interesting papers are collected in a continuing sequence of laser beam shaping conferences.[2-5]

12.2 SYSTEM CONSIDERATIONS

Beam shaping design can be approached using either physical or geometrical optics. However, there are general constraints and limits that result from a physical optics formulation of the problem. In particular, there is an uncertainty principle relation that defines necessary conditions for obtaining a good solution to the beam shaping problem.

12.2.1 PHYSICAL OPTICS CONSIDERATIONS

Most beam shaping problems can be adequately formulated in terms of Fresnel diffraction theory.[6] An extensive treatment of both vector and scalar diffraction theory can be found in the books by Stamnes[7] and Walther.[8] The concept of field mapping is applicable to beam integrators as well as field mappers. The basic field mapping problem can be expressed in terms of the Fresnel integral with the phase function explicitly defined as

$$U(x_0, y_0) = \frac{\exp(ikz)}{i\lambda z} \int \int U(x_1, y_1) \exp i\psi(x_1, y_1)$$
$$\times \exp\left\{\frac{ik}{2z}\left[(x_0 - x_1)^2 - (y_0 - y_1)^2\right]\right\} dx_1\, dy_1$$

(12.1)

where:

$k = 2\pi/\lambda$, and $U(x_1, y_1)$ is the complex representation of the input beam
$\psi(x_1, y_1)$ is the phase function representing the lossless beam shaping element
$U(x_0, y_0)$ is the shaped complex field in the output plane at distance z

By expanding the last exponential in the integrand of Equation 12.1 and including the remaining quadratic phase function in the beam shaping element, ψ, one can express the beam shaping problem as a Fourier transform (Fraunhofer integral) as

$$U(x_0, y_0) = \frac{\exp(ikz)}{i\lambda z} \exp\left[x_0^2 + y_0^2\right] \int \int U(x_1, y_1) \exp i\psi(x_1, y_1)$$
$$\times \exp\left[-i\frac{2\pi}{\lambda z}(x_0 x_1 + y_0 y_1)\right] dx_1\, dy_1$$

(12.2)

where ψ in this equation differs from that of Equation 12.1 by a quadratic phase factor. In terms of either of these two equations, the beam shaping problem is to determine the phase function, ψ, when $U(x_1, y_1)$ and the magnitude of $U(x_0, y_0)$ are specified. This is equivalent to simultaneously specifying the magnitude of a function and the magnitude of its Fourier transform. Closed-form beam shaping solutions based on a Fourier transform optical system are given in Section 12.3.2.

The essence of the beam shaping problem is illustrated in Figure 12.3. In this figure, an input beam of rays, in this case with a Gaussian distribution, is incident on a plane representing the optical system. The output rays are bent so as to come to a uniform distribution in the output plane. This is the Gaussian-to-flattop beam shaping problem. A wavefront can be computed by noting that in the geometrical optics, approximation rays are normal to the wavefront. Once the wavefront is determined, one can then establish a phase function that would produce the shaped beam. Implementing the phase function may be a complicated optical design process, and may be best achieved with multiple optical surfaces (elements).

In fact, the simple process outlined above can give a direct solution to a beam shaping problem if the phase function can be realized by an optically thin phase element. However, determining the optical phase element is generally a complicated problem. Additional complications arise if the thin element approximation is not applicable, or if both phase and irradiance profiles are specified. The ray bending depicted in Figure 12.3 is monotonic. Nonmonotonic ray bending schemes are also possible and some may offer advantage. A simple example of a nonmonotonic system would be a ray bending scheme that folds the upper half of the input into the lower half of the output and conversely. In fact, the bending might have a random component. It is easy, if one is not careful, to arrive at bending schemes that are not realizable, in that they do not, even approximately, satisfy Maxwell's equations. Further, certain ray bending schemes may fall outside the realm of geometrical optics.

In the end, what can be obtained for a given beam shaping problem is limited by physical optics (electromagnetic theory). The simplest result of this theory is an uncertainty principle that depends upon input beam size and wavelength, and output profile size and distance. This result is closely related to diffraction limits on imaging optical systems.

12.2.2　THE UNCERTAINTY PRINCIPLE AND β

The uncertainty principle of quantum mechanics, or equivalently the timebandwidth-product inequality associated with signal processing, can be applied to the beam shaping problem. The uncertainty principle is a constraint on the lower limit of the product of the root-mean-square width of a function and its root-mean-square bandwidth,[9,10]

$$\Delta_x \Delta_y \geq \frac{1}{4\pi} \tag{12.3}$$

Applying the uncertainty principle to the beam shaping problem of Equations 12.1 or 12.2, one obtains a parameter β of the form[1]

$$\beta = C\frac{r_0 y_0}{\lambda z} \qquad (12.4)$$

where:

r_0 is the input beam half-width
y_0 is the output beam half-width
C is a constant that depends on the exact definition of beam widths

If the diffraction-limited minimum beam size is approximated by $\lambda z/2y_0$, Equation 12.4 can be interpreted as a multiple of the desired beam size divided by the diffraction-limited beam size. As will be discussed in the following sections, a good field mapping solution to Equations 12.1 or 12.2 is obtainable if it is suitably large. Also, and not unrelated, is the parameter involved in the stationary phase solution for Equations 12.1 or 12.2.[1]

12.2.3 STATIONARY PHASE AND GEOMETRICAL OPTICS

Solutions to the beam shaping problem defined by Equation 12.2 can be obtained by application of the method of stationary phase. Stamnes[7] provides an extensive discussion of the method of stationary phase and its application to diffraction problems. Walther[8] applies the method of stationary phase to the wave theory of lenses. A detailed stationary phase treatment of the beam shaping problem is given in Reference 1 of Chapter 2.

In one dimension, the method of stationary phase[11,12] gives an asymptotic approximation to integrals of the form

$$I(\beta) = \int_a^b f(x)e^{i\beta\phi(x)}\mathrm{d}x \qquad (12.5)$$

where β is a dimensionless parameter.

As discussed previously, it will be seen that the term β arising from the stationary phase solution of Equation 12.5 has the same form as the term β derived from uncertainty principal considerations. The first term in the asymptotic phase approximation to the integral in Equation 12.5 is given by

$$I_c(\beta) \sim e^{i\{\beta\phi(c)+\mu\pi/4\}}f(c)\left[\frac{2\pi}{\beta|\phi''(c)|}\right]1/2 \qquad (12.6)$$

where primes denote derivatives,

$$\mu = \mathrm{sign}\ \phi''(c) \qquad (12.7)$$

c is a simple stationary point defined by

$$\Phi'(c) = 0, \ \Phi''(c) \neq 0 \qquad (12.8)$$

Equation 12.6 is commonly referred to as *the stationary phase formula*. Similar results are obtained in two dimensions with $\Phi''(c)$ replaced by the Hessian matrix for Φ.

The essence of the beam shaping problem is to equate $|I_c(\beta)|^2$ with the desired irradiance in the output plane. That is, the magnitude squared of the right-hand side of Equation 12.6 is equal to the desired output irradiance. Using this condition with Equation 12.8 leads to a second-order differential equation for the beam shaping phase function $\Phi(x)$. The details of obtaining the explicit form of the differential equation from Equations 12.6 and 12.8 are rather tedious.[1] In general, care must be taken with respect to the absolute value of Φ'' in the denominator of Equation 12.6. This condition requires that the phase $\Phi(x)$ is a convex function, a function whose second derivative is either positive or negative everywhere. This turns out not to be a problem for the case of mapping a Gaussian into a rect function. When the phase of the phase function is not a convex function over the entire input beam, the problem becomes more difficult. In this case, a solution to the problem can be obtained by dividing the input beam into regions over which the phase function is convex, and combining the solutions for these regions in a seamless manner. In practice, this may be a very difficult problem.

In two dimensions, the general form of the equation to be solved is

$$F(\omega_x, \omega_y) = \frac{1}{2\pi} \int_{-\infty}^{\infty} \int_{-\infty}^{\infty} f(\xi, \eta) \exp[i(\beta\phi(\xi, \eta) - \xi\omega_x - \eta\omega_y)]d\xi d\eta \qquad (12.9)$$

where:

$\xi = x/ri$ and $\eta = y/ri$ are normalized input variables with ri defining the length scale

$\omega_x = x/r_0$ and $\omega_y = y/r_0$ are normalized output variables in Fourier transform plane with r_0 defining the length scale

The stationary phase solution improves asymptotically with increasing dimensionless parameter $\beta = 27\pi r_i R_0/\lambda z$, where λ is the optical wavelength and z is defined following Equation 12.1. The geometrical optic approximation is obtained in the large β limit. Note that this is the same form as β in Equation 12.4.

The stationary phase evaluation of integrals of the type given by Equation 12.6 generally leads to second-order partial differential equation for the phase function Φ. The resulting partial differential equation can then be solved for Φ, subject to an energy boundary condition determined by Parseval's theorem. The partial differential equation reduces to a second-order ordinary differential equation for the separable and circularly symmetric problem. The optical element is then designed to realize $\beta\Phi$, where β scales the solution to the desired geometry.

12.3 FIELD MAPPERS

Field mapping is basic to all beam shaping. We define field mapping as the design of a set of optical elements that map a specified input optical field into a desired output optical field. Frequently, it is only the magnitude of the output optical field that is specified or constrained.

12.3.1 TECHNIQUES

The most general statement of the lossless beam shaping problem is to determine the phase function that minimizes the mean-square error between the desired output and the output produced by the phase function. In terms of Equation 12.2, the problem is to minimize the integral

$$R = \int \left| U(x_0, y_0) - \bar{U}(x_0, y_0) \right|^2 \, dx_0 \, dy_0 \tag{12.10}$$

where $\bar{U}$ is the desired output optical field.

Frequently, it is the magnitude of the output field that is specified. In this case, it would be desirable to minimize an integral of the form

$$R = \int \left| \left| U(x_0, y_0) \right|^2 - \left| \bar{U}(x_0, y_0) \right| \right|^2 \, dx_0 \, dy_0 \tag{12.11}$$

Unfortunately, it appears to be very difficult to obtain direct solutions to Equations 12.10 or 12.11. The solution for a special case of Equation 12.10 is given in Reference 1 of Chapter 3.

As discussed in Section 12.2.3 and in the next section, solutions for the beam shaping problem can be obtained by applying the method of stationary phase to the diffraction integral defining the beam shaping problem. This approach can give closed-form solutions to the problem in some cases. When this is not the case, solutions may be obtained by numerical evaluation of the stationary phase formula.

Another approach to solving the beam shaping problem is to use geometrical optics techniques. This is sometimes referred to as the geometrical transformation method. The essence of this approach is illustrated in Figure 12.3 (see Section 12.3.3). Generally, this approach uses the eikonal and conservation of energy along a ray bundle to obtain a differential equation for the lens surfaces. As in the application of the method of stationary phase, closed form or numerical solutions can be obtained for the resulting differential equation.

Analytical or closed-form solutions are desirable in that they provide insight into the problem. They can also make explicit the solution dependence on system parameters, for example, the variables in Equation 12.4. However, there can be beam shaping design problems that are difficult to attack analytically. In this case, one approach is to apply an iterative or search type algorithm to the problem.[13-18] Algorithms that have been applied to beam shaping include the Gerchberg–Saxton algorithm, Yang-Gu amplitude-phase retrieval and iterative algorithm (YG algorithm), input–output algorithm, simulated annealing, and genetic algorithms.

The authors would like to caution the reader that the availability of an algorithm alone does not solve the problem. It is always good to understand the basic physical constraints on the problem before attempting to apply an algorithm. For example, Churin[17] points out that there are serious convergence problems with the Gerchberg–Saxton technique, especially for diffraction-limited target sizes. The diffraction-limited case is clearly associated with a low β value and one should not expect any algorithm or technique to produce a good result. Generally, in applying an algorithm to design a beam shaping element, one must be careful to adequately constrain, but not overconstrain the problem. Also, any numerical approach requires a sampling of the input and output space. One must be careful to adequately sample in both domains. Dresel et al.[14] and Brauer et al.[19] note that phase values of adjacent object points are not independent. If this fact is neglected, speckle can occur in the output field. Requiring smooth phase functions with strong coupling between neighboring regions can eliminate speckle patterns.

In general, the application of an algorithm will require the computation of a diffraction integral or a detailed ray trace. Standard optical software can be used for these computations. Also, the application of an algorithm will typically give a solution that is a complex function of the form $e^{i\psi}$. This is not the same as determining ψ. The function $e^{i\psi}$ readily yields ψ modulo 2π. It may be desirable to apply phase unwrapping[20] to the solution to get a continuous phase. Phase unwrapping is much harder in dimensions greater than one. For separable problems, one-dimensional phase unwrapping is appropriate.

It would not be practical to go into the details of these algorithms in this chapter. We will discuss the Gerchberg–Saxton algorithm because it is a relatively simple iterative algorithm that clearly illustrates the application of algorithms to the beam shaping problem. We will also discuss the application of popular optimization algorithms to the beam shaping problem.

12.3.2 Gerchberg–Saxton Algorithm

Lossless beam shaping is related to the phase retrieval problem of image recovery or reconstruction, in that both problems are involved with determining the phase associated with a Fourier transform magnitude. The Gerchberg–Saxton algorithm and other algorithms have been investigated in detail with respect to the image recovery problem.[21] The Gerchberg–Saxton algorithm specialized to the problem of lossless mapping of an input beam in to an output beam with a specified magnitude is shown in Figure 12.5. The algorithm is started by assuming a beam shaping phase ψ_0. The initial phase, ψ_0, is an estimate or guess. It can be obtained from an approximate solution to the problem, from a solution to a similar problem, or as a random guess. The input function is then Fourier transformed, and the resulting magnitude is replaced by the output constraint $|G|$. This result is then inversely transformed

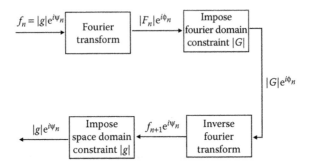

FIGURE 12.5 Diagram of the Gerchberg–Saxton algorithm for lossless mapping of the input beam to a beam with specified irradiance.

and the input beam function is imposed on the result. At this point, one has an estimate, ψ_1, of the beam shaping phase. The iteration is repeated resulting in an estimate ψ_n. Hopefully, the algorithm will converge to a result that is a reasonable approximation to the desired output $|G|$.

The form of the Gerchberg–Saxton algorithm illustrated in Figure 12.5 is also known as the error-reduction algorithm. Dainty and Fienup (Reference 27 of Chapter 7.) discuss the error-reduction algorithm in detail. They point out that it can be viewed equivalently as the method of successive approximations, as a form of steepest-decent gradient search, or as a projection onto sets in Hilbert space. In terms of a Hilbert space projection, the Fourier transform magnitude constraint is a projection onto a nonconvex set, and thus convergence is not guaranteed. They give proof that the algorithm *converges* in the weak sense that the squared error cannot increase with increasing iterations. It is the authors' observation, based on discussions with others who have tried the algorithm, that the results are mixed, with good results obtained only part of the time.

12.3.3 Optimization Algorithms

Optimization-based techniques are particularly useful where the complexity of the problem makes a closed-form solution difficult, if not impossible. The shaping of a Gaussian input beam into a triangular profile would fall into this category. The main issue lies in reducing the beam shaping problem to the evaluation of a single number, or merit, function that can be optimized by the computer. Traditional lens design programs are oriented toward imaging systems, where points on an object surface are brought to focus as points on an image surface. Beam shaping problems involve the redistribution of a beam, where image-based merit functions do not apply. Beam shaping problems, therefore, require specialized merit functions. Once such a merit function is developed, an appropriate optimization routine can be applied to the problem.

A general merit function based on conservation of energy has been effective in solving a variety of beam shaping problems.[22–24] It is beyond the scope of this chapter to treat this method in detail, but an outline will be presented for radially symmetric problems. Assuming that the incoming beam is represented as a set of N rays distributed equally across the input surface of the system:

$$\rho_i = \left(\frac{r}{N}\right)i, \quad i = 0,\ldots,N \tag{12.12}$$

where:
  ρ_i is the distance from the optical axis for each incoming ray
  r is the system radius for the incoming beam

The energy $\sigma(\rho_i)$ entering a differential input ring between ρ_i and ρ_{i-1} can be balanced with the energy $u(P_i)$ leaving a differential output ring between P_i and P_{i-1}, which results in:

$$u(P_i) = \sigma(\rho_i)\left(\frac{\cos\left(\theta_i^{\text{in}}\right)\rho_i(\rho_i - \rho_{i-1})\cos\left(\chi_i^{\text{out}}\right)}{\cos\left(\theta_i^{\text{out}}\right)P_i\left(P_i - P_{i-1}\right)}\right) \tag{12.13}$$

where:
  θ_i^{in} is the angle between the input surface normal and the direction of an
      incoming ray
  θ_i^{out} is the angle between the output surface normal and the direction of an
      outgoing ray
  χ_i^{out} is the angle between the output surface normal and the optical axis for a
      particular ray location
  P_i is the radial distance between the output ray and the optical axis

These angles and radii can be obtained from a ray trace in most optical design packages. For a Gaussian input distribution, the differential input energy would be

$$\sigma(\rho_i) = \exp\left(-\frac{\rho_i^2}{\rho_N^2}\right) \tag{12.14}$$

where ρ_N is the waist radius.
  A merit function for a uniform output profile could look like:

$$M = \sqrt{\frac{1}{N}\sum_{i=1}^{N}\left(u(P_i) - \bar{u}\right)^2}; \quad \bar{u} = \frac{1}{N}\sum_{i=1}^{N}u(P_i) \tag{12.15}$$

This merit function, which is to be minimized, is the root sum of squares (RSS) of the energy variations in the differential output rings. In practice, an additional term in the merit function would also be required to limit the size of the output profile to what was desired.[22-24]

With a merit function defined, an optimization routine must be employed. For many years, local optimization routines, such as damped least squares, have been used for optics problems. Damped least squares optimization is equivalent to trying to find the lowest valley in a region by always heading downhill from the starting point. If the starting point is in the basin of a high valley, the lowest point of that valley will be found, but it would not be the lowest valley in the region. In practice, this is not strictly true—damped least squares methods can jump over small hills in the solution space—but in general these methods are very dependent upon the starting point to find a solution. In traditional lens design of imaging systems, a body of wisdom has accumulated that aids the designer in selecting a good starting point. In the design of beam shaping optics, however, there are currently no good design rules to aid the designer. Most of the problem involves finding appropriate aspheric or diffractive surfaces to perform the shaping task. For solving this kind of problem, global optimization techniques are particularly interesting. Ideally, these techniques would find the lowest valley in a region, or the global minimum in a problem.

Three global optimization techniques have received the most attention in the literature: genetic algorithms, simulated annealing, and the Tabu search. It is probably too early to say whether one of these global methods is superior to the other, or whether any of them will frequently find the best solution. The complexity of the problems prohibits the manual search of solution space to confirm the performance of the algorithm. However, genetic algorithms have been used to obtain satisfactory solutions to a number of beam shaping problems.[22] The merit function described above was evaluated tracing $N = 200$ rays with a commercially available software package, and the results were supplied to an external genetic algorithm. Genetic algorithms operate analogously to biological evolution, where the attributes of promising designs are combined, and random changes are introduced to avoid stagnation in the solution.

The approach described above is very general, and can be extended to nonsymmetric problems. In the following section, we describe a simplified approach that exploits the global optimization capabilities found in several commercial lens design programs. The method involves the construction of a merit function using a minimum number of rays for computational speed. Rather than being evenly spaced, the rays are distributed with a density that matches the incoming beam profile. The merit function computes how well the ray density (and the wavefront if required) matches what is desired at the output plane. This custom merit function is then called an optimization operand by the lens design program, which uses its internal optimization routines. For radially symmetric problems and for separable problems, such as the shaping of a circular beam into a rectangular beam, this simplified approach appears to work well.[25]

In practice, this simplified approach involves the tracing of approximately 20 rays that are distributed across the input surface that corresponds to the desired input profile. For 20 rays, 1/20th of the incoming beam energy would be contained in a differential ring with an inner radius at ray i and an outer radius at ray $i + 1$. After shaping, the rays are distributed across the output surface with a density that corresponds to the desire output profile. A merit function can be constructed that finds the RSS of the deviation for each output ray from its target location. As this merit function approaches zero, the optical system is transforming the input beam into the desired output beam. The optical shaping surface is represented as a polynomial surface, with only even terms being required for symmetric problems. The global optimization algorithm in the optical design program attempts to find the best values of these polynomial coefficients.

As an example of this process, consider a case where a Gaussian input beam must be shaped to provide uniform, circular irradiation upon a planar surface. This problem can readily be solved analytically, but it illustrates the simplified optimization method. The problem is radially symmetric, so only one axis needs to be computed. For the input ray distribution, we need to calculate the positions of the rays that result in equal energy in the rings between them. The input energy $E(\rho_1)$ contained inside a circle with radius ρ_i is

$$E(\rho_i) = \int_0^{2\pi} \int_0^{\rho_i} \rho e^{-2\rho^2/\rho_0^2} d\rho \, d\theta \qquad (12.16)$$

where ρ_0 is the radius at $1/e^2$ of the peak energy.

This equation can be manipulated to find the radial position ρ_i corresponding to a given fraction of input energy:

$$\rho_i = \sqrt{-\frac{\rho_0^2}{2} \ln\left(1 - \frac{E_i}{E_\infty}\right)} \qquad (12.17)$$

where:

E_∞ is the total energy of the beam

E_i/E_∞ is the fraction of energy contained within the ith radius

The fraction E_i/E_∞ is set to $1/N$, $2/N$, $(N-1)/N$ to find the input ray positions ρ_i. These are then normalized to the entrance pupil radius for the optical design program.

On the output side, one might at first think that the rays just need to be distributed at equal increments out to the target radius. This is incorrect, however, because the problem must again be considered in terms of rings of equal energy. For an output ray at radius P_i, the fractional output energy within that radius is given by

$$u(P_i) = \frac{\pi P_i^2}{\pi P_N^2} \qquad (12.18)$$

where N is the total number of rays.

This equation is solved for $u(P_1) = 1/N$, $2/N$, N/N to find the output ray positions.

Figure 12.6 illustrates a macro for the ZEMAX optical design program that computes the merit function for this problem. A robust merit function should also detect

```
! Macro to determine merit function for beam shaping for circular target shapes.

! VEC1(n) are positions for incoming rays, normalized to pupil.
! VEC2(n) are target positions for exiting rays in lens units.
! Number is the total number of rays to trace.
! Size is desired radius at the target in lens units.
! Final is the target surface number.
! Error is the RSS of the deviations of the rays from desired targets.
! Ro is the Gaussian beam radius in lens units.
! Pupil is the entrance pupil radius.

! Initialize variables:
Error = 0
Final = NSUR()
Number = 20
Size = 1.5
Ro = 3
Pupil = AVAL() / 2
! AVAL() returns the entrance pupil diameter from the lens file.

! This loop populates the incoming (normalized to pupil) and outgoing (lens
! units) ray positions:
FOR n = 1, Number - 1, 1
  VEC1(n) = SQRT((-1/2 * Ro * Ro * LOGE(1 - n/Number))) / Pupil
  VEC2(n) = SQRT(Size * Size * n / Number)
NEXT n

! The ray for 100% input energy is defined as twice the beam radius.
VEC1(Number) = 2 * Ro / Pupil
VEC2(Number) = Size

! This loop traces the rays and computes the RSS target position error. The ray
! error is multiplied by 100 to make the RSS value easier to visually track
! during optimization. If the ray misses the final surface, the error
! contribution is made large.
FOR n = 1, Number, 1
  RAYTRACE 0,0,0,VEC1(n)
  IF (0 == RAYE)
    Error = Error + POWR(100*(RAYY(Final) - VEC2(n)), 2)
  ELSE
    Error = Error + POWR(1000 * VEC2(n), 2)
  ENDIF
NEXT n

OPTRETURN 0 = Error
```

FIGURE 12.6 Merit function macro for transforming a Gaussian beam to a uniform circular target spot. (From Holswade, S.C. and Dickey, F.M., Laser beam shaping via conventional design software, *Proc. SPIE*, 4443, 36–46, 2001. With permission.)

cases where rays miss the output surface, and it should make their contribution to the merit function correspondingly large. For a target surface such as a sphere, rays could easily be directed beyond the target aperture during the optimization process. The merit function of Figure 12.6 addresses this possibility.

This merit function was used for a problem where a Gaussian input beam with waist radius of 3 mm was transformed into a uniform circular target spot with a radius of 1.5 mm. This optical system is illustrated in Figure 12.7. The beam shaping surface was represented as a 12th-order polynomial function of r, and 20 rays were used. The coefficients of this surface were set to zero (flat plate) at the beginning of optimization, using a 550 MHz processor. The system was first optimized using the standard damped least squares algorithm. Within a few minutes, it had found a lens profile that was not too far from the final solution. It is recommended to always start with damped least squares optimization to verify the design parameters and to

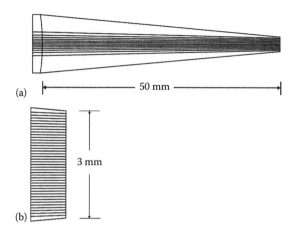

(a)

(b)

50 mm

3 mm

FIGURE 12.7 (a) Plan view of a beam shaping optical system that converts a Gaussian beam into a uniform, circular target spot. (b) Magnified view of the final 1 mm of the target zone. Note that Zemax internally computes the ray positions for this plot such that the target rays are evenly spaced. (From Holswade, S.C. and Dickey, F.M., Laser beam shaping via conventional design software, *Proc. SPIE*, 4443, 36–46, 2001. With permission.)

debug the merit function. The ZEMAX *Hammer* algorithm, a global search method that begins with the current solution, was then employed to refine the merit function to a small value. It was left running overnight, although it came fairly close to the final value within approximately an hour. The resulting target spot is shown in Figure 12.8. A larger number of rays helped smooth the region near the target edges, but at the cost of longer optimizations.

For the case of shaping Gaussian beams into rectangular geometries, a separable approach is employed. The system is solved for each axis independently, and the shaping element profile is the sum of these two solutions. Planar, rather than circular, symmetry now applies for each axis. Unfortunately, the computation of the incoming ray positions for these cases is not straightforward for the simplified mathematical functions found in optical design macro languages. The input energy $E(x_i)$ contained inside a ray at position x_i is

$$E(x_i) = \int_0^{x_i} e^{-2x^2/\rho_0^2} dx \qquad (12.19)$$

where ρ_0 is the radius at $1/e^2$ of the peak energy.

This integration requires the use of error functions, which are not found in optics macro languages, and which only slowly converge in series expansions. It is best to just solve this equation numerically for the normalized ray positions and then manually insert them into the merit function. For the separable case with planar, uniform target geometry, the output ray positions are simple fractions of the desired target

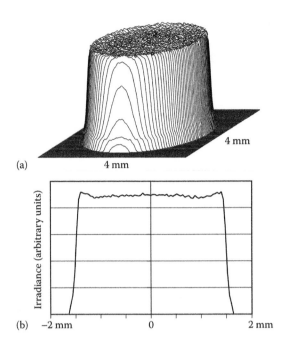

FIGURE 12.8   (a) Surface plot of the target spot for the system in Figure 12.7. (b) Cross section through the center of this surface. (From Holswade, S.C. and Dickey, F.M., Laser beam shaping via conventional design software, *Proc. SPIE,* 4443, 36–46, 2001. With permission.)

size on that axis. Figure 12.9 shows a merit function that uses 20 rays to solve this problem. This merit function was used for a problem where a Gaussian input beam with waist radius of 3 mm was transformed into a uniform square target spot with a 3 mm width. The beam shaping surface was represented as an eighth-order polynomial function of x and y, and 20 rays were used. Beginning with a flat plate surface, the y-axis was optimized using the damped least squares routine. Within a few minutes, it stopped, although the solution was not as close as for the circular case. The ZEMAX *Hammer* algorithm was again employed to refine the merit function to a small value. It was left running overnight. These y-axis coefficients were then copied to the corresponding x-axis coefficients. The resulting target spot is shown in Figure 12.10.

For shaping problems where a collimated output (planar wavefront) is needed, at least two shaping surfaces are required. The first surface would create the desired profile at the second surface, and the second surface would redirect the rays to be collimated with the optical axis. The merit function would then contain an additional term that would be the RSS of the deviations of the ray output angles from the optical axis.[22,25] Note that for successful collimation, the collimating element typically requires a much higher order surface than the shaping element. Also, diffraction effects must be considered when the beam propagates any appreciable distance.

Optimization-based techniques are powerful once the beam shaping problem is defined in terms that a computer can manipulate. For cases where a known solution

```
! Macro to determine merit function for beam shaping to rectangular target
shapes.

! VEC1(n) are incoming ray positions normalized to the Gaussian beam radius.
! Other variables are the same as the circular case macro.

! Initialize variables:
Error = 0
Final = NSUR()
Number = 20
Size = 1.5
Ro = 3
Pupil = AVAL() / 2

! The ray for 100% input energy is defined as twice the beam radius.
VEC1(1) = 0.0313534
VEC1(2) = 0.0628307
VEC1(3) = 0.0945592
VEC1(4) = 0.126674
VEC1(5) = 0.15932
VEC1(6) = 0.19266
VEC1(7) = 0.226881
VEC1(8) = 0.2622
VEC1(9) = 0.29888
VEC1(10) = 0.337245
VEC1(11) = 0.377708
VEC1(12) = 0.420811
VEC1(13) = 0.467295
VEC1(14) = 0.518217
VEC1(15) = 0.575175
VEC1(16) = 0.640776
VEC1(17) = 0.719766
VEC1(18) = 0.822427
VEC1(19) = 0.979982
VEC1(Number) = 2

! This loop populates outgoing (lens units) ray positions:
FOR n = 1, Number, 1
  VEC2(n) = Size * n / Number
NEXT n

! This loop traces the rays and computes the RSS target position error:
! If the ray misses the final surface, the error contribution is made large.
! Note that the RAYTRACE normalizes the input rays to the pupil.
FOR n = 1, Number, 1
  RAYTRACE 0,0,0,VEC1(n)*Ro/Pupil
  IF (0 == RAYE)
    Error = Error + POWR(100*(RAYY(Final) - VEC2(n)), 2)
  ELSE
    Error = Error + POWR(1000 * VEC2(n), 2)
  ENDIF
RAYY(Final), " ", RAYE
NEXT n

OPTRETURN 0 = Error
```

FIGURE 12.9 Merit function macro (single axis) for the separable problem of transforming a Gaussian beam into a rectangular target spot. (From: Holswade, S.C. and Dickey, F.M., Laser beam shaping via conventional design software, *Proc. SPIE*, 4443, 36–46, 2001. With permission.)

already exists, it should be used in lieu of reworking the problem using a computer. However, many problems do not have published solutions, and optimization techniques are useful in seeking working solutions without performing a mathematical derivation. An example would be the shaping of a Gaussian input beam to provide uniform irradiance on the surface of a cylinder. With modification to the desired

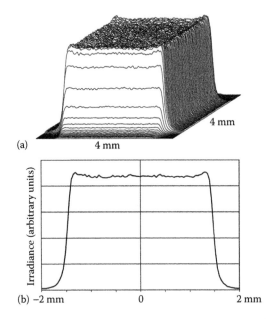

(a)   4 mm

(b) −2 mm   0   2 mm

FIGURE 12.10   (a) Surface plot of the target spot for a desired uniform square profile. (b) Cross section through the center of this surface. (From Holswade, S.C. and Dickey, F.M., Laser beam shaping via conventional design software, *Proc. SPIE*, 4443, 36–46, 2001. With permission.)

output ray locations for the curved axis, this problem could be solved using the separable merit function of Figure 12.9. Another example would be the shaping of a source that did not have a simple input profile, such as an incoherent LED source. It must be stressed that the optimization techniques discussed here apply only for geometric beam shaping problems; diffraction effects are not considered. As discussed in Section 12.6.2, the designer should ensure that diffraction can be ignored before using a geometric technique.

Finally, iterative methods using physical optics principles are being studied to perform beam transformations into complex geometries.[26,27] An example would be the transformation of a Gaussian input beam into the number "2." These methods are also being studied to design beam shaping systems that are tolerant of input beam variations.[28]

12.3.4   CLOSED-FORM SOLUTIONS

The beam shaping problem described by Equation 12.2 can be directly implemented by the system shown in Figure 12.11.[1,29,30] In the figure, the last two elements comprise the beam shaping system; the first two elements are a beam expanding telescope. The beam shaping system consists of a shaping element (phase function ψ) and a Fourier transform (focusing) lens. The beam expanding telescope, which may or may not be necessary, provides a means of increasing by increasing the input beam diameter. It is interesting to note that Harger[31] may have been the first to discuss the problem of determining the phase modulation that relates the magnitude of a

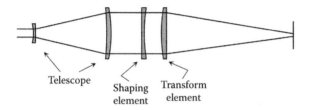

FIGURE 12.11 Beam shaping system implementing Equation 12.2. (From Dickey, F.M. and Holswade, S.C., Gaussian laser beam profile shaping, *Opt. Eng.* 35, 3285–3295, 1996. With permission.)

function and the magnitude of its Fourier transform. Although his interest was radar signal processing, Harger's approach is the essence of the beam shaping problem discussed in the following paragraphs.

Using the method of stationary phase, Romero and Dickey[29] have obtained solutions for converting Gaussian beams into uniform profiles with both rectangular and circular cross sections. In these solutions, the phase 1*i* in Equations 12.1 and 12.2 is given by $\psi = \beta\Phi$, where Φ is a general solution defined below. For a circular Gaussian beam input, the problem of turning a Gaussian beam into a flattop beam with rectangular cross section is separable. That is, the solution is the product of two one-dimensional solutions. β and $\Phi(\xi)$ are thus calculated for each dimension. The phase element will then produce the sum of these phases $(\beta_x\Phi_x(x) + \beta_y\Phi_y(y))$. The corresponding one-dimensional solution for Φ is

$$\phi(\xi) = \frac{\sqrt{\pi}}{2}\,\xi\,\mathrm{erf}(\xi) + \frac{1}{2}\exp(-\xi^2) - \frac{1}{2} \tag{12.20}$$

where

$$\xi = \frac{\sqrt{2x}}{r_0} \quad \text{or} \quad \xi = \frac{\sqrt{2y}}{r_0}$$

$r_0 = 1/e^2$ radius of the incoming Gaussian beam.

The solution for the problem of turning a circular Gaussian beam into a flattop beam with circular cross section is

$$\phi(\xi) = \frac{\sqrt{\pi}}{2}\int_0^\xi \sqrt{1 - \exp(-\rho^2)}\,d\rho \tag{12.21}$$

where

$$\xi = \frac{\sqrt{2r}}{r_0}$$

r is the radial distance from the optical axis.

As previously mentioned, the quality of these solutions depends strongly on the parameter β. For the two solutions given in Equations 12.20 and 12.21, β is given by

$$\beta = \frac{2\sqrt{2\pi}\, r_0 y_0}{f\lambda} \qquad (12.22)$$

where:

   $r_0 = 1/e^2$ radius of incoming Gaussian beam
   $y_0 = $ half-width of desired spot size (the radius for a circular spot, or half the width of a square or rectangular spot)

The effect of β on the quality of the solution for the problem of mapping a Gaussian beam into a flattop beam with a rectangular cross section is illustrated in Figure 12.12. In the figure, we give simulation results for a shaped beam profile with a rectangular cross section, with β values of 4, 8, and 16.

The beam shaping configuration discussed is very general. Solutions for different profiles and cross sections can be obtained using the method of stationary phase. Also, the phase element can be designed using optimization algorithms and the Gerchberg–Saxton algorithm. There are several properties associated with the

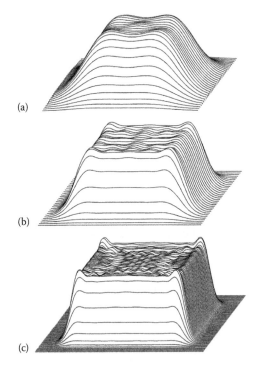

(a)

(b)

(c)

FIGURE 12.12 Simulated shaped beam with square cross section. (a) $\beta = 4$, (b) $\beta = 8$, (c) $\beta = 16$. (From Dickey, F.M. and Holswade, S.C., Gaussian laser beam profile shaping, *Opt. Eng.* 35, 3285–3295, 1996. With permission.)

lossless beam shaping configuration shown in Figure 12.11 that are important to system design considerations. We will list them here, noting that the details are provided in the references.

- Element spacing: assuming the validity of the Fresnel integral, the spacing between the phase element and the Fourier transform lens is not critical.
- Single element design: the phase element and the Fourier transform lens can be combined as one element.
- Scaling: the Fourier transform lens focal length may be changed to scale the output spot size without changing β.
- Positive/negative phase: the sign of the beam shaping element phase, ψ, can be changed without changing the output beam profile (irradiance). It does, however, change properties of the beam before and after the output plane. In one case, the beam goes through a quasi-focus before the output (focal) plane; in the other case, the beam goes through a quasi-focus after the output plane.
- Quadratic phase correction: the solutions given in Equations 12.20 and 12.21 were derived assuming a plane wave (uniform phase) input beam. Small quadratic phase deviations associated with a diverging input correspond to a small shift in the output plane with a proportional scaling of the shaped beam size.
- Collimation: the output beam can be collimated using a conjugate phase plate in the output plane that cancels any nonuniform phase component.

Unlike beam integrator techniques, field mappers are typically sensitive to alignment errors and variations in the input beam size.[1] The effects of a $0.1r_0$ decentering of the input Gaussian beam, with respect to the phase element in the system corresponding to Figure 12.12, is illustrated in Figure 12.13. Decentering with respect to the transform lens is less significant for the irradiance. This is a result of the fact that the effect of translating a function is the addition of a phase factor in the Fourier transform of the function. The effects of a 10% increase/decrease in the input beam size with respect to the design beam size are illustrated in Figure 12.14. The profile

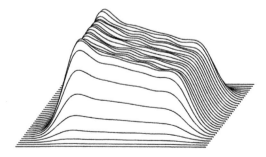

FIGURE 12.13 Input beam decentered from optical axis by $0.1r_0$, $\beta = 8$. (From Dickey, F.M. and Holswade, S.C., Gaussian laser beam profile shaping, *Opt. Eng.* 35, 3285–3295, 1996. With permission.)

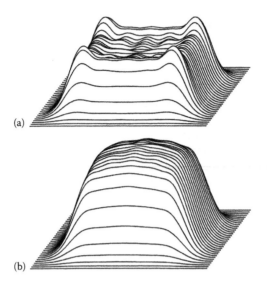

FIGURE 12.14 Effects of deviations in input beam size from design values: (a) 10% increase, (b) 10% decrease. (From Dickey, F.M. and Holswade, S.C., Gaussian laser beam profile shaping, *Opt. Eng.* 35, 3285–3295, 1996. With permission.)

of the beam also varies as one goes away from the design output plane. It is usually desirable to check the performance of a given design with respect to variations of tolerances in alignment and design parameters.

12.3.5 GEOMETRICAL OPTICS SOLUTIONS

In the limit of large β, the stationary phase solutions discussed in the previous section are equivalent to geometrical optics solutions in that they can be obtained by bending rays as suggested in Figure 12.3. However, the geometrical approximation is not valid unless β is sufficiently large. For small β, diffraction effects will dominate the system performance. A purely geometrical optics approach does not reveal the effects of low β. Further, the solutions in the previous section are obtained for the optical system in Figure 12.11 and can frequently be implemented using the thin lens approximation. This approach has the major advantages of the scaling property mentioned in the previous section and the use of fabrication/design techniques associated with thin elements, such as gray-scale lithography or diffractive optics.

When a single element design is of interest and/or the thin element approximation is not applicable, the beam shaping optics can be designed directly using geometrical optics. The geometrical optics design problem is illustrated in Figure 12.15. In the figure, r is the input coordinate, R is the output coordinate, and $z(r)$ is the function describing the surface of the beam shaping element. For an element with one plane surface, as shown in the figure, we can set $z(0) = 0$. The geometrical optics beam shaping design problem described in Figure 12.15 can be solved using the energy balance condition and Snell's law (the energy balance condition being the conservation of energy along a ray bundle between any two surfaces).

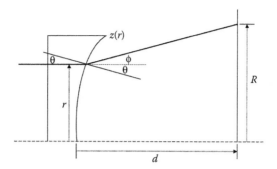

FIGURE 12.15 Geometrical optics diagram for a flattop design.

To illustrate this approach, we will apply the energy balance condition and Snell's law to determine the shape of the element surface in Figure 12.15 for a circular Gaussian input beam and a circular flattop output beam. The input beam irradiance is given by

$$\sigma(r) = e^{-2(r/r0)2} \qquad (12.23)$$

For this problem, the energy balance condition is given by

$$\int_0^{2\pi} d\theta \int_0^r \sigma(r)r\, dr = \int_0^{2\pi} d\theta \int_0^R \Sigma(R)R\, dR \qquad (12.24)$$

where $\Sigma(R)$ is the output irradiance defined by

$$\sum(R) = \begin{cases} K, & R \le R_{max} \\ 0, & R > R_{max} \end{cases} \qquad (12.25)$$

Equation 12.24 can be integrated to give

$$R = \left\{ \frac{r_0^2}{2K} \left[1 - e^{2r^2/r_0^2} \right] \right\}^{1/2} \qquad (12.26)$$

where

$$K = \frac{r_0^2}{2R_{max}^2} \left[1 - e^{2r_{max}^2/r_0^2} \right] \qquad (12.27)$$

r_{max} is the working aperture of the shaping element.

The relation between r and R described by Equations 12.26 and 12.27 define optical rays from the shaping element to the output plane.

Snell's law for the ray shown in Figure 12.15 can be written as

$$n_0 \sin(\theta + \Phi) = n \sin\theta \qquad (12.28)$$

or equivalently

$$\tan\theta = \frac{\sin f}{(n/n_0)} - \cos\phi \qquad (12.29)$$

Noting that $(dz(r)/dr) = \tan\theta$, we can obtain a differential equation for the surface $z(r)$ as

$$\frac{dz(r)}{dr} = z'(r) = \frac{R-r}{(n/n_0)\sqrt{(R-r)^2 + (d-z(r))^2} - d + z(r)} \qquad (12.30)$$

Equation 12.30 can be easily solved using many of the available math software packages.

To illustrate the solution, consider the design of a beam shaper with the following parameters: $r_0 = 5$ mm, $r_{max} = 10$ mm, $R_{max} = 0.02$ mm, $d = 100$ mm, $n = 1.5$, and $n_0 = 1$. Note, in this case, $\beta = 8$. The second surface of the element shown in Figure 12.15 is obtained for these parameters by solving Equation 12.30. The solution in graphical form is shown in Figure 12.16. A calculation, including diffraction effects, of the output for this system is shown in Figure 12.17.

Shealy[32] and Evans and Shealy[22] have treated the geometrical optics approach to the design of laser beam shapers in detail. In fact, Equations 12.23–12.30 are either taken from or are based on the treatment by Shealy.[32] In addition to the energy balance condition and the application of Snell's law, they include a constant

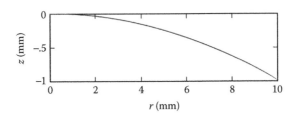

FIGURE 12.16 Solution for the second surface of the element in Figure 12.15 for an input beam diameter of 10 mm and an a 0.02-mm-diameter output flattop corresponding to $\beta = 8$.

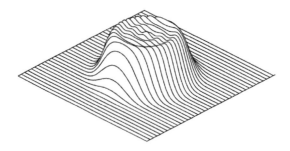

FIGURE 12.17 Calculated diffraction field in the output plane corresponding to the surface in Figure 12.16.

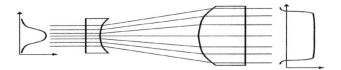

FIGURE 12.18 Two lens system for large, uniform, collimated beams.

optical path length condition. This condition is invoked when it is desirable to produce a uniform irradiance beam that is also collimated. A minimum of two elements is needed to produce collimated beams. A configuration has been suggested by Rhodes and Shealy[33] that is especially applicable to the production of relatively large collimated beams with a uniform irradiance profile. The basic concept is illustrated in Figure 12.18. In the figure, the second surface of the first lens directs the incident rays so that they are uniformly distributed at the first surface of the second lens. That surface then redirects the rays so that they are collimated. A general theory of this two lens beam shaping system is detailed by Shealy[32] and Evans and Shealy.[22] Their approach is geometrical optics, which assumes a large β. The design approach starts with the eikonal and invokes conservation of energy along a ray bundle between the two surfaces. Special attention is given to constant optical path length designs that give a collimated (minimum divergence) output. The result is a differential equation for the lens surface. They also develop similar methods for two-lens systems using gradient-index (GRIN) glasses. Ichihashi and Furuya[34] have applied this design approach to the problem of beam shaping for printed wiring board laser drilling. Hoffnagle and Jefferson[35] have specialized the design to the condition that both elements in Figure 12.18 are planoconvex.

It is interesting to note that the first lens in Figure 12.18 effectively accomplishes the same function as the combination of the shaping element and transform lens in Figures 12.11 or 12.15. The main difference is that the output surface for the system in Figures 12.11 or 12.15 is assumed planar, while the output surface of the first lens in Figure 12.18 is the first surface of the second lens.

12.4 BEAM INTEGRATORS

A multiaperture integrator system basically consists of two components: (1) a subaperture array consisting of one or more lenslet arrays, which segments the entrance pupil or cross section of the beam into an array of beamlets and applies a phase aberration to each beamlet, and (2) a beam integrator or focusing element, which overlaps the beamlets from each subaperture at the target plane. The target is located at the focal point of the primary focusing element, where the chief rays of each subaperture intersect. Laser beam integrators were introduced by Ream in 1979.[36] Beam integration first appeared in a refereed journal in 1986.[37] The first detailed analysis of beam integrators was published by Dickey and O'Neil.[38] Brown et al.[39] provide an extensive treatment of laser beam integrators.

Beam integrators can be loosely divided into two categories: diffracting and imaging. A simple diffracting beam integrator (also called a nonimaging integrator) is

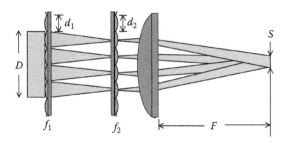

FIGURE 12.19 Basic configuration for the imaging integrator. (From Dickey, F.M. and Holswade, S.C. 2000. *Laser Beam Shaping: Theory and Techniques.* Marcel Dekker, New York. With permission.)

illustrated in Figure 12.4, consisting of a single lenslet array and a positive primary lens. The target irradiance is the sum of defocused diffraction spots (point spread functions) of an on-axis object point at infinity (assuming a collimated input wavefront). The diffracting beam integrator is based on the assumption that the output is the superposition of the diffraction fields of the beamlet apertures. The diffraction field is obtained using the Fresnel integral. If the beam is not spatially coherent over each beamlet aperture, a more complicated integral is required, and, generally, one would not be able to obtain a reasonable replica of the lenslet aperture. For example, a spatially incoherent field is approximated by a Lambertian source of relative large extent that would not produce overlapping irradiance distributions at the output plane.

Figure 12.19 illustrates an imaging multiaperture beam integrator. This type of integrator is especially appropriate for sources with low spatial coherence. From a ray optics perspective, these sources produce a wavefront incident over a range of field angles on the lenslet apertures. The first lenslet array segments the beam as before and focuses the beamlets onto a second lenslet array. That is, each lenslet in the first array is designed to confine the incident optical radiation within the corresponding aperture in the second array. A second lenslet array, separated from the first by a distance equal to the focal length of the secondary lenslets, together with the primary focusing lens, forms a real image of the subapertures of the first lenslet array on the target plane. The primary lens overlaps these subaperture images at the target to form one integrated image of the subapertures of the first array element. Reimaging the lenslet apertures mitigates the diffraction effects of the integrator in Figure 12.4. Imaging integrators are more complicated than diffracting integrators in that they require a second lenslet array with an associated alignment sensitivity. Diffracting integrators are thus more frequently the integrators of choice.

There are four major assumptions in the development of diffracting beam integrators. They are as follows:

1. The input beam amplitude (or equivalently irradiance) is approximately uniform over each subaperture. This allows for the output to be the superposition (average) of the diffraction patterns of the beamlet defining apertures. It is expected that small deviations will average out in the output plane. That is, the errors associated with a particular aperture will not dominate.

2. The phase across each subaperture is uniform. The discussion in Assumption 1 applies in this case also. For example, a linear phase across a subaperture results in a redirection of the beamlets, causing a misalignment in the output.
3. The input beam divergence does not vary significantly with time. Generally, an input beam divergence will result in a nonoverlapping of the beamlets in the target plane. This can be corrected in many cases with correction optics in the input beam. However, a time-varying divergence would negate the possibility of correction.
4. The input beam field should be spatially coherent over each subaperture. This is inherent in Assumption 1 since the diffraction patterns are assumed to be described by a Fresnel integral.

The imaging integrator does not require Assumption 4 as it does not necessarily require that the output pattern be described by a diffraction integral.

The basic problem for the diffracting integrator is that each lenslet then maps a uniform input intensity into a uniform output intensity *via* the Fourier transform. It can be shown[40] that the desired lenslet phase function is

$$\Phi\left(\frac{x}{d}\right) = \beta\left(\frac{x}{d}\right)^2 \tag{12.31}$$

This quadratic phase factor describes a thin lens. Again the solution includes the parameter β that is a measure of the quality of the solution. The parameter, for this case, is given by

$$\beta = \frac{\pi d S}{\lambda F} \tag{12.32}$$

where d, S, and F are defined in Figure 12.4.

Note that β is a dimensionless constant and, as previously discussed, is related to the mathematical uncertainty principle. Increasing β decreases the effects of diffraction in the output.

Using paraxial geometrical optics, it can be shown that the spot size S on the target is equal to the focal length F of the primary lens divided by the f number of the subaperture lens,

$$S = \frac{F}{f/d} \tag{12.33}$$

This result is also obtained using diffraction theory and Fourier optics.

In addition to the diffraction effects discussed above with respect to β, multiaperture beam integrators generally exhibit interference effects. They are effectively multiple beam interferometers. The coherence theory of multiaperture beam integrators is developed in Reference 1 of Chapter 7. Depending on the degree of spatial coherence of the source, the output irradiance will contain an interference (speckle) component. For these conditions, the integrated irradiance of the coherent component is adequately described by

$$I(x,y) = \left| \sum_{0,0}^{M,N} A_{mn} \exp\left\{ i[k(\alpha_m x + \beta_{n} y) + \theta_{mn}] \right\} \right|^2 |F(x,y)|^2 \qquad (12.34)$$

where:

α_m and β_n are the direction cosines associated with each beamlet

θ_{mn} is the phase of the beamlet

A_{mn} is the amplitude of the beamlet field and the function $F(x,y)$ is the diffraction integral of the beamlet-limiting aperture

$F(x,y)$ is the Fourier transform of the lenslet phase function for the optical configuration in Figure 12.4

The first factor in Equation 12.34 describes the averaging and interference effects of the integrator. The interference effect is a result of the sum of linear (in x and y) phase terms, which can be viewed as a Fourier series. The spatial period for the resulting interference pattern is given by

$$\text{Period} = \frac{\lambda F}{d} \qquad (12.35)$$

It should be noted that when the coherence between the beamlets is negligible, Equation 12.34 reduces to

$$I(x,y) = \sum_{0,0}^{M,N} |A_{mn}|^2 |F(x,y)|^2 \qquad (12.36)$$

which is the diffraction pattern of a single lenslet aperture. The results for a beam integrator applied to a beam with a degree of spatial coherence and moderate β is illustrated in Figure 12.20.[39,41] The irradiance fluctuations in the figure are a combination of the interference pattern and diffraction. Results very close to an ideal flat-top can be obtained for large β and incoherent sources such as some excimer lasers.

Another approach to beam integration is the channel integrator. As first proposed, a channel integrator is a reflective cylinder with a rectangular crosssection.[42] Other cross sections are possible. The channel integrator concept can be easily explained in one dimension. These ideas can then be extended to two dimensions.

A one-dimensional schematic of the channel integrator is shown in Figure 12.21. In the figure, the two solid lines labeled 1 and 2 represent the channel integrator. The input beam is focused, with focal length F, on the center of the front face of the integrator. The integrator aperture size is S. To understand the integrator, consider the

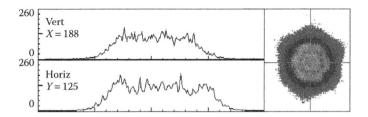

FIGURE 12.20 Measured profile at the focal plane of a diffracting beam integrator illustrating the effects of a partially coherent source. (From Dickey, F.M. and Holswade, S.C. 2000. *Laser Beam Shaping: Theory and Techniques.* Marcel Dekker, New York. With permission.)

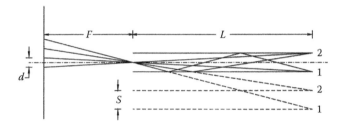

FIGURE 12.21 Schematic of the channel integrator.

input rays centered on the optical axis that are bounded by an aperture of width d. This bundle of rays just fills the output aperture of the integrator. The ray bundle above, with the same aperture width d, will also fill the output aperture after a reflection. This can be readily visualized by repeating virtual integrator walls, shown by the dashed lines. These virtual walls define the reflections and the input ray bundles that fill the output aperture. It can be seen that successive apertures with size d in the input are geometrically projected with inversion on the output aperture. Thus, this configuration is equivalent to the multiaperture arrays discussed above. Although they are not addressing *laser* beam shaping, Chen et al. provide interesting analysis that is applicable to channel integrators.[43]

The above argument can readily be extended to channel integrators with square or rectangular cross sections. It appears that channel integrators can be made with cross sections that can both tile a plane and meet the condition that each edge is a line of reflection. This clearly eliminates circular (or elliptical) cross sections. Using a different argument, channel integrators with circular cross sections can also be eliminated by considering the diagram that results from rotating Figure 12.21 about the optical axis. In this case, one can show that there are concentric rings that are mapped onto the ring defining the outer aperture of the integrator. One can also show that there are interspersed rings that map to a point at the center of the output aperture. The result is that the input irradiance mapping is not uniform over the output aperture. A complete analysis of the channel integrator with respect to possible aperture shapes would involve tiling and group theory, and is beyond the scope of this chapter.

For the channel integrator, the equation corresponding to Equation 12.33 relating the integrator (input) aperture size to the output spot size is given by

$$d = \frac{FS}{L} \qquad (12.37)$$

Equation 12.35 for the interference pattern period is also valid in this case since it depends on the angle between adjacent beamlets. The parameter β (as derived in Reference 1 of Chapter 3, Appendix A) is not directly applicable for the channel integrator. It is suggested that the related Fresnel number[44] be used for β in this case. The Fresnel number for the channel integrator is

$$N_F = \frac{d^2 L}{\lambda F (L + F)} \qquad (12.38)$$

This equation for the Fresnel number is obtained by including the phase function representing the lens in the Fresnel integral defining the propagation of the beam. Requiring a large Fresnel number implies that L should be large or F should be small.

As discussed above, the channel integrator is equivalent to the multiaperture beam integrator comprising refractive lenslet arrays or multiaperture reflective systems. There are a couple of disadvantages associated with channel integrators. One major disadvantage is that they tend to be lossy due to the multiple reflections involved. Another disadvantage is the complexity of needing, in most cases, to add a second lens to relay (image) the output pattern onto the working surface. There is a possibility of eliminating the first lens by tapering the channel integrator. One advantage of the channel integrator is the high-power handling capability. This is a result of the fact that they can be made of metal structures suitable for cooling.

12.5   DIFFUSIVE BEAM SHAPERS

The design and characteristics of diffractive diffusers, also called diffuser beam shapers, are discussed by Brown.[45] Diffuser beam shapers are essentially field mappers. They are designed to diffract the incident beam into the desired irradiance distribution with a built-in speckle (or random) pattern. The basic design procedure is: (1) multiply the desired irradiance pattern (magnitude) by a random function (speckle pattern), (2) inverse Fourier transform this result to get the input field, and (3) binarize the phase of the inverse transform function to define the beam shaping diffuser. The technique is illustrated in Figure 12.22.

Diffuser beam shapers generally offer the advantage of being much more tolerant to alignment errors than conventional field mappers. Although we class them as field mappers, their speckle and alignment tolerance properties are more like beam integrators. Perhaps they can be viewed as beam integrators with lenslet aperture size approaching zero and with the phase varying from lenslet to lenslet.

Diffractive diffusers are effectively random phase gratings. Generally, they are far field gratings that produce the desired output in the far field of the diffuser. This

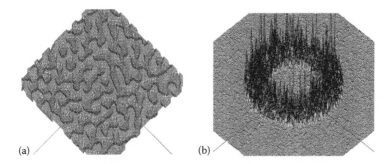

(a)   (b)

FIGURE 12.22 Simulated diffuser beam shaping. (a) A portion of the binary (2 level) phase structure for the ring diffuser and (b) simulated plot of the ring diffuser. This simulation used a Gaussian input beam of 0.5 mm diameter. A spherical phase curvature was then applied to simulate a lens with a focal length of 10.0 mm. Using scalar wave theory, the field was propagated 10.0 mm to the focal plane of the lens. (From Brown, D.M., Dickey, F.M., and Weichman, L.S., Multi-aperture beam integration systems, Chapter 7, Ref. 1.)

FIGURE 12.23 Illustration of a diffractive diffuser designed to produce the image of Einstein. Note, the image character in maintained as the beam propagates. (Courtesy of D.R. Brown, MEMS Optical, Inc.)

is especially true for the above design procedure since it involves a Fourier transform relation between input and output. Due to the small size of the grating structure, the beam is in the far field almost immediately beyond the optic. Since the diffuser is obtained from the Fourier transform (spectrum) of the desired output, the beam will continue to expand with a predefined angular divergence, while maintaining the character (shape) of the desired output/image.

Diffractive diffusers are especially useful for producing complicated patterns such as logos, alignment patterns, and art. An example is shown in Figure 12.23. The figure shows a representation of the input Gaussian beam and a representation of the diffractive diffuser designed to produce a likeness of Albert Einstein. This is followed by experimental results at two distances from the diffuser.

12.6   CHOOSING A SHAPING TECHNIQUE

Of the three basic types of beam shaping, aperturing is the simplest to discuss. This technique is easy to implement and low-cost, but it suffers from large power losses. It is often used as a quick way to produce a relatively uniform profile to test if beam shaping will enhance an application. If so, then a more advanced technique can be pursued.

The other two types of laser beam shaping techniques, field mapping and beam integration, are low loss or lossless. Field mappers work only for beams with a known field distribution, such as single-mode beams, and they are generally highly sensitive to alignment and beam dimensions. Integrators work for both coherent and multimode beams, where the input field distribution may not be known, and they are much less sensitive to alignment and beam size. However, interference effects are a problem with integrators, especially for coherent beams. This categorization into mappers and integrators is independent of the method, such as diffraction theory or geometrical optics, used to design the beam shaping optics.

Before comparing various methods of mapping and integrating beams, however, some thought must be given to the method of measuring beam shaping performance. Misunderstandings about how closely a given technique matches what is desired can lead to poor choices of methods.

12.6.1 Metrics for Beam Shaping-Picking a Figure of Merit

There can be several pitfalls associated with determining how well a particular shaping technique matches desired results. Understanding the various figures of merit, or metrics, is thus important. The most common metric is probably the mean square difference between the desired and shaped beam profile. Other examples of metrics are the integrated absolute difference, the maximum absolute difference, and the peak irradiance divided by the average irradiance. Nonanalytical criteria include *ad hoc* experimental performance and *it looks good*. If a metric has an established history for a given application, it is best, where practicable, to use the same metric for design and analysis.

The mean square difference is appealing because it is analytically very tractable and has an energy or power interpretation. The integrated absolute difference is a viable criterion but it is less tractable analytically. It can readily, however, be employed with numerical techniques. The absolute difference can cause problems. Two functions could have a large absolute difference over a small region, with only a small variation in their mean square difference.

The discussion so far has been concerned with the energy or amplitude of the optical field. In addition, there may be simultaneous constraints on the error in the phase of the optical field. The phase of the optical field is generally not important in energy-intensive applications such as material processing, cutting, and welding. The phase of the shaped beam is likely to be important in applications involving optical signal processing, interferometry, holography, and interference lithography. Any of the metrics discussed above can be applied to the phase of the shaped beam; however, the energy interpretation of the mean square difference is not associated with the phase of the optical field.

In describing a physical problem mathematically, such as beam shaping, care must be taken that the mathematical description matches what is desired. For example, a profile with the lowest mean square difference with respect to a uniform profile may not look the *best* to the designer who compares it to other possibilities. Profiles tend to look more uniform as they get larger than the desired target size. However, their mean square difference would be large since they would have sizeable amounts of energy beyond the bounds of the desired target size. On the other hand, a beam with a large mean square difference from what is desired may work perfectly for the application. For example, in a material removal application, a fine structure in the beam would not be resolvable by the physical process of material ablation, even though it would mathematically result in a high mean square difference with respect to a uniform profile. In designing such a system, it would make sense to smooth the computed profile before computing the mean square difference, since the physical situation is essentially averaging the beam as well.

12.6.2 A Selection Process

Like many optical problems, there is no single beam shaping method that addresses all situations. The nature of the input beam, the system geometry, and the quality of the desired output beam all affect the choice of technique. In considering a beam shaping application using single-mode Gaussian beams, it is first advisable to calculate the parameter:

$$\beta = \frac{2\sqrt{2\pi}r_0 y_0}{f\lambda} \tag{12.39}$$

where:
  λ is the wavelength
  r_0 is the radius at the $1/e^2$ point of the input beam
  y_0 is half-width of the desired output dimension
  f is the focal length of the focusing optic or the working distance from the optical
       system to the target plane for systems without a defined focusing optic

For beams or output spots that do not have circular symmetry, approximate heights and widths can be used. This parameter sets limits on the quality of the solution available. For simple output geometries such as circles and rectangles, some rules of thumb can be defined: if $\beta < 4$, a beam shaping system will not produce acceptable results. For $4 < \beta < 32$, diffraction effects are significant and should be included in the development of the beam shaping system. For $\beta > 32$, diffraction effects should not significantly degrade the overall shape of the output beam. However, interference effects will still be present for multifaceted or diffusing approaches. These may or may not be an issue for a particular application. For more complicated output geometries, β may have to be significantly higher to produce acceptable results. There are no good general rules for these cases.

The particular beam shaping approach depends on the nature of the input beam and the desired output shape. If the desired output shape is a complicated, nonsymmetric pattern, then the diffuser approach is probably a good option. For relatively simple output shapes, such as the circles, rectangles, and squares used in material processing, there are several other approaches. Some of these depend on the quality of the input beam. If the input beam is multimode, with significant irradiance variations across its profile, then the multiaperture approach is most appropriate. This approach reduces the effect of input beam variations, but at the expense of interference effects in the output profile. The multiaperture approach also works well when the input beam size changes over time or between sources. The diffuser approach can also work in this situation.

For applications where the input beam has a well-characterized profile, the nature of the output beam can suggest an approach. If the output pattern must be projected onto a nonplanar surface, geometric techniques work well, as long as β is sufficiently large. The geometrical approach also works well for producing a shaped beam with uniform phase. These beams will propagate for a reasonable distance with the desired profile. Using most other techniques, the profile will degrade beyond the target plane.

Often the input beam is single-mode with a Gaussian profile, and the desired output profile is either circular or rectangular. In these cases, the diffraction-based approach works particularly well. This technique produces an output profile free of interference effects. The profile degradation due to diffraction is also predicted by the solution, and the technique suggests ways to reduce this degradation to an acceptable level. Finally, the solution is known and can be easily scaled to any situation.

As long as the input profile is known, optimization-based techniques have the potential to work well for a number of desired output patterns or beams, including nonsymmetric profiles. These techniques also have the advantage of a minimum of mathematical development to arrive at a solution. In addition, the basic algorithms, once implemented, remain the same for designing additional output patterns.

ACKNOWLEDGMENTS

The authors thank R. N. Shagam of Sandia National Laboratories for the diffraction calculation and the plot of Figure 12.17.

REFERENCES

1. Dickey, F.M., and Holswade, S.C., *Laser Beam Shaping: Theory and Techniques,* Marcel Dekker, New York, 2000.
2. Dickey, F.M., and Holswade, S.C., eds., *Laser Beam Shaping, Proc. SPIE,* 4095, 2000.
3. Dickey, F.M., Holswade, S.C., and Shealy, D.L., eds., *Laser Beam Shaping II, Proc. SPIE,* 4443, 2001.
4. Dickey, F.M., Holswade, S.C., and Shealy, D.L., eds., *Laser Beam Shaping III, Proc. SPIE,* 4770, 2002.
5. Dickey, F.M. and Shealy, D.L., eds., *Laser Beam Shaping IV, Proc. SPIE,* 5175, 2003.
6. Goodman, J.W., *Introduction to Fourier Optics,* McGraw-Hill, New York, chap. 5, 1968.
7. Stamnes, J.J., *Waves in Focal Regions,* IOP Publishing, England, 1986.
8. Walther, A., *The Ray and Wave Theory of Lenses,* Cambridge University Press, England, 1995.
9. Franks, L.E., *Signal Theory,* Prentice-Hall, New Jersey, 1969.
10. Bracewell, R.N. *The Fourier Transform and its Applications,* McGraw-Hill, New York, 1978.
11. Bleistein, N., and Handelsman, R.A., *Asymptotic Expansion of Integrals,* Dover, New York, 1986.

12. Bleistein, N. *Mathematical Methods for Wave Phenomena,* Academic Press, Orlando, 1984.
13. Tan, X., Gu, B.-Y., Yang, G.-Z., and Dong, B.Z., Diffractive phase elements for beam shaping: a new design approach, *Appl. Opt.,* 34, 1314–1320, 1995.
14. Dresel, T., Beyerlein, M., and Schwider, J., Design and fabrication of computer-generated beam-shaping holograms, *Appl. Opt.,* 35, 4615–4621, 1996.
15. Dresel, T., Beyerlein, M., and Schwider, J., Design of computer-generated beam-shaping holograms by iterative finite-element mesh adaption, *Appl. Opt.,* 35, 6865–6873, 1996.
16. Li, Q., Gao, H., Dong, Y., Shen, Z., and Wang, Q., Investigation of diffractive optical element for shaping a Gaussian beam into a ring-shaped pattern, *Opt. Laser Technol.,* 30, 511–514, 1998.
17. Churin, E.G., Diffraction-limited laser beam shaping by use of computer-generated holograms with dislocations, *Opt. Lett.,* 24, 620–621, 1999.
18. Liu, J. and Gu, B.-Y., Laser beam shaping with polarization-selective diffractive phase elements, *Appl. Opt.,* 39, 3089–3092, 2000.
19. Brauer, R., Wyrowski, F., and Bryngdahl, O., Diffusers in digital holography, *J. Opt. Soc. Am. A,* 8, 572–578, 1991.
20. Ghiglia, D.C. and Pritt, M.D., *Two-dimensional Phase Unwrapping: Theory, Algorithms, and Software,* Wiley, New York, 1998.
21. Stark, H., *Image Recovery: Theory and Application,* Academic Press, Orlando, 1987.
22. Evans, N.C., and Shealy, D.L., Optimization-based techniques for laser shaping optics, Chap. 5, Ref. 1.
23. Evans, N.C., and Shealy, D.L., Design and optimization of an irradiance profile-shaping system with a genetic algorithm method, *Appl. Opt.,* 37, 5216–5221, 1998.
24. Evans, N.C., and Shealy, D.L., Design of a gradient-index beam shaping system via a genetic algorithm optimization method, *Proc. SPIE,* 4095, 26–39, 2000.
25. Holswade, S.C., and Dickey, F.M., Laser beam shaping via conventional design software, *Proc. SPIE,* 4443, 36–46, 2001.
26. Schimmel, H., and Wyrowski, F., Amplitude matching concept for design of wave-transforming systems, *Proc. SPIE,* 4092, 26–37, 2000.
27. Schimmel, H., and Wyrowski, F., Amplitude matching strategy for wave-optical design of monofunctional systems, *J. Mod. Opt.,* 47(13), 2295–2321, 2000.
28. Bühling, S., and Wyrowski, F., Solving tolerancing and 3D beam shaping problems by multifunctional wave optical design, *Proc. SPIE,* 4092, 48–59, 2000.
29. Romero, L.A., and Dickey, F.M., Lossless laser beam shaping, *J. Opt. Soc. Am. A,* 13, 751–760, 1996.
30. Dickey, F.M., and Holswade, S.C., Gaussian laser beam profile shaping, *Opt. Eng.,* 35, 3285–3295, 1996.
31. Harger, R.O., *Synthetic Aperture Radar Systems: Theory and Design,* Academic Press, Orlando, 1970.
32. Shealy, D.L., Geometrical methods, Chap. 4, Ref. 1.
33. Rhodes, P.W., and Shealy, D.L., Refractive optical systems for irradiance redistribution of collimated radiation: Their design and analysis, *Appl. Opt.,* 19, 3545–3553, 1980.
34. Ichihashi, K., and Furuya, N., Development of the optical system for PWB (Printed Wiring Board) laser drilling. Laser Microfabrication—ICALEO, D-36, 2000.
35. Hoffnagle, J.A., and Jefferson, C.M., Design and performance of a refractive optical system that converts a Gaussian to flattop beam, *Appl. Opt.,* 39, 5488–5499, 2000.
36. Ream, S.L., A convex beam integrator, *Laser Focus,* 15, 68–71, 1979.
37. Deng, X., Liang, S., Chen, Z., Yu, W., and Ma, R., Uniform illumination of large targets using a lens array, *Appl. Opt.,* 25, 377–381, 1986.

38. Dickey, F.M., and O'Neil, B.D., Multifaceted laser beam integrators, general formulation and design concepts, *Opt. Eng.,* 27, 999–1007, 1988.
39. Brown, D.M., Dickey, F.M., and Weichman, L.S., Multi-aperture beam integration systems, Chap. 7, Ref. 1.
40. Dickey, F.M., and Holswade, S.C., Gaussian beam shaping: Diffraction theory and design, Appendix A of Chap. 3, Ref. 1.
41. Weichman, L.S., Dickey, F.M., and Shagam, R.N., Beam shaping element for compact fiber injection systems, *Proc. SPIE,* 3929, 176–184, 2000.
42. Geary, J.M., Channel integrator for laser beam uniformity on target, *Opt. Eng.,* 27, 972–977, 1988.
43. Chen, M.M., Berkowitz-Mattuck, J.B., and Glaser, P.E., The use of a kaleidoscope to obtain uniform flux over a large area in a solar or arc imaging furnace, *Appl. Opt.,* 2, 265–271, 1963.
44. Saleh, B.E.A., and Teich, M.C., *Fundamentals of Photonics,* Wiley, New York, 1991.
45. Brown, D.R., Beam shaping with diffractive diffusers, Chap. 6, Ref. 1.

Index

Note: Page numbers followed by f and t refer to figures and tables, respectively.

Printed and bound by CPI Group (UK) Ltd, Croydon, CR0 4YY

01/11/2024

01782617-0015